Practical
Solar Energy
Technology

Practical Solar Energy Technology

Martin L. Greenwald

Montclair State College
Upper Montclair, New Jersey

Thomas K. McHugh

Heliotherm
Media, Pennsylvania

Prentice-Hall, Inc., Englewood Cliffs, New Jersey 07632

Library of Congress Cataloging in Publication Data

Greenwald, Martin L., (date)
    Practical solar energy technology.

    Includes index.
    1. Solar energy. I. McHugh, Thomas K.,
(date). II. Title.
TJ810.G74 1985   621.47   84-23721
ISBN 0-13-693979-1

Editorial/production supervision: *Kathleen M. Lafferty*
Cover design: *20/20 Services, Inc.*
Cover illustration: *Craig Rouse*
Manufacturing buyer: *Anthony Caruso*

Printed in the United States of America

10 9 8 7 6 5 4 3 2 1

ISBN 0-13-693979-1   01

Prentice-Hall International, Inc., *London*
Prentice-Hall of Australia Pty. Limited, *Sydney*
Editora Prentice-Hall do Brasil, Ltda., *Rio de Janeiro*
Prentice-Hall Canada Inc., *Toronto*
Prentice-Hall Hispanoamericana, S.A., *Mexico*
Prentice-Hall of India Private Limited, *New Delhi*
Prentice-Hall of Japan, Inc., *Tokyo*
Prentice-Hall of Southeast Asia Pte. Ltd., *Singapore*
Whitehall Books Limited, *Wellington, New Zealand*

To David and Emily

It is the next generation
for whom solar
holds so much promise.

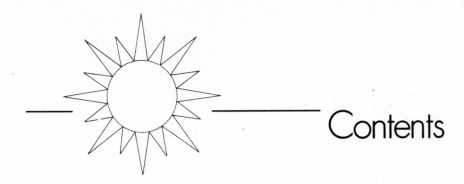

Contents

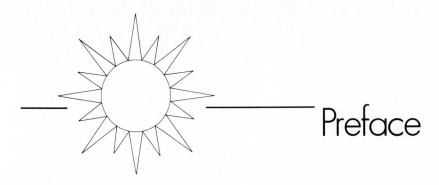

Preface

Practical Solar Energy Technology is a vocational/technical and college text that focuses on the design, installation, and maintenance of solar energy systems. The areas covered encompass solar domestic hot water, space heating, and swimming pool heating systems. It can be easily understood by the layperson with little technical background. State-of-the-art flat-plate collector technology and the latest solid-state electronic control devices employed within the industry are described in detail. Various collector and system designs are examined so as to give the reader a well-rounded knowledge of systems and applications. The outstanding features of this text are:

1. The fundamentals of plumbing and heating technology as applied to the design and operation of various types of hot water and space heating systems are examined. This information is directed primarily at residential and commercial applications powered by fossil fuels. Since most solar energy systems interface or work side by side with fossil-fuel backup heating systems, a basic understanding of the operational principles of such system technology is a prerequisite to an understanding of solar system technology.

2. Easily understood methods for accurately undertaking heat loss analysis of a variety of residential and commercial structures to determine the Btu requirements for solar system sizing are presented. Included are suggestions for implementing a variety of energy conservation measures to minimize heat loss.

3. The methodology necessary for properly sizing solar domestic hot water, space heating, and swimming pool heating systems, with an eye toward maximizing solar system output while minimizing system expense is introduced.

4. Methods of determining solar system performance, cost-benefit analysis, and return on investment are presented. This analysis includes the effects of inflation, maintenance, interest rates, insurance costs, and any other factors that the consumer wishes to use to influence the analysis scenario. Third-party financing, a relatively new concept in solar system purchasing, is also discussed.

5. Comprehensive information relating to solar system installation, maintenance, and troubleshooting procedures is presented. Heretofore, this information has been available only to commercial system installers and manufacturers. The "how to" aspects of these areas are examined in depth. The emphasis is on time-tested installation, system check out, and troubleshooting procedures.

Our society faces critical choices concerning future energy sources. These choices will help to determine the quality of life in the emerging postindustrial, highly technological society of the future. The sun is a source of enormous and usable power. Implementing solar technology on an ever-broadening scale can help to ensure a better future for us all, with a minimum legacy of waste and depleted resources for future generations.

Acknowledgments

The authors wish to acknowledge the assistance of many people who generously gave of their time and abilities in this endeavor. First and foremost our thanks to Craig Rouse, a skilled solar engineer and draftsman whose talented illustrations grace this text throughout. Also, to Mark A. Butler, solar engineer and draftsman, who assisted with illustrations. Next, to Ed Westlake, Sr., president of Heliotherm Division of Westlake Plastics Company, who generously allowed the authors access to the solar engineering experience gained by his corporation throughout the past 10 years. Also, to the following people for their editorial and technical input: Dr. Arthur W. Earl, Professor Emeritus, Department of Industrial Studies, Montclair State College, New Jersey; and Dr. William Makofske, Professor of Environmental Studies and Physics, Ramapo College of New Jersey. And finally, to our wives, Barbara and Olga, who pushed aside work and home schedules, which allowed us the necessary freedom to complete this manuscript.

Martin L. Greenwald
Thomas K. McHugh

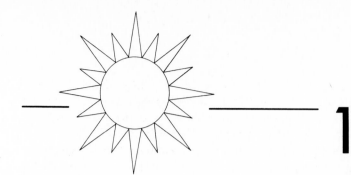

1

Solar Energy in Perspective

ENERGY USE THROUGH THE AGES

The study of technology is essentially an investigation as to how humankind progressed from being gatherers to being producers. Tools, the extension of the human body, multiply by a vast magnitude our ability to do work. This lessens the expenditure of energy needed to accomplish specific tasks. The use of energy mirrors the use of technology and is an integral part of it. The history of human energy use, beginning with using what was readily available without further processing and progressing to the production of complex energy systems and fuel sources, is truly staggering when measured in terms of both power output and fuel consumption.

For centuries, human beings used power sources that were readily available in nature: wind, water, wood, and coal. Wind has long been used to provide power for sailing ships, pumping water, and grinding grain. The earliest known windmills were built in Persia (Figure 1.1). They were known as "sail windmills" because the blades resemble sails used on wind-powered sailing ships. These machines are well suited to mechanical power generation due to slow rotational velocities. During high winds, the sails must be taken in to prevent overspeeding. This prevents damage due to centrifugal forces that are built up as a result of excessive rotational speeds. Machines similar in design to the sail windmill are still in use today and can easily be fabricated from readily available materials. Windmills were common in Europe during the Middle Ages, and their use was widespread through the close of the nineteenth century. The prairie windmill (Figure 1.2) is still a common sight in many parts of the United States, utilized to pump water for a variety of agricultural and household purposes. The use of wind for powering sailing vessels and providing mechanical power needed to pump water and grind grain was widespread through the middle of the nineteenth century. At that time, the development of the steam engine signaled the beginning of mechanized society and ushered in the industrial revolution.

Wood has been used for centuries as a primary energy source and accounted for almost 90% of the fuel consumed in the United States by 1850; coal accounted for the remaining 10%. However, extensive cutting of the forests necessary to sustain wood consumption at that time began both to raise the price of wood and increase the transportation distances.

For these reasons, the use of coal began to rise. By the early twentieth century, coal was supplying nearly 80% of residential and industrial fuel demand, especially in the growing iron and steel industries. Coal was to remain king of fuel well into the twentieth century. Its use began to decline, however, as oil and gas production increased. Crude oil had been discovered in 400 B.C.; however, little practical use had been found for it. The market for crude oil began to develop after its discovery in America in the mid-nineteenth century. Although the market development for crude oil was slow at first, oil would become the dominant fuel source within 75 years of its discovery.

FIGURE 1.1   Mediterranean sail windmill.

FIGURE 1.2   Prairie windmill, United States. These machines have been in use in the past 100 years to pump water for agricultural and residential purposes.

ENERGY CONSUMPTION PATTERNS

At the present time, the industry and technology of industrialized nations is inescapably wedded to the use of fossil fuels: coal, oil, and natural gas—sources of energy that are both finite and diminishing. Society is highly dependent on oil and gas within the areas of agriculture, medicine, pharmaceuticals, and transportation. Present energy consumption within the United States is approximately 78 quads per year.* World consumption is approximately 220 quads per year. The approximate primary energy mix of the United States is illustrated in Figure 1.3. Of the total primary energy consumed within the United States, just under half is petroleum. Daily import figures for petroleum are somewhat erratic, ranging from a

*A quad is a commonly used measure of energy production: 1 quadrillion Btu. One quad is roughly equivalent to 180 million barrels of crude oil, 239 billion kilowatthours of electricity, or 42 million tons of coal.

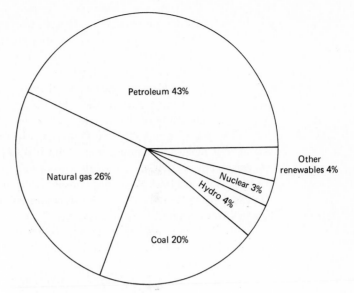

FIGURE 1.3 Approximate energy mix of the United States, 1980. (Adapted from data supplied by the New York State Energy Office.)

high of almost 50% to less than 20%. Although successful efforts in energy conservation have dramatically lowered oil import figures, the long-term stability of reduced imported oil is open to question.

Examination of the fuels that are used to generate electricity in the United States also reveals an overwhelming dependence on fossil fuels. The fuel mix used to generate electricity is illustrated in Figure 1.4. Whereas petroleum is the dominant fuel source within the United States, oil takes second place to coal in electrical generation. Coal continues to grow in popularity as a fuel for electrical generation as the price of other fuels continues to rise more quickly than the price of coal.

The use of hydroelectric energy, perhaps the most environmentally sound and one of

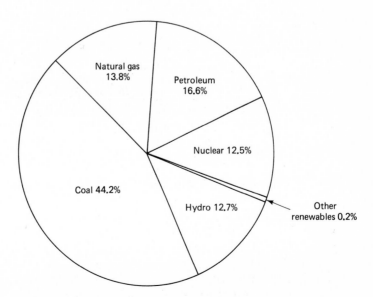

FIGURE 1.4 Primary energy used to generate electricity. Coal continues to be king as far as generating electricity is concerned. Its use is rising as other fuels become more expensive. (Data from New York State Energy Office.)

the most cost-effective of all primary energy sources, is not projected to increase significantly in the future. Although there exist a number of abandoned small-scale hydroelectric facilities in the northeast, most of the locations capable of large-scale energy production have been developed. The reactivation of small-scale facilities is not expected to contribute significantly to overall energy production.

The likelihood of increased use of nuclear energy for electrical generation is presently clouded, due in large part to the poor economics of building and operating nuclear power plants and disposing of the radioactive waste.

The use of renewable energy systems has risen dramatically during the past 10 years. Of all the renewable technologies, solid-fuel heating accounts for the major portion of this increase. The rebirth of solid-fuel heating has made significant inroads into the marketplace of the more traditional sources of home heating fuels, notably fuel oil and natural and liquefied petroleum (LP) gas. The trends in solid-fuel heating are moving away from radiant heaters toward central heating systems.

While some analysts predict a second coal age, one must remember that society paid dearly in human and environmental terms for the first one. During the approximately 100 years in which coal has been in active use as a fuel source in the United States, well over 1 million miners have been disabled, additional tens of thousands have died, and unknown numbers continue to suffer from diseases related to the extraction of "black gold." It appears that the future of coal use is heavily dependent on the availability of petroleum and gas, environmental restrictions relating to mining operations and land reclamation procedures, and prices of other fuels relative to coal.

Although there remains a great deal of uncertainty as to the future availability of fossil fuels, an equally important factor for future availability is the trend of increasing rather than static energy consumption patterns. Energy consumption in the United States is allocated among five major sectors. The existing energy use within each sector, with approximate percentages of total consumption, is illustrated in Figure 1.5. The electric utility industry has historically been the largest single user of energy within the United States, dwarfing other sectors of the economy by comparison. Prior to the advent of the energy crisis in the early 1970s, total U.S. energy consumption was projected to increase at an annual rate in excess of 7%. This figure was based on past energy consumption and growth patterns. However, as a result of problems arising from fuel availability plus successful conservation efforts under-

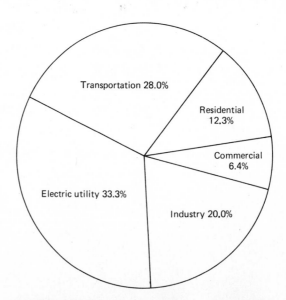

FIGURE 1.5 Percent of energy use by sector. Electric utilities and transportation comprise almost two-thirds of our energy usage. (Data from New York State Energy Office.)

taken throughout the 1970s, this increase was reduced to approximately 3% per year. As admirable as the achievement of this lowered growth figure is, assuming an optimistic projection of fossil fuel availability, Americans are still left with a questionable amount of readily available reserves of affordable fossil fuels.

An uncompromising fact of nature reveals that each barrel of oil extracted from the earth makes each successive barrel more expensive and difficult to retrieve. Given the unstable nature of politics and economics which influence the extraction and use of primary energy sources, educated guesses and computer-assisted scenarios as to future fuel availability tend to be difficult at best, and often inaccurate.

Before any renewable energy technology can be applied, an examination must be made as to what the nature of the energy end-use consumption pattern is. End-use energy consumption patterns within the residential, commercial, industrial, and transportation sectors are illustrated in Figure 1.6. Of the four sectors listed in Figure 1.6, the residential area can be most directly influenced by the individual person, in terms of total energy consumed and the type of fuel used to perform specific tasks. Figure 1.6 reveals the energy consumption patterns of an average home located in the northeastern United States. Although costs of

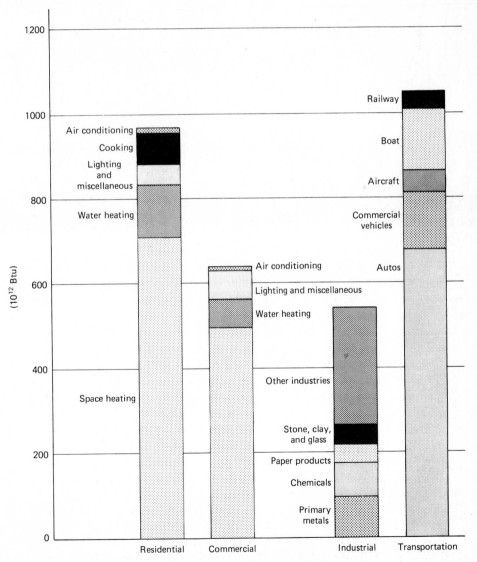

FIGURE 1.6 Energy end-use consumption, 1979. (Data from New York State Energy Office.)

heating and cooling homes vary widely over the United States, space and water heating often account for nearly 80% of overall residential energy expenditures.

The operation of common electric appliances other than electric space and water heating accounts for a relatively small proportion of the energy dollars spent by the average household. One traditional problem associated with home energy costs has always been the mismatch of the primary energy source and end use. For example, in using electricity for space heating, a high-grade energy source, electricity, is wastefully directed through a resistance heating element in a unit baseboard heater to raise the temperature of a room from 30°F to 70°F. Thus fuel burning in a utility power plant in excess of 1000°F is utilized to raise the interior temperature level of a dwelling by 40°F. This represents a substantial mismatch of primary energy source and end use. Similar correlations can be made for other fuels and energy sources, but the result is the same: inefficient use of the primary fuel source, making it unavailable for other, more efficient uses. For example, electricity that is used for interior space heating is unavailable for use in electric motors or communications equipment, both of which make highly efficient use of electrical energy. Low-grade heat sources, such as solar collectors, are ideally suited for end-use tasks such as space and water heating, which do not require high temperatures for successful operation. It is important to note that most solar space heating and domestic hot water installations are not sized to provide 100% of the required heat or hot water, for to do so would make the installation far too expensive to ever be economical. Rather, solar systems are commonly designed to provide sizable fractions, generally between 60 and 80%, of the space heating or hot water requirements. Sizing solar systems in this way maximizes system output and minimizes cost. Most solar installations rely on a standard hot water heater or central heating unit for backup energy during cloudy periods or when space heat and hot water demands are in excess of solar system capability.

Prior to examining the design features of solar systems which allow their interface with existing space heating, hot water, and pool systems, we will examine more closely what solar energy is and some of the factors that influence its availability.

THE SUN AS A SOURCE OF POWER

All the energy that strikes the surface of the earth from outer space can be classified either as *direct solar energy* (from the earth's sun) or *nonsolar energy*. Nonsolar energy comes in such forms as radiant energy from other stars, meteors passing within range of the earth's gravitational pull, and background cosmic radiation. Nonsolar energy accounts for a relatively small percentage of the energy that reaches the earth. Research continues into the origins and explanation of nonsolar energy. The vast majority of energy falling on our planet comes from the sun. This energy takes the form of electromagnetic energy: oscillating and vibrating electrical and magnetic fields.

Electromagnetic Energy

Electromagnetic energy has two principal features: frequency and wavelength. *Frequency* is the number of times per second that a particular type of energy form repeats a specific type of behavior, and is illustrated in Figure 1.7. This phenomenon is similar to the pitch of notes from a guitar string. As the frequency with which the string vibrates is increased, the note becomes higher and the time interval between successive oscillations decreases. As the frequency of the electromagnetic oscillations increase, the energy level of the wave increases. Thus high-frequency x-rays and gamma rays can penetrate many substances, including human tissue, whereas low-frequency radio waves are merely reflected from most surfaces.

Frequency also influences the second characteristic of electromagnetic energy: wavelength. *Wavelength* is the distance between repetitions of wave shapes, illustrated in Figure 1.8. Note that as the frequency increases, the wavelength decreases. The nature of the source of the electromagnetic waves determines the frequency of the oscillations. Thus different atomic structures exhibit different frequencies of vibration by which they are characterized. This wide variety of electromagnetic energy can be broken down into a spectrum: an array of characteristics that are associated with energy of different frequencies and wavelengths.

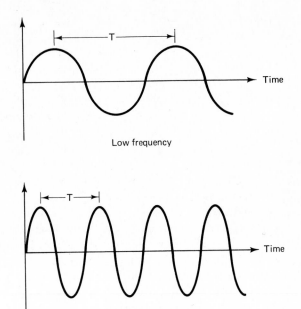

FIGURE 1.7 Wave frequency. Frequency is the number of repetitions each second that occur from a given material. These repetitions are sometimes referred to as oscillations.

The electromagnetic spectrum is illustrated in Figure 1.9. Note that visible light occupies a relatively small portion of the spectrum.

The Solar Constant

The amount of electromagnetic energy emitted by the earth's sun is enormous and radiates out from the sun equally in all directions. The earth captures a relatively small por-

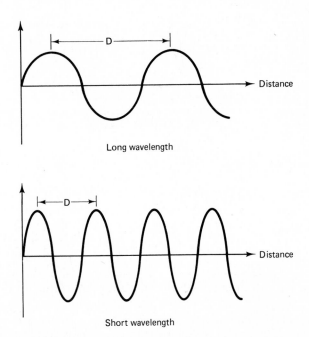

FIGURE 1.8 Wavelength. Wavelength is the distance between exact repetitions of shape, measured in meters.

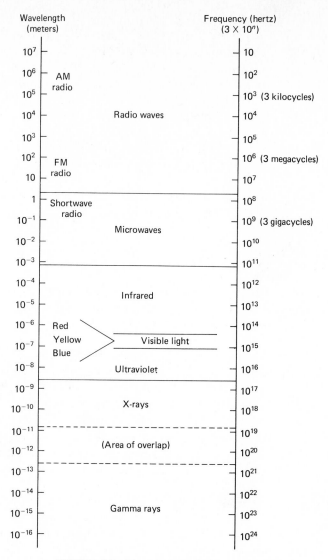

FIGURE 1.9 The electromagnetic spectrum.

tion of this energy. The energy produced by the sun is the result of a complex series of nuclear reactions known as *fusion*. In the nuclear fusion process, hydrogen atoms within the sun combine to form helium. These reactions take place at extremely high temperatures, releasing vast quantities of energy into space. This energy must travel approximately 93 million miles before it reaches the outer surface of the earth's atmosphere. At this point, the energy intensity is referred to as the *solar constant*. The solar constant refers to the measurement of the amount of energy falling on an imaginary area of space directly outside the earth's atmosphere. The amount of energy available at this point in space is 428 Btu per hour per square foot. Not all of this energy reaches the surface of the earth, however. The atmosphere of the earth greatly modifies this incoming energy, constantly absorbing, dispersing, and reflecting it as it travels, causing all the climactic characteristics with which we are familiar: rain, wind, and global temperature fluctuations. In addition to atmospheric effects, a percentage of the incoming solar energy is directly reflected back into outer space without affecting the earth's atmosphere. This energy is reflected off clouds, atmospheric dust particles, the polar ice caps, and the surface of the earth's land masses. This reflection is known as *albedo*. The earth's albedo, measured by satellites, has been shown to be approximately 30% of the solar constant. Thus solar energy can follow one of three pathways as it approaches our planet: It can be directly reflected back into outer space as a function of the earth's albedo; it can be absorbed within the atmosphere to power climatic conditions; and it can be absorbed on the surfaces of the earth, powering the biological processes of photosynthesis. Following these three energy pathways creates a very delicate balance within the earth's ecosystem. Changes that affect either the albedo or the makeup of the atmosphere can have far-reaching effects on life on our planet. Such changes could alter prevailing weather and temperature patterns, as well as the level of the oceans, for example. One of the great challenges for both present and future generations will be to maintain the delicate balance of the ecosystem that has evolved during the past several million years in order to ensure the future survival of our planet.

Understanding the nature of solar energy is part of the background necessary for proper solar system utilization. In addition, there are basic laws of physics that must be understood as well: laws that govern all energy systems. A discussion of these basic principles and laws of physics follows.

FUNDAMENTALS OF ENERGY CONVERSION AND HEAT TRANSFER

Laws of Thermodynamics

Humankind has progressed from using relatively unsophisticated energy sources to using complex energy systems that are vast both in the magnitude of the energy they produce and in the fuel they consume. No matter how simple or complex the energy system, it is subject to specific laws of physics that govern its behavior and energy availability. The laws of physics that relate most directly to energy use and production, known as the *laws of thermodynamics,* explain the behavior of all energy forms in which heat is produced.

The First Law of Thermodynamics

The first law of thermodynamics considers the *quantity* of available energy within a system and is sometimes referred to as the *principle of conservation of energy.* It states that the total amount of energy in a closed system is constant. Although energy can be changed from one form into another, the net quantity of energy available remains the same. In this process, energy is neither created nor destroyed, but merely changed from one form to another. Consider, for example, the automobile as an energy system; application of the first law of thermodynamics states that the automobile will transform gasoline into usable mechanical energy. Heat will be given off as a by-product of moving the automobile either forward or backward. In this process, no energy has been consumed. It has merely changed in form. The gasoline has been fed into an internal combustion engine, and the chemical energy potential within the gasoline has been changed into mechanical energy. Burning the gasoline produces an explosive force that rotates the crankshaft of the automobile engine, and this rotating motion is transmitted to the wheels of the vehicle. Heat has been emitted to the atmosphere as a by-product of this energy transformation process. The automobile has consumed nothing; it has merely changed the energy that is available from one form to another.

The Second Law of Thermodynamics and Entropy

The second law of thermodynamics considers the *quality* of energy that is available in any system and introduces the term *entropy.* The second law states that heat, when left to itself, always flows from hot to cold. To change this natural direction and make heat flow from cold to hot requires that additional work be applied to the system and waste heat will always be produced as a by-product of the work performed. As a result of this waste heat, less useful energy is available at the conclusion of the task than was available at the start. No system, regardless of the complexity of its structure or the genius of its design, can be 100% efficient. The efficiency of the modern automobile illustrates this principle most clearly. The thermodynamic efficiency of a modern internal combustion engine is approximately 25%; 75% of the energy available in the fuel is lost to the cooling system of the engine or is used in overcoming internal friction between moving parts. If one considers the additional efficiency losses inherent in the drive train and transmission, and the frictional loss through the tires, the overall efficiency of a modern automobile is approximately 10%. In practical terms this means that 10 cents of every dollar spent for fuel is actually used to push the car forward; the remaining 90 cents is used to overcome heat and friction losses. The concept of entropy relates to the phenomenon of waste heat and available energy.

Entropy is a term that describes randomness in a physical system. The entire universe, from an energy point of view, can be considered as a landscape made up of peaks and valleys (Figure 1.10). The peaks represent the existence of high-quality energy, such as crude oil and natural gas, and the valleys represent low-quality energy, in essence waste heat. Human beings are constantly using up the peaks of available high-quality energy to perform a variety of tasks, and thus filling in the low-quality energy valleys. The valleys represent mainly waste heat, energy that is unavailable for further use. Waste heat is randomly spread out through the universe. Thus as we consume high-quality energy sources, the entropy or randomness of a system increases as a result. The second law of thermodynamics can be expressed in terms of entropy in the following way:

> The entropy of a system will always increase or remain constant, but can never decrease.

To gain a fuller understanding of the concept of entropy, let's examine some situations in which entropy changes as a result of work applied to a system. Assume that a farmer is going to plant a small field of corn. The field in its natural state has numerous rocks which are randomly distributed across its surface. In order to plant the field, the rocks must be removed and placed along the hedge row. Whether all the rocks are moved by hand or by machine, energy must be expended. If a bulldozer is used to remove the rocks, gasoline, a low-entropy, high-quality fuel, is burned in the internal combustion engine of the bulldozer. The gasoline, in turn, produces waste heat. The waste heat from the engine quickly dissipates into the atmosphere and once dissipated is no longer available for use. Clearing the field by hand would make no difference in terms of entropy. Food must be consumed to provide energy for the manual labor required to clear the field. The food consumed requires energy and fertilizer to make it grow. Waste heat is given off as a result of the oxidation of the carbohydrates and sugars within the body as the muscles work to remove the rocks. Although we decrease the entropy or randomness of the field by removing the scattered rocks and placing them uniformly along one boundary wall, we actually create net increasing entropy for the system and environment as a result of our work, whether done by hand or by machine.

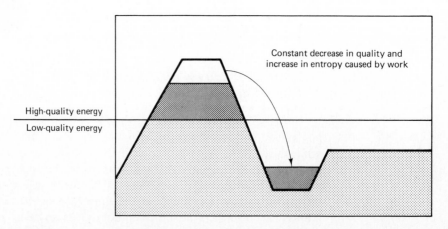

FIGURE 1.10 Energy availability. Humankind is constantly consuming the high-quality energy which is limited in supply.

Close examination of this principle reveals that the entropy of the universe is constantly increasing. In the far-distant future, when the universe reaches the highest state of entropy, all energy will be one homogeneous mixture, and all mass will be at the same temperature, a few degrees above absolute zero. With no available energy, work will be impossible and life as we know it will cease to exist. This point is referred to as the *heat death* of the universe. Although total entropy is many billions of years in the future and no concern at the present time, the concept of entropy is valuable, for it illustrates that each gallon of oil or each cubic foot of natural gas consumed makes each remaining gallon or cubic foot that much more valuable. As the consumption of high-quality energy sources continues at ever-increasing rates, entropy increases and so do the costs of remaining low-entropy fuel sources.

The laws of thermodynamics explain the basic nature of heat and outline the overall efficiency limits that govern most of the processes with which we are familiar. Heat moves within energy systems in a number of distinct ways, and it is now time to examine some of these methods.

Radiation, Conduction, and Convection

Heat is transferred from one area, or object, to another in three fundamentally distinct ways, known as radiation, conduction, and convection. Although each of these methods is mutually exclusive, an operating solar system often utilizes all three methods simultaneously and a basic understanding of each is required for the proper sizing and design of a solar heating system. Figure 1.11 illustrates the areas within a solar heating system where each of the three types of heat transfer occur.

Radiation

Radiation (Figure 1.12) is the transfer of heat by the transmission of electromagnetic energy. Radiant energy moves through space in a manner similar to waves moving over the

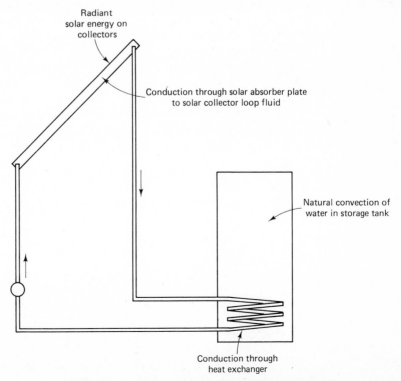

FIGURE 1.11 Forms of heat transfer in a solar system.

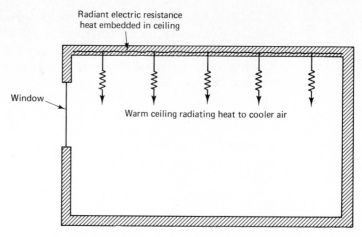

FIGURE 1.12   Transfer of heat by radiation.

surface of a pond after a stone has been dropped into the water. The amount of radiant heat emitted from a surface depends not only on the temperature of the object, but on the nature of the surface of the object as well. White, reflective, and shiny surfaces are poor absorbers of radiant heat. Also, they are poor emitters of radiant heat because the heat tends to be reflected back into the object with the shiny surface. On the other hand, black, dull, or nonreflective surfaces are good absorbers and emitters of radiant energy.

Solar collectors gather radiant energy. To absorb the maximum amount of this radiant energy, solar collectors should have a dark, nonreflective surface. A solar storage tank, which is to be placed in a cool unheated basement area, should have a white reflective surface that tends not to emit the heat contained in the solar-heated water.

Conduction

The transfer of heat by conduction requires that two objects be in direct contact with one another (Figure 1.13). Heat will flow from the hotter object to the cooler object at the point where the surfaces of the two objects touch. Objects that are good conductors of electricity, such as copper and silver, are also excellent conductors of heat because many excess electrons are available within the atomic structure of these materials to transport heat. Heat is sometimes referred to as random kinetic energy. The nature of conduction can be illustrated by the action that takes place when one plays a game of pool. When the cue ball collides with other balls on the table, the energy in the cue ball is transferred to the other balls through random collisions on the table. The amount of energy transferred during these collisions depends on the force and angle of the striking cue ball, and the number of other balls on the table. Similarly, the speed at which heat is transferred through any material is depen-

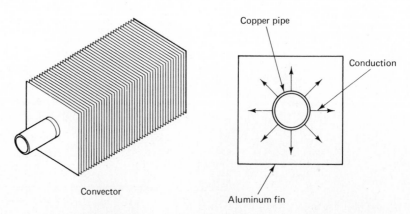

FIGURE 1.13   Transfer by heat by conduction.

dent on the *thermal conductivity* of that substance. Materials differ significantly in their thermal conductivity. They may be classified as *insulators* (such as urethane foam, fiberglass, or other materials which are poor conductors of heat and therefore good insulators) or as *conductors* (such as copper or aluminum). The quantity of heat that flows through any material is dependent on its thermal conductivity, time, the surface area of the material, the thickness of the material, and the temperature difference between the materials in contact. The general equation which describes the flow of heat by conduction is

$$Q = KAT\frac{\Delta t}{d} \tag{1.1}$$

where Q = amount of heat conducted
K = thermal conductivity constant given for that material
A = area of the conducting surface
T = time of conductance
Δt = temperature difference between the two surfaces in contact
d = thickness of the conducting material

The thermal conductivity, K, can be expressed in Btu per inch per square foot per hour per degree Fahrenheit (Btu/in. $\cdot$ ft^2 $\cdot$ hr $\cdot$ °F). K is different for each substance. It has become common to speak in terms of a material's R value instead of its K value. *R values* are the mathematical reciprocals of K values and they describe the ability of a material to resist the flow of heat. A material with an R value of 10 will resist the flow of heat twice as effectively as a material with an R value of only 5. K values, on the other hand, describe the ability of a material to conduct heat rather than to retard it. The heat conduction equation, using R values instead of K values, is stated as follows:

$$Q = \frac{AT\,\Delta t}{R \text{ for thickness considered}} \tag{1.2}$$

Since we normally consider heat conduction in Btu per hour, and R is already stated in terms of hours, it can be said that

$$Q = \frac{A\,\Delta t}{R} \tag{1.3}$$

To illustrate the use of equation (1.3) for performing a heat loss analysis in a typical home, let's use a wall with a total surface area of 100 ft^2. The interior room temperature is 70°F and the outside air temperature is 0°F. The wall, insulated with 3½ in. of fiberglass, has a total R value of 15. Using equation (1.3) the wall will have a heat loss to the outside air of 467 Btu per hour, calculated as follows:

$$100 \text{ ft}^2 \times (70°F - 0°F) \times {}^1/_{15} = 467 \text{ Btu per hour}$$

Table 1.1 lists the R values of common insulating and building materials. As shown in Figure 1.11, heat transfer by conduction is crucial to the efficient operation of a solar heating system. The radiant solar energy that strikes the solar collectors must be conducted into the fluid within the collector for efficient removal of the energy from the absorber of the collector. The materials chosen for the construction of the solar collector absorber plates should be excellent conductors of heat. Another critical factor for efficient heat conduction within a solar system is the proper design and construction of the heat exchangers. These devices, located either within the solar storage tank or in external units, enable quick transfer of heat from the collector-loop fluid to the storage water.

Convection

Convection is a method of transferring heat through the movement of a fluid. The fluid can be either a liquid or a gas. One of two types of convection is used in most heating

Table 1.1 *R* VALUES OF COMMON INSULATION
AND BUILDING MATERIALS

Material Description	R Value, per Inch
Fiberglass insulation	3.7
Cellulose insulation	3.7
Vermiculite	3.5
Polystyrene	4.0
Urethane foam	7.5
Building brick	0.20
Wood siding	0.85
Concrete block	0.20
Asphalt shingles	0.40 (common thickness)
Wood shingles	0.94 (common thickness)
Insulating sheathing board	2.64
Gypsum board (sheetrock)	0.45 (standard ½ in.)
Plywood	0.63 (standard ½ in.)
Plaster	0.20
Cement	0.20
Ceiling tile	1.20 (standard ½ in.)
Solid wood door	2.00 (standard 1¾ in.)
Combination storm windows	1.80

systems: *natural convection* or *forced convection*. An example of natural convection is the flow of air created by radiators in a home heating system. As air molecules within the room are heated by coming into contact with the hot radiators, the air becomes less dense and rises to the ceiling. Cooler air from below moves toward the radiator to replace the warm air that has risen. Thus a natural convection airflow is established as the warm air is constantly rising and the cool air is moving in below to take its place (Figure 1.14).

A natural convection system is also established within the solar water heater in Figure 1.11. The collector-loop fluid inside the heat exchanger transfers heat gathered from the sun by conduction to the potable water surrounding the heat exchanger at the bottom of the tank. This warm water rises to the top of the tank and cooler water falls to the bottom to replace it. In the examples of the room radiator and storage tank, movement of air and water is restricted to natural convection. More efficient heat transfer can take place, however, if

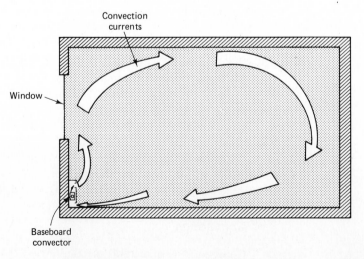

FIGURE 1.14 Transfer of heat by convection.

the convection is forced, involving movement of the fluid to the desired point in a heating system using a fan, pump, or similar device. This use of circulating pumps in a solar collector loop to move the fluid within the system, and fans to move heated air through a duct system in a warm air heating application, are examples of forced convection. Forced convection increases the efficiency of the heat transfer process because the fluids give up their heat to the surrounding cooler surfaces more rapidly than by natural means. This reduces the temperatures of the fluids in the system, resulting in larger temperature differentials. The larger the temperature differential between the solar absorber plate and the collector fluid within, the greater will be the exchange and transfer of heat between these two surfaces.

The solar engineer should design a system that utilizes heat transfer techniques that keep the collectors operating at as cool a temperature as possible. The cooler the solar collectors are relative to the temperature of the ambient air, the more efficient will be the collection and transfer of energy. Solar collectors that operate at excessively high temperatures indicate that much of the heat is being lost to the atmosphere rather than being transferred to storage. Proper convection, conduction, and radiation characteristics in a solar system will allow the collectors to operate in the range 5 to 10°F warmer than the storage medium of the system.

UNITS OF ENERGY: THE BTU

In accordance with the laws of thermodynamics, energy can be neither created nor destroyed, merely changed from one form to another. It is this convertible nature of energy that allows us to determine which are the most economical fuels to use in order to achieve specific objectives. For example, a gallon of fuel oil can be compared to a *therm* (100 ft^3) of natural gas on an energy content basis. If the same unit of energy is used for both fuels and the cost of each of the fuels is known, the cost per unit of energy comparison can be determined. Prior to undertaking such an analysis, one must understand the basic unit of energy, the Btu.

Water, the most common substance on the surface of our planet, provides a logical basis for measuring units of energy. A Btu, or British thermal unit, combines two methods of measurement, temperature and weight. One Btu is the amount of energy required to raise the temperature of 1 pound of water 1 degree Fahrenheit. Conversely, 1 Btu is released from 1 pound of water when it cools 1 degree Fahrenheit.

The Btu is the standard unit of energy used by architects, heating contractors, builders, and heating appliance manufacturers. It is mandatory that people who wish to become familiar with solar energy systems be knowledgeable about Btu figures. The United States is undergoing metrication, a change from the British system of measurement to the international metric system. The metric counterpart of the Btu is the calorie. One calorie is the amount of heat necessary to raise 1 kilogram of water 1 degree Celsius. One calorie equals 3.97 Btu. Therefore, the metric system can also be used to determine heat energy.

This basic unit of energy facilitates an understanding of conversion tables that enable cost comparisons between various fuels to be made. When undertaking such analyses, the conversion factors in Table 1.2 should be used. Let's assume that the homeowner is presently using electricity for space heating, with a cost per kilowatthour (kWh) of 10 cents. The homeowner is trying to decide whether or not to install a central wood heating boiler in the home. Wood is available at $125 per cord. The current delivered price of electricity is $29.30 per million Btu. The cost of the same amount of energy supplied by a wood boiler would be $10.41 assuming a delivery of mixed hardwoods, a savings of approximately 60% on the costs of heating the home using wood rather than electric resistance heating. Similar calculations can be made for other types of fuels and for domestic hot water production in addition to space heating. It should be noted that the figures shown do not include the purchase or installation costs of the various appliances, which should be added to the fuel costs to derive figures that more appropriately represent the full cost of each method investigated. The conversion table is useful in making any type of energy decision and is of primary importance when designing new or retrofitted solar systems.

Table 1.2   CONVERSION FACTORS AND BTU EQUIVALENTS OF COMMON FUELS

Fuel Type	Description
Natural gas	One therm = 100,000 Btu = 100 ft^2 Assumed efficiency of unit = 75% $/million Btu = 13.33 × $/therm
Fuel oil	138,000 Btu/gallon Assumed efficiency of unit = 65% $/million Btu = 11.15 × $/gallon
LP gas (propane)	93,000 Btu/gallon Assumed efficiency of unit = 75% $/million Btu = 14.34 × $/gallon
Electricity	3412 Btu/kilowatthour (kWh) Assumed efficiency of unit = 95% $/million Btu = 293 × ¢/kWh
Mixed hardwoods	24 million Btu/cord One cord = 128 ft^3/90 ft^3 wood Assumed efficiency of unit = 50% $/million Btu = $/cord ÷ 12
Mixed softwoods	15 million Btu/cord One cord = 128 ft^3/90 ft^3 wood Assumed efficiency of unit = 50% $/million Btu = $/cord ÷ 7.5
Coal	12,500 Btu/pound Assumed efficiency of unit = 60% $/million Btu = $/ton ÷ 15

SOURCE: *Adapted from* Heating with Coal *by John W. Bartok, Garden Way Publishing Co., Charlotte, Vt., 1980.*

THE SOLAR FUTURE

Increasing use of solar energy systems will most certainly bring about a decrease in the use of fossil fuels. Due to the finite nature of fossil-fuel resources, solar energy will take on an increasingly important role as a transitional energy resource. Fossil fuels will be diverted away from space heating and domestic hot water heating and applied to more important purposes: those of agricultural, medicine, and pharmaceuticals production.

Solar energy, used to provide domestic hot water, space heating, and swimming pool heating in residential and commercial applications, will continue to grow. The solar market is in its infancy, and it seems likely that market penetration will continue to be directly linked to the price and availability of oil, natural gas, solid fuels, and tax incentives for some time to come.

Domestic hot water systems will be the most widely used solar technology. Operating on a yearly basis, the economics of these systems are excellent. In some areas of the United States, peak-electricity rates are in excess of 20.0 cents/kWh. The *payback* period for a solar domestic hot water system, the time it takes for the money that a person saves by virtue of installation of the solar system to recover initial system cost, can be as little as 2 to 3 years, with a return on investment of close to 50%. Manufactured from off-the-shelf plumbing components and time-proven flat-plate collectors, well-designed domestic hot water systems can be relied on to provide a majority of the domestic hot water requirements for the homes and buildings in which they are installed.

Furthermore, interfacing solar hot water systems with existing fossil-fuel hot water systems in most residential and commercial buildings is relatively simple. As a result of these factors, the use of solar hot water systems can be expected to grow considerably within the foreseeable future.

Tapping solar energy for residential and commerical space heating is slightly more complicated than domestic hot water production. Engineering of solar space heating systems requires knowledge of hydronics and heat exchange technology if the system is to be interfaced successfully with the existing fossil-fuel heating systems. Considerable experience has been gained during the recent past relating to the design and operation of solar space heating systems. Solar space heating systems cost considerably more than their domestic hot water counterparts due to the increased number of solar collectors and other system components required. Backup domestic hot water systems are relatively inexpensive, but backup space heating systems are not. Paying for two heating systems, one solar and one backup, can place considerable economic burdens on the home or business owner. In addition, most space heating systems are in use for only 5 to 8 months per year. Because protracted use of any energy system prolongs the payback and reduces the return on investment, most consumers find it desirable to combine solar space heating with either domestic hot water or pool heating systems in order to make practical use of the solar system on a year-round basis whenever possible.

One must remember that space heating accounts for the majority of energy dollar outlays in many areas of the country regardless of the fuel or system used to provide the heat. For this reason, solar space heating systems will continue to grow in popularity and in market penetration, albeit at a slower pace than their domestic hot water system counterparts.

In many instances, due to the modularized nature of solar system components, domestic hot water systems can often be upgraded in size and complexity to take on space heating functions. When this is borne in mind, the investment in a solar domestic hot water system becomes a down payment on a larger, more complex system with vastly increased capabilities. Much will need to be done to provide consumers and installers with upgraded engineering skills, and better training will undoubtedly be available within the solar industry in the coming years.

The use of solar swimming pool heating systems has grown dramatically during the past few years, and this growth can be expected to continue. Impressive improvements have been made in these systems during this time, resulting in solar pool heating systems that are both inexpensive and reliable. Pool owners can choose between self-contained solar units for heating only the swimming pool, and systems that contain space heating capabilities as well. Self-contained solar pool heating systems are relatively inexpensive and generally employ unglazed and uninsulated solar collectors. Solar pool heating systems, depending on geographical location, are designed either to extend the swimming season during the summer months, as in the northeastern United States, or to heat the pool on a year-round basis, as in California or Florida. Pool heating systems dependent on fossil fuels such as oil, propane, or natural gas consume large quantities of energy and can be expensive to operate. Solar pool heating systems can replace a major portion of this energy. Insulating blankets are generally used in addition to the solar system to prevent evaporation and radiation of heat from the swimming pool to the atmosphere during inclement weather and evening hours.

The political implications of energy production and consumption are numerous. As solar power begins to assume an increasing portion of America's energy requirements, political complications are bound to increase. The energy fabric of society is such that the future energy mix will probably be made up of a number of energy sources working in harmony, rather than the one or two dominant fuel systems experienced during the past 75 years. The increasing use of renewable energy sources should provide downward pressure on traditional fuel pricing structures. Also, as the depletion of traditional fossil fuel resources continues, solar and other renewable energy systems will become necessities rather than alternatives. Through planning and education, rational energy choices can be made that will result in an energy future with which we can all live.

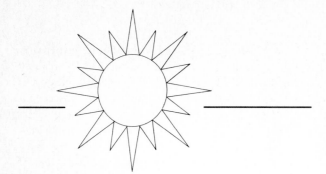

2
Solar Collector Design Fundamentals

The flat-plate collector is the most widely used and economical method for collecting solar energy. Solar collectors are designed to heat either air or liquid. This heated fluid then transfers the collected solar energy to storage for later use to provide either domestic hot water, process heat, or space heating. The operating temperature range of these systems is between 40 and 200°F. The amount of storage and the number of collectors used in each system varies, based on both technical and economic considerations. Since the flat-plate collector forms the core of any operating solar system, we will now examine the design and construction features of these devices.

BASIC COLLECTOR COMPONENTS

The basic components of a solar flat-plate collector are illustrated in Figure 2.1. Shown are a supporting box, a transparent glazing cover, an absorber plate, and the collector insulation. These components are designed and arranged so as to provide for the greatest amount of solar energy absorption while minimizing heat loss from the unit. A well-designed collector must operate efficiently over a wide range of weather and temperature conditions, yielding working fluids in the range 80 to 180°F. The given site and specific heating application will sometimes dictate the precise configuration of each of the collector components. Regardless of component modifications from one unit to another, all collectors must be designed to overcome heat loss associated with high-temperature operation in order to yield a cost-effective installation.

COLLECTOR HEAT LOSS

Reducing heat loss in a solar collector is a complex problem, since all the components within the collector box are prone to lose heat at the elevated temperature ranges usually encountered in collector operation.

Heat loss from the absorber plate occurs as it heats up under the influence of incoming solar radiation. As the temperature of the absorber increases, heat that is not transferred to the circulating fluid within the absorber flow passages (either air or liquid) tends to be lost by conduction to the collector box components. These losses occur due to insufficient or unbalanced fluid flow within the circulating passages of the absorber plate. In addition, improper tube spacing on the absorber can cause hot spots. Losses from the absorber can occur due to shadows cast by the collector frame, effectively cooling the absorber plate perimeter. In addition, as the temperature of the absorber increases, greater amounts of energy are lost by conduction and convection to air over the absorber, and by reradiation of infra-

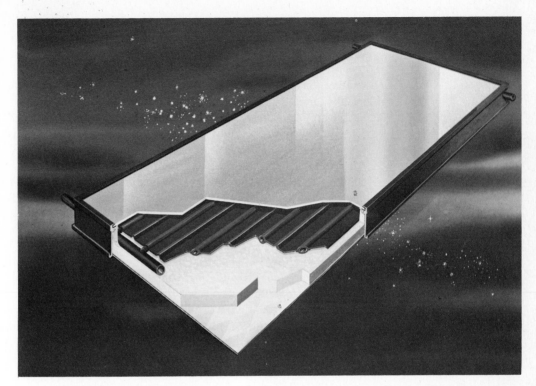

FIGURE 2.1 Flat-plate collector cross section. (Courtesy of American Solar King.)

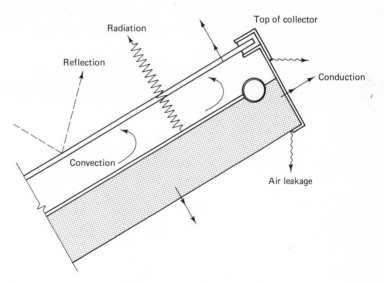

FIGURE 2.2 Collector heat loss.

red energy from the absorber surface. Figure 2.2 illustrates the variety of ways in which heat is lost from a standard flat-plate collector. Each subcomponent in the collector is designed to minimize the heat losses illustrated and will be examined next in greater detail.

COLLECTOR FRAME: MATERIALS OF CONSTRUCTION

The collector box provides the necessary support structure for all the collector components. The frame must be capable of supporting the weight of the finished collector assembly with no deformation of shape, and to withstand winds in excess of 100 mph. Most commercial collector frames are fabricated from extruded aluminum. Aluminum is an ideal material for this purpose: It is strong, corrosion resistant, and not affected by elevated temperatures within the operating range of most flat-plate collectors. Aluminum can be extruded into a variety of complex shapes. Figure 2.3 illustrates the extrusion process.

Aluminum ingots are fed through a series of dies which are the negative shape of the required finished stock. After extruding the shape, the aluminum is straightened and cut to lengths for ease in handling.

Many homemade collector boxes are fabricated from wood, for obvious economic reasons; however, wood is not a suitable material for the majority of solar collector applications. Reasons for this are: lack of resistance to decay over extended periods of time, relatively low temperatures of ignition, and lack of dimensional stability under conditions of constantly changing temperature and humidity. Although lumber is available that is pressure treated with chemicals to resist corrosion and decay, the other disadvantages remain. Steel is sometimes used for collector boxes, but it will oxidize over long periods of time. Note from Figure 2.1 that the design of the collector box extrusion incorporates space for insulation, as well as support for the absorber plate and cover glazing. These extrusions should be designed to allow the absorber plate to be mounted and fully supported within the enclosure, yet not be in direct thermal contact with the outer frame of the collector box. In poorly designed mounting extrusions, the absorber plate is in direct contact with the outer collector frame, resulting in significant heat loss from the absorber (Figure 2.4).

Although most flat-plate collectors incorporate one glazing layer, the extrusions are easily modified to accommodate additional layers if called for. Just as the absorber is insulated from direct contact with the outer collector frame, the glazing, too, should not come into direct contact with the frame. This is accomplished by use of special gaskets made from ethylene propylene diene monomer (EPDM) rubber.

Backing plate

The backing sheet or plate serves a number of functions in the collector: It provides rigid support for the collector insulation, helps to seal out moisture and debris, allows for ease in handling during installation, and provides for a pleasing aesthetic appearance of the panel should the back of the unit be visible after installation (which is often the case when collectors are installed on roof racks). Aluminum sheet is generally used for this purpose.

FIGURE 2.3 Extruding aluminum collector components. (Courtesy of Warner Mfg.)

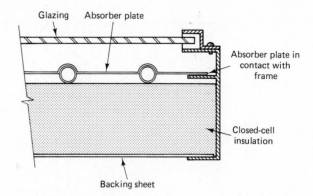

FIGURE 2.4 Direct contact of absorber plate with collector frame. This type of collector construction allows for large heat losses from the absorber plate through the collector frame.

Fasteners

All fastening devices used in conjunction with other collector components should be made from stainless steel. The majority of these fasteners are used to secure the glazing battens to the collector box and for the absorber plate and corner miter fastening.

Mounting flanges

The collector usually incorporates into the frame design a method for securing the collector to the roof or to rack supports. An integral mounting flange which is part of the frame extrusion is illustrated in Figure 2.5. A modification of the integral mounting flange is the use of auxiliary mounting clips that fasten to mounting studs inserted into recesses in the collector frame extrusion (Figure 2.6). This type of mounting system allows for the use of many different types of interchangeable hold-down clips. Also the position of these clips is adjustable, a feature not usually found with integral mounting flanges. Both flush roof mounting and elevated installation procedures can be followed for either type of mounting arrangement illustrated in Figures 2.5 and 2.6.

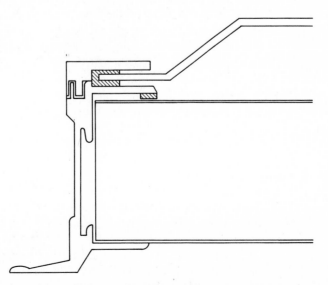

FIGURE 2.5 Integral mounting flange designed into collector extrusion. (Courtesy of Heliotherm.)

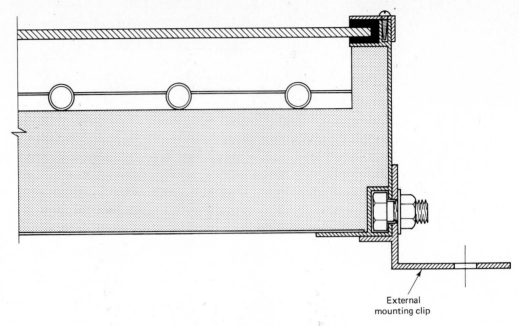

FIGURE 2.6 External mounting clips.

ABSORBER PLATE   The single most important component of the flat-plate collector is the absorber plate. This plate must be designed so as to simultaneously absorb incoming solar radiation and transfer this heat to the fluid circulating within the collector loop. The overall efficiency of the absorber is determined by the methods of manufacture and the materials used in its construction and the absorption and emission characteristics of the absorber surface. There are several different methods used to design and construct absorber plates.

Absorber Plate Configurations

Three of the principal types of absorber plate configurations are illustrated in Figure 2.7. The most efficient configuration of a solar absorber plate is the integral flow tube pattern. Construction of this type of absorber involves fusing two sheets of metal in which the flow tube pattern has been coated with a resist. After fusing, the flow pattern in pneumatically inflated. This results in the flow tubes being totally encapsulated between the two fused sheets of metal and a high degree of contact between the surface area of the absorber and fluid flow channels within. Also, integral flow absorbers possess great strength and durability, as well as the ability to perform under a wide range of temperature extremes without being subject to significant deformation. Intricate flow patterns can be achieved with this type of design, resulting in very high operating efficiencies.

The roll-form type of absorber, illustrated in Figure 2.8, is in widespread production. This fabrication technique involves a sheet of metal that is rolled and clamped around the absorber riser tubes. The space between the absorber sheet and the riser tube is usually filled with a high-temperature solder or similar heat-conducting material. Operating with a straight linear flow pattern, the thermal contact between riser and absorber in these systems is often over 120° and results in efficient heat transfer from absorber surface to heat transfer fluid. The use of dissimilar metals for the absorber sheet and flow tubes should be avoided due to corrosion that would result from electrolytic decomposition at the junction of the two metals.

The solder bond absorber configuration is the most easily constructed of all types of absorber plates, the factor which has lead to this type of design being widely used among do-it-yourselfers. The major drawback in this type of absorber plate configuration is the rel-

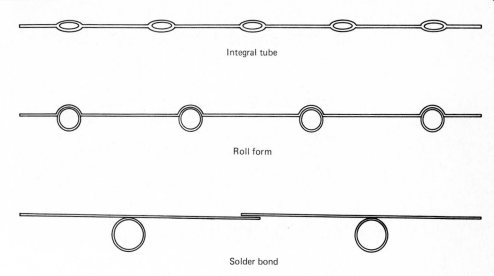

Integral tube

Roll form

Solder bond

FIGURE 2.7 Three principal types of absorber plate configuration.

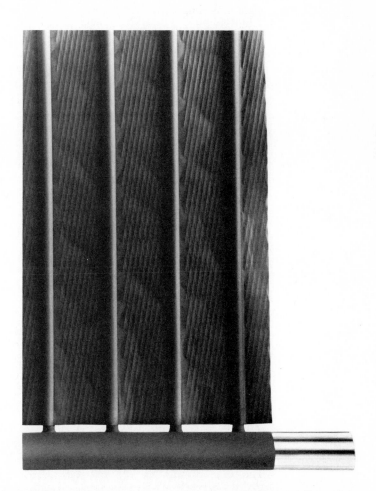

FIGURE 2.8 Roll-formed absorber plate. (Courtesy of Terra-Light Corp.)

atively poor thermal contact between absorber plate and flow tube passage, resulting in lower operating efficiencies than the other two configurations described.

The trickle-type absorber uses an absorber manufactured in a corrugated pattern. These corrugations are similar in appearance to corrugated aluminum and galvanized steel roofing materials. This shape forms external water flow passages in the depressions of the corrugations. As the water flows down the absorber, it picks up heat. A trickle-down collector system, while relatively inexpensive to construct, contains several inefficiencies in its mode of operation. Since free water is trapped between the absorber plate and the collector glazing, fogging of the interior glazing surface can occur. This happens due to the cyclical evaporation and condensation of the water from the absorber surface. Also, the life of the absorber plate can be severely restricted by accelerated corrosion if not constructed properly.

Materials of Construction

There are several different materials used in the construction of solar absorber plates: copper, steel, aluminum, EPDM rubber, and copolymer plastics. Of these, copper is used most frequently. Care must be exercised in the choice of absorber plate materials for compatibility of the solar system with existing plumbing components. For example, all-aluminum absorber plates must be isolated from the copper plumbing system in the home, since electrolytic corrosion will take place between these two highly active metals. In this instance, external heat exchangers must be used for transfer of the heat from the collector-fluid loop to the solar storage tanks.

Ethylene propylene diene monomer (EPDM) rubber absorber plates, while comparatively benign with respect to electrolytic corrosion, have poorer heat transfer capabilities than do either copper or aluminum. EPDM rubber absorbers are used most extensively in swimming pool heating applications, illustrated in Figure 2.9.

Recent developments in plastics polymer technology have witnessed the introduction of solar absorber panels made from these materials. These absorbers find their widest application in swimming pool collectors (Figure 2.10).

Copper is the most widely used of all materials for the manufacture of solar collector absorber plates. Its excellent heat conductivity, ease of forming, and compatibility with the existing plumbing and heating systems are some of the reasons for its widespread use.

FIGURE 2.9   EPDM rubber absorber plate. (Courtesy of Bio-Energy Systems.) (Besicorp.)

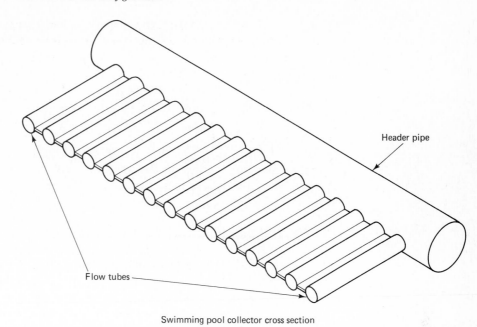

Header pipe

Flow tubes

Swimming pool collector cross section

FIGURE 2.10 Typical copolymer plastic absorber plate configuration. (Courtesy of Sun-Glo Solar, Ltd.)

FLUID FLOW PATTERNS AND CONFIGURATIONS

Liquid flow pattern arrangements used in flat-plate collectors are classified as either series or parallel. Each of these flow arrangements is illustrated in Figure 2.11. Of the two, the parallel flow configuration offers the greatest system efficiency. The length of travel of all fluid within the collector is short and of equal distance. The length of flow in a series configuration, however, is longer than in the parallel configuration. This longer flow path results in greater frictional and pressure-drop losses within the absorber, necessitating the use of larger circulators to keep the fluid moving at the required flow rate. In addition, a series flow pat-

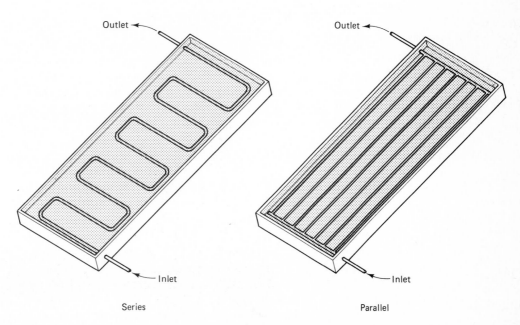

Outlet

Outlet

Inlet

Inlet

Series

Parallel

FIGURE 2.11 Series and parallel flow patterns.

Table 2.1 TUBE SPACING FOR MAXIMUM EFFICIENCY

Collector Plate Thickness (in.)	Collector Plate Material Tube Spacing (in.)		
	Copper	Aluminum	Steel
0.004	3.5	3.0	1.6
0.008	4.5	3.7	2.4
0.012	4.9	4.1	2.6
0.020	5.7	5.1	3.0
0.040	6.3	5.9	3.7
0.060		6.5	4.1
0.080		7.1	4.5
0.120			5.1

SOURCE: *Adapted from* Solar Energy Handbook *by AMETEK. Copyright 1979 by the authors. Reprinted with the permission of the publisher, Chilton Book Co., Radnor, Pa.*

tern can result in uneven heating on the surface of the absorber plate, resulting in reduced heat transfer efficiency.

The spacing of the absorber plate flow tubes is dependent on many factors: the diameter of the flow passages, the thickness of the absorber plate, the thermal conductivity of the absorber plate, and the insulation characteristics of the collector. Table 2.1 lists the most advantageous spacing arrangement for fluid flow tubes based on the thickness of the absorber plate material.

SELECTIVE COATINGS

Figure 2.2 illustrates the susceptibility of flat-plate collectors to the three basic types of heat loss: conduction, convection, and radiation. Proper design of the collector box, insulation materials, and glazing characteristics can help to reduce conduction and convective loss from the collector box. Radiation losses, however, are minimized by special coatings that are applied to the surface of the absorber plate.

The passive appearance of an operating solar collector belies the highly complex nature of the heat transfer reactions that take place on and within the absorber plate. In looking at how some of these reactions occur, let's assume that no fluid is circulating within the absorber and that the collector is in a stagnating condition. As incoming solar radiation strikes the surface of the absorber, infrared heat is produced. The temperature of the absorber plate begins to rise in response to the heat production until a state of equilibrium is reached. At this point the amount of energy absorbed by the collector is equal to the collector heat loss to the atmosphere. The temperature at equilibrium remains fairly constant, since the incoming and outgoing energy of the absorber plate is in balance. Although stagnation temperatures vary from one collector to another based on absorber configuration and collector efficiency, temperatures of between 400–450°F are not uncommon. The reradiation of energy to the atmosphere is referred to as *thermal emissivity*. The lower the emissivity of the absorber plate, the higher will be the operating temperature of the collector. It is desirable, therefore, to construct absorber plates from materials that possess low emissivity rates and high solar absorptance. The problem, however, is that most materials that possess low rates of emissivity are also relatively poor solar absorbers.

The most successful and economically feasible solution to this correlation of low emissivity and low absorptance has been the development of selective coatings that are applied to the surface of the absorber. Materials such as black chrome and black nickel are electroplated onto the absorber plate. These materials have solar absorptances in the area of

Table 2.2 SELECTIVE SURFACE MATERIALS

Material	Developer	Solar Absorptance	Thermal Emittance
Semiconductor lead dioxide (on copper)[a]	AMETEK	0.99	0.25
Black chromium on bright nickel (on aluminum or steel)[b]	Olympic	0.95	0.10
	Honeywell	0.977	0.19
	NASA-Lewis	0.93	0.12
Black nickel[b] (on nickel or copper)[c] (on galvanized iron)[c]	Honeywell	0.90	0.08
	Tabor	0.90	0.05
	Tabor	0.89	0.12
Copper oxide (on copper)[c]		0.89	0.17

[a]*U.S. Patent No. 3.958.554, Dr. Ferenc Schmidt; assigned to AMETEK, Inc.*

[b]*R. L. Lincoln, D. K. Deardorff, and R. Blickensderfer. "Development of ZrOxNy Films for Solar Absorbers," Society of Photooptical Instrumentation Engineers. Vol. 68 (1975), p. 161.*

[c]*H. Tabor. "Selective Surfaces for Solar Collectors," Chap. IV of* Low Temperature Engineering Application of Solar Energy, *ASHRAE, New York.*

SOURCE: Adapted from Solar Energy Handbook *by AMETEK. Copyright 1979 by the authors. Reprinted with the permission of the publisher, Chilton Book Co., Radnor, Pa.*

0.90 or higher with accompanying thermal emissivity rates of approximately 0.10. Table 2.2 describes the absorptance and emissivity levels of commonly used selective absorber surfaces.

In addition to electrodeposited methods for applying selective absorber surfaces, special black paints are available for absorber surface treatment (Figure 2.12). Although these paints do not equal the performance of electrodeposited selective surfaces with regard to absorptance and emittance, they do achieve solar absorptance rates in excess of 93% with emissivity of between 20 and 50%. Also, paints are much less costly to apply to the absorber than are electrodeposited surfaces. The biggest market for selective surface spray paints is in the do-it-yourself hobbyist application. Care must be exercised in the selection of paint coatings for this purpose, for if the paint has not been specifically formulated for solar collector applications, peeling and discoloration with accompanying degradation in performance will result.

COLLECTOR GLAZING

Solar collectors are almost always manufactured with one or more transparent covers. The only general exceptions to this are swimming pool collectors manufactured from rubber and plastic polymers which are in service primarily during the summer months. The purpose of the cover or glazing is to reduce heat losses from the absorber plate and collector box. At times, more than one layer of glazing may be used, depending on the environment and operating conditions in a specific application. A guide to the selection of the number of cover plates used in a solar collector is given in Table 2.3. It should be noted that while additional layers of glazing will reduce convective heat losses, the added layers also reduce the amount of collectible solar energy due to internal reflection and transmission losses.

Therefore, care should be exercised in selecting the number of glazing layers incorporated into the collector design. As a general rule, flat-plate collectors for domestic hot water and space heating applications are designed with one layer of transparent glazing material in conjunction with a selective surface absorber plate.

FIGURE 2.12   Selective surface spray paint. (Courtesy of Dampney Co., Inc.)

Materials of Construction

Glass, fiberglass compounds, and acrylic plastics are commonly used as collector glazing materials. Of these, glass is the most widely used. Ordinary window glass is generally unsuitable for solar glazing applications due to its high iron content, resulting in significant loss of transmission capability. Maximum transmission is often limited to between 80 and 85% of available sunlight with window glass. To overcome this deficiency, low-iron glass, referred to as "water white glass" is employed. (To distinguish between water white glass and ordinary window glass, view the sheet of glass from the edge: Water white glass will be clear, window glass will be green.) Water white glass possesses transmission values of between 88 and 92%. Also, the glass is usually tempered for increased strength and durability.

One type of plastic used for collector glazing is polymethyl methacrylate, known as Plexiglas.* Plexiglas is an acrylic that can be shaped when heated and possesses a solar transmittance of 92%. Two types, Plexiglas G and Plexiglas K, will neither discolor nor deteriorate when exposed to sunlight. A thermoformed acrylic collector glazing is illustrated in Figure 2.13.

Fiberglass-reinforced plastics and polymer-based materials have been in use for several years in the solar industry. Applications for these materials vary from greenhouse glazings to solar collector glazings. Transmission values for these materials exceed 85%, and they are generally less expensive per square foot than either glass or acrylics. All polycarbonate

*Plexiglas is a trademark of Rohm and Haas Corporation.

Table 2.3 GUIDE TO THE SELECTION OF NUMBER OF TRANSPARENT COVER PLATES

Collection Temperature above Ambient Temperature $(t_c - t_a)$	Typical Applications	Optimum Number of Cover Plates	
		Black-Painted Absorber $\epsilon = 0.9$ or 0.95	Selective Absorber $\epsilon = 0.2$
−5 to +5°C (−10 to +10°F)	Heat source for heat pump Heating of swimming pools in summer Air heating for drying	None	None
5 to 35°C (10 to 60°F)	Summer water heating Heating of swimming pools in water Air heating for drying Solar distillation Space heating in nonfreezing climates	1	1
35 to 55°C (60 to 100°F)	Winter water heating Winter space heating	2	1
55 to 80°C (100 to 150°F)	Summer air conditioning Steam production in summer Refrigeration Cooking by boiling	3	2

FIGURE 2.13 Molded Plexiglas solar glazing. (Courtesy of Heliotherm.)

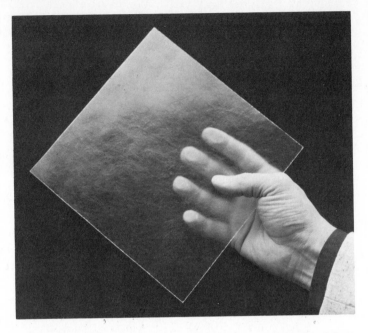

FIGURE 2.14 Fiber-reinforced plastic glazing. (Courtesy of Solar Components Corp.)

glazings must be stabilized for ultraviolet light to delay deterioration at least a few years. A typical polymer-based glazing material is illustrated in Figure 2.14.

Gasketting

Glass, plastic, and fiberglass compounds expand when heated and contract upon cooling. Since the operating cycles of all flat-plate collectors alternate through heating and cooling phases, the constant expansion and contraction of the glazing must be taken into account when designing the collector box and glazing support structure. EPDM rubber gaskets are fitted to the perimeter of the glazing material to seal it into the collector frame. This material is flexible enough to permit the glazing to expand and contract and thus prevent stress buildup. These gaskets should be inspected occasionally to assure integrity of the material since the constant expansion and contraction of the glazing will eventually cause gasket failure. Gasket failure is less common with acrylic glazing than with glass, since the plastic will flex at elevated temperatures rather than undergoing linear expansion with subsequent abrading of the gasket. Figure 2.15 illustrates a typical gasket for collector glazing manufactured from EPDM rubber.

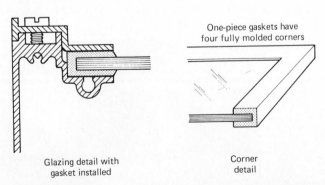

FIGURE 2.15 EPDM rubber gasketing. (Courtesy of Pawling Rubber Corp.)

Most collector designs feature glazing hold-downs that can easily be removed for inspection of the edge gasket and glazing replacement (Figure 2.16).

INSULATION MATERIALS AND METHODS

Heat losses from the back and sides of a collector frame are minimized by the incorporation of a variety of thermal insulating materials. Two of the most widely used materials for this purpose are fiberglass and foam-type insulation. Although there are many types of insulating materials available, few can perform adequately given the operating characteristics of the flat-plate collector. For successful solar collector operation, insulating materials must be able to do the following:

1. Operate under extremely high temperatures (in excess of 400°F under stagnation), without outgassing, igniting, or chemically degrading.
2. Be relatively impervious to moisture. Absorption of moisture by an insulating material radically reduces the R value of the material. Insulation with closed cell and unicellular chemical structures do not readily absorb moisture.
3. Be cost-effective relative to the price of the finished solar collector.

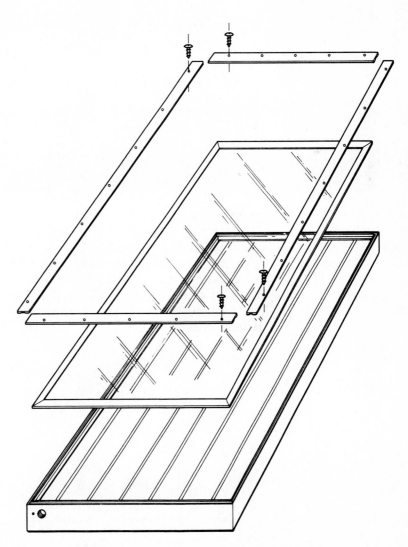

FIGURE 2.16   Hold-down batten for collector glazing.

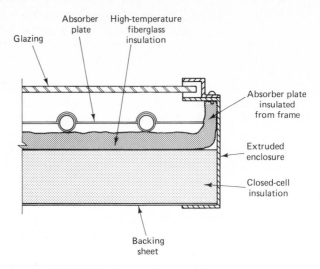

FIGURE 2.17   Insulation of the collector.

Figure 2.17 illustrates a method of insulating the absorber plate within the collector box. Note in Figure 2.17 that insulation is placed on the sides of the collector frame as well as beneath the absorber. This arrangement greatly reduces heat loss from the total perimeter of the absorber, in addition to directly beneath it.

Urethane foam-type insulation material is supplied in rigid sheet form. The sheets are cut to size and placed beneath the absorber plate, as in Figure 2.18.

In some instances, collector manufacturers use a combination of foam and fiberglass insulation in the collector box. The long-term durability of both of these materials is quite good, resulting in low conduction and convection losses from the absorber plate and collector box, if the insulating materials are properly installed. The effectiveness of insulation, however, depends upon relatively airtight construction within the collector box to minimize heat losses due to air infiltration and moisture condensation.

The characteristics of isocyanurate and fiberglass foam insulation are listed in Table 2.4.

FIGURE 2.18   Isocyanurate foam insulation in a solar collector.

Table 2.4 CHARACTERISTICS OF POLYISOCYANURATE AND FIBERGLASS INSULATION

	Material	
Property	Isocyanurate	Fiberglass
Density	2.0 lb/ft^3	0.6–1.0 lb/ft^3
Closed cell content	90%	—
Thermal conductivity (*K* factor)	0.16–0.17 Btu-in./ft^2-hr-°F	Varies with density
Thermal resistance (*R* value) per inch of thickness at 75°F	6.2–5.8 unfaced or sprayed 7.7–7.1 (impermeable skin)	3.16 (batt) 2.20 (loose fill)
Water vapor permeability	2–3 perm-in.	100% perm-in.
Water absorption	Negligible	< 1% by weight
Capillarity	None	None
Fire resistance Flame spread Fuel contributed Smoke developed	Combustible 25 5 55–2000	Noncombustible 15–20 5–15 0–20
Toxicity	Produces CO when burned	Some toxic fumes due to binder combustion
Effect of age Dimensional stability	 0–12% change	 None (batt) Settling (loose fill)
Thermal performance	0.11 new 0.17 aged 300 days	None
Fire resistance	None	None
Degradation due to: Temperature Cycling Animal Moisture Fungal/bacterial Weathering	 Above 250°F Not known None Not known Does not promote growth None	 None below 180°F None None None Does not promote growth None
Corrosiveness	None	None
Odor	None	None

SOURCE: *Adapted from* Insulating Techniques and Estimating Handbook, *Frank R. Walker Publishing Co., Chicago, 1980.*

AIR-COOLED COLLECTOR DESIGNS

Several solar energy systems rely on air-cooled rather than liquid-cooled collectors for heat transfer. Air-cooled collectors are simpler in design than their liquid-cooled counterparts. There are distinct advantages and disadvantages associated with the use of air- versus liquid-cooled collectors. Air-cooled collectors are far less complicated than liquid-cooled collectors in design and fabrication. No integral liquid flow passages are required which greatly reduce the cost of the absorber plates. Many different types of absorber configurations are available for air-cooled applications since heat transfer from collector plate to the passing air takes place by both convection and conduction. Air tends to stratify when heated, producing what is referred to as *laminar flow*. This stratification is unacceptable since the distinct layers of air tend to insulate and hinder rather than accelerate heat transfer. To help alleviate the tendency toward stratification, air-cooled collectors are designed to introduce turbulence into the airflow within the collector. This helps to assure even heating of the air as it traverses the absorber surface. Several common types of airflow absorber configurations are illustrated in Figure 2.19. Note that all the airflow configurations are engineered to increase available surface area of the absorber plate to the passing air. Maximizing absorber surface area helps to keep the absorber plate at a uniform temperature during operation, while facilitating maximum heat transfer.

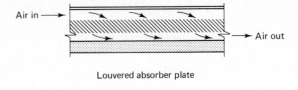

Louvered absorber plate

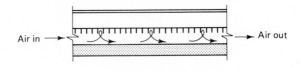

Fins on absorber plate

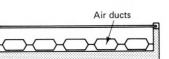

Integral air ducts in absorber plate

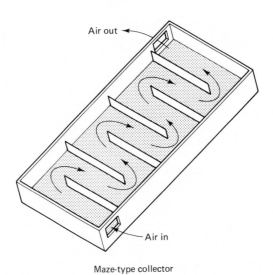

Maze-type collector

FIGURE 2.19 Air-cooled flat-plate collector flow configurations.

There are two major disadvantages associated with the use of air-cooled collectors in comparison to liquid-cooled systems. The first is the lack of an efficient mechanism to provide for domestic hot water. Since domestic hot water is the most cost-effective of residential and commercial solar applications, this shortcoming limits the widespread use of hot air collectors for many space heating systems, as well as small domestic hot water installations.

The second disadvantage associated with air-cooled flat-plate collectors is the use of air, rather than a liquid, as the heat transfer medium. The specific heat of air is approximately 0.25, while that of water is 1.0. Thus approximately four times the weight of air must be moved through a collector to transfer an equivalent amount of heat in a water-cooled design. For this reason, many hot air solar systems use no storage at all, their useful operation occurring between the hours of 10 A.M. and 3 P.M. Those systems that do incorporate storage often rely on large areas of crushed rock or phase-change materials to store the collected energy. Such storage systems radically increase system costs over systems that forgo storage.

It should be kept in mind, however, that air-cooled heating systems can prove to be more cost-effective than liquid-cooled designs in many instances. Low-temperature hot air heating requirements such as supplemental room heating, crop and grain drying applications and similar agricultural tasks, as well as residential and commercial hot air heating with limited domestic hot water demands are well-matched end uses for solar hot air collector systems.

FLAT-PLATE COLLECTOR EFFICIENCY

The efficiency of a solar flat-plate collector is usually the most important feature examined by a prospective system purchaser. Since the economic advantages of solar energy rely on the ability of the system to deliver the maximum amount of incoming solar radiation to storage, collector efficiency is of primary importance in determining the overall cost-effectiveness of the system.

The practice of specifying a single efficiency figure for any solar collector is difficult and often misleading to the consumer. Since the operating environment of the solar collector is constantly changing, collector efficiency will change as well. For example, the *instantaneous efficiency* of a flat-plate collector at solar noon on a clear summer day will be higher than the instantaneous efficiency of that same collector at 4 P.M. or 9 A.M. on the same day. In addition to daily changes in instantaneous efficiency, seasonal variation have a large impact on collector performance.

Regardless of operating environment, the overall efficiency of any flat-plate collector can be plotted in order to describe its performance under a wide array of operating parameters.

The Efficiency Curve

In its most basic terms, the *efficiency* of a solar collector can be defined as the ratio of the amount of usable energy delivered by the collector to the amount of solar energy incident on the collector surface, written as

$$\eta = \frac{\text{usable energy collector output}}{\text{solar energy input}}$$

where η is the collector efficiency.

When taking into account all the factors that influence a flat-plate collector, ranging from weather conditions to the integrity of its construction, the efficiency equation for a solar collector is as follows:

$$\eta = F_R\,(\tau\alpha) - F_R U_L \frac{T_i - T_a}{I} \tag{2.1}$$

where η = instantaneous collector efficiency
$\quad F_R$ = collector heat removal factor
$\quad U_L$ = overall heat loss coefficient of the collector
$\quad \tau\alpha$ = glazing transmission and absorptance of absorber plate
$\quad T_i$ = temperature of inlet fluid
$\quad T_a$ = ambient air temperature
$\quad I$ = solar radiation incident on the collector

With this information, we can now plot a typical efficiency curve for a flat-plate collector. Equation (2.1) yields a point on a graph which is the instantaneous efficiency based on the inputs to the efficiency equation. If we take several different operating conditions as inputs into our equation, we end up with a series of points that can be connected with a straight line to yield an efficiency curve for the collector. A typical efficiency curve for a flat-

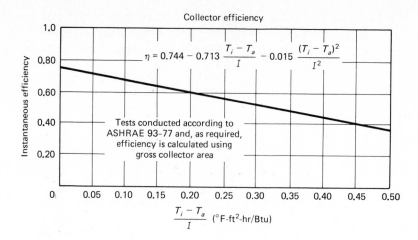

FIGURE 2.20 Efficiency curve for typical liquid-cooled flat-plate collector. (Courtesy of Heliotherm.)

plate collector with an all-copper absorber plate is illustrated in Figure 2.20. The vertical axis of the graph represents the collector instantaneous efficiency, η. The horizontal axis represents the component of the efficiency equation known as the fluid parameter, $(T_i - T_a)/I$. The intersection of the efficiency line with the vertical axis of the graph represents the collector efficiency when fluid temperature entering the collector is equal to the ambient air temperature. This point is referred to as the *intercept*. It is this figure, the intercept, which is most often supplied by collector manufacturers describing overall collector performance. We can see from the graph, however, that the intercept efficiency level of the collector is achieved only when $T_i = T_a$. What this tells us, therefore, is that solar flat-plate collectors operate most efficiently at low fluid temperatures. The instantaneous efficiency at the y intercept is a prediction of the efficiency that the collector is able to achieve under optimum operating conditions only. We will discuss shortly the methods for determining efficiency intercept values from other than extrapolated instantaneous efficiency curves.

Note that the efficiency curve slopes downward, approaching the horizontal axis of the graph as the fluid parameter, $(T_i - T_a)/I$, increases in value. The efficiency of the collector reaches zero under stagnation when fluid circulation through the collector has ceased and no energy is being removed from the absorber plate. A shallow slope of the curve indicates relatively little loss in efficiency as the inlet fluid temperature increases or as the ambient temperature decreases, and is indicative of good insulation and collector construction.

Efficiency measurements for flat-plate collectors are conducted by independent testing laboratories under highly controlled conditions. Most of the efficiency tests currently conducted by these laboratories follow the guidelines established by ASHRAE 93–77.* A typical testing facility for flat-plate collectors is illustrated in Figure 2.21.

The usefulness of the theoretical efficiency curve is somewhat limited in practical applications. Although it can predict the performance of a collector under a set of specified operating conditions, it cannot predict the long-term performance of the collector in an operating system. Extensive field monitoring of the system is necessary if actual long-term efficiency and performance are to be determined. However, computer programs have been in use for years to predict overall monthly and yearly solar contributions. These programs rely on the characteristics of the collector efficiency curve to predict performance. In Chapter 8 we explain the use and accuracy of the most widely used computer program, the *f*-chart.

*ASHRAE 93–77, available from ASHRAE, 1791 Tullie Circle NE, Atlanta, GA 30329.

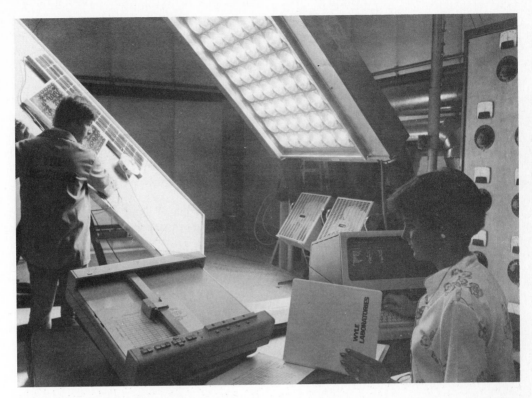

FIGURE 2.21 Test facility for flat-plate collectors. (Courtesy of Wyle Laboratories Scientific Services and Systems Group.)

Factors that Affect Collector Efficiency*

The collector intercept figure is widely used for an overall description of collector efficiency. Because there are a number of different methods used to determine the intercept, the components of this figure should be closely examined. The intercept value found on collector efficiency curves is an extrapolation of the instantaneous efficiency points that have been extended through the vertical axis of the efficiency graph. As such, these points are merely theoretical assumptions of maximum collector performance. However, the intercept can be calculated from the various operating efficiencies of the critical collector components. These calculations can then be used to verify manufacturers' claims of intercept values that are represented on published efficiency curves. The main components that determine the collector intercept are: net aperture to gross collector surface area, optical transmission of the collector glazing material, solar absorptance of the absorber plate, thermal conduction efficiency at the absorber plate/fluid flow tube junction, thermal conduction of the bond between absorber plate and fluid flow passage, and efficiency of fluid flow through the absorber plate. Each of these characteristics will be examined next in more detail.

Net Aperture Area

Most collectors incorporate a supporting frame and glazing hold-down system that reduce the overall outside dimensions of the solar collector to what is known as the *net effective aperture area,* the actual surface area through which incoming solar radiation can reach the absorber plate. For example, a solar collector measuring 3 ft × 8 ft, utilizing a 1-in. alu-

*This section is adapted from "Applications of Solar Energy for Heating and Cooling of Buildings," ASHRAE GRP 170, New York, 1977.

minum glazing hold-down around the perimeter, contains 1.8 ft² of hold-down material. This material reduces the 24.0-ft² total surface area of the collector to a net aperture of 22.2 ft². Some systems for measuring collector efficiency incorporate the net aperture of the collector in the efficiency testing procedure, whereas others utilize the gross area dimensions. When comparing the performance figures for two different collectors of equal dimensions, it should be ascertained if both collectors were tested using the same aperture measurements. Typically, the ratio of gross area to net effective aperture rarely exceeds 0.93. Using the example of a 1-in. hold-down on a 3 ft × 8 ft collector, the ratio of net to gross aperture area is 0.925.

Optical Transmission of Glazing Material

The primary barrier to be penetrated by the incoming solar radiation on its way to the collector absorber plate is the glazing material. Low-iron-tempered glass has an optical transmission of approximately 91%, while the transmission of clear acrylic is approximately 92%. The major portion of the solar radiation lost in both the glass and acrylic is due to reflection from both the interior and exterior surfaces of the glazing. A small amount of radiation is absorbed by the material itself. Reflection losses are held to a minimum when the incoming light is perpendicular to the plane of the glazing. As the angle of incidence either increases or decreases from the perpendicular plane, reflection losses from the glazing will increase. Thus transmission values above 92% are difficult to achieve. Transmission values are typically found to be between 80 and 92% in most flat-plate collectors, depending on the glazing material.

Solar Absorptance

We have examined how the use of selective surfaces on absorber plates increases solar absorptance while holding emissivity levels to a minimum. The best commercially available selective coatings yield absorptance values of approximately 95%. This figure will decrease if the selective coatings are either too thick, poorly bonded to the absorber plate, or poor in quality. The reader is referred to Table 2.2 for specific absorptance and emission values for various selective coatings. The absorptance of electrodeposited selective surfaces such as black chrome can be assumed to be within the range 0.91 to 0.96, with 0.95 being an average value.

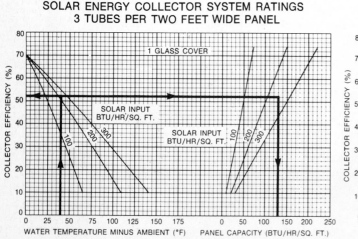

FIGURE 2.22   Effect of tube spacing on collector efficiency. (Courtesy of Revere Solar and Architectural Products.)

Thermal Conduction Efficiency at the Absorber Plate/Fluid Flow Junction

The thermal conduction efficiency of the absorber plate at the junction of the fluid flow passages is a measurement of the amount of energy that is actually transferred to the fluid conduits from the absorber plate surface. This figure represents the net energy available for transfer to the collector fluid loop. The efficiency of this transfer process depends on the following: the material used in construction of the absorber plate, thickness of absorber plate material, the type of construction employed in absorber plate manufacture, and the spacing of the fluid flow passages. As the number of flow passages within the absorber increases, the percentage of wetted surface area of the absorber plate increases as well, allowing for greater heat transfer. Figure 2.22 illustrates the effect that tube spacing has on collector performance and efficiency. Given a tube spacing of approximately 2 in. on-center, one can expect thermal conductivity efficiency between 0.90 and 0.98, with 0.95 being typical in a well-designed flat-plate absorber.

Thermal Conduction of Bond between Absorber Plate and Flow Passage

The fluid flow tubes must be properly bonded to the absorber plate in order to eliminate conduction losses at this junction. Absorber plates with integral flow passages in which no bonding is required offer almost ideal bond conductance properties. If the bonding techniques in roll-formed absorber plates offer direct metal-to-metal contact over a wide surface area, heat transfer across the bonded area will be very high. Home-built collectors in which copper tubing is soldered to a flat metal absorber plate usually suffer significant transmission losses at this critical point. Well-designed absorber plates usually have fluid bonding efficiencies of approximately 95%.

Fluid Flow Efficiency

The efficiency with which heat is transferred to the circulating collector fluid depends on the temperature difference between the absorber plate and fluid, the percentage of wetted surface area of the absorber, and the flow characteristics of the fluid within the absorber flow passages. The highest percentage of wetted surface area occurs with integral flow, roll-bond-type absorber plates, where absorber to fluid efficiencies can be as high as 0.98. Typical roll form absorber plates have fluid flow efficiencies of approximately 0.95. The use of fluids other than water within the absorber require higher flow rates to prevent reduced heat transfer due to lower specific heat values. Fluid flow through the collector should be uniform throughout all the risers in the absorber to help eliminate hot spots. The absorber manifolds should be sized so that they are large enough to allow for balanced and equal flow rates throughout the solar collector, whether it be a two-collector domestic hot water system or a 30-collector space heating system.

Calculation of the Intercept

Once all the efficiency characteristics of a specific flat-plate solar collector have been determined, the theoretical intercept can be determined. For example, using a well-constructed flat-plate collector as an illustration of this procedure, we will assign typical efficiency values to each of the parameters discussed. A theoretical intercept can then be derived as follows:

The collector in question is a single-glazed flat-plate collector with black chrome selective absorber, having the following component efficiencies:

Net aperture to gross surface area   0.92

Optical transmission of glazing	0.92
Absorptance of absorber plate	0.95
Absorber plate conduction efficiency	0.95
Bond conduction efficiency	0.95
Absorber fluid efficiency	0.95

Therefore,

$$\text{intercept} = 0.69$$

It can be seen from the calculation above that even a good collector design that features high-quality material and component configurations can yield only a theoretical intercept of 0.69. Claims of collector intercepts greater than 0.75 should be examined carefully to ascertain the validity of the testing procedures employed to derive the values claimed. Conversely, an intercept value below 0.60 indicates that there is need for improvement in the design of the collector.

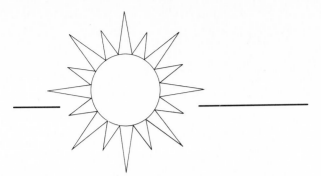

3

Conventional Domestic Hot Water Heating Systems

The cost of providing domestic hot water has risen dramatically during the past few years. The standard hot water heater can in certain instances be the most expensive appliance to operate in the home. Until recently, relatively few improvements have been directed toward the technology of residential and commercial water heating systems. Interestingly, solar water heating units were common in many parts of the United States during the early part of the twentieth century prior to the advent of inexpensive fossil fuels. Solar water heating remains a long-term inexpensive technique for providing most of the hot water used in residences and commercial establishments. Fundamentally an adaptation of standard water heating technology, solar systems substitute direct and diffuse solar energy for the electricity, gas, or oil customarily used. To gain a full understanding of the solar domestic hot water system, one must examine conventional hot water heating system designs. A typical domestic hot water system is illustrated in Figure 3.1. Cold water enters the building either from the local water mains or from a separate drilled well. Entering water pressures vary between 20 and 60 psi in most areas. The entering cold water is immediately channeled in two directions: one line enters the hot water heater, the other travels directly to each cold water outlet: showers, sinks, commodes, and washing machines. Every system should be designed to enable all water lines to be completely drained so that the water system can be completely shut down if necessary. Within this framework, let's examine the various types of hot water heaters in use today.

Electric Heaters

One of the most common types of domestic hot water heaters is the electric hot water heater (Figure 3.2). Electric hot water heaters are available in a variety of standard sizes that include 30-, 40-, 50-, 60-, 80-, and 120-gallon capacities. A rule of thumb for sizing hot water heaters is to assume that each person in the household will consume approximately 20 gallons of hot water per day; thus a family of four will typically consume an average of 80 gallons of hot water per day. The standard hot water heater contains two heating elements: one at the top and one at the bottom of the tank (Figure 3.3). The heating elements, insulated from the shell of the tank to prevent electrical short circuits and shocks, are made from met-

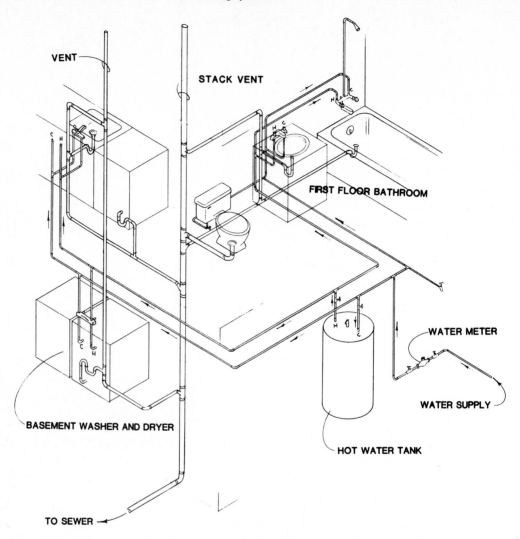

VENT

STACK VENT

FIRST FLOOR BATHROOM

WATER METER

WATER SUPPLY

HOT WATER TANK

BASEMENT WASHER AND DRYER

TO SEWER

FIGURE 3.1   Typical residential plumbing system.

als with a high resistance to electricity. When an electrical current is passed through these elements, virtually all the electricity is converted into heat. The heat is transferred directly to the water surrounding the elements. Cold water enters the bottom of the tank and as it is heated rises to the top of the tank, from where it is piped throughout the home.

Water temperature within the tank is controlled by two thermostats, each thermostat controlling one of the heating elements (Figure 3.4). Each thermostat is placed in contact with the outside of the water heater's pressure vessel and wired as illustrated. As the water temperature drops inside the tank, the bimetallic elements contract, closing the electrical circuit to the heating element. During ordinary use, the heating element at the bottom of the tank is responsible for the majority of the water heating load. This element will stay on until the water temperature within the tank satisfies the thermostat setting. As the water is heated, it rises to the top of the tank. During standby conditions, or when domestic hot water consumption is at a minimum, the upper heating element does not ordinarily switch on, since there is a sufficient supply of hot water in the top of the tank to satisfy the top thermostat setting. When hot water consumption is high, however, cooler water may reach the upper portion of the tank in sufficient quantities to activate the upper heating element.

When the upper heating element is operating, the lower element automatically disconnects. This switching sequence prevents overloading of the electrical system. Most electric hot water heaters also contain an upper-limit control. This control will automatically shut off the hot water heater at a predetermined temperature, usually between 180 and 200°F.

FIGURE 3.2 Electric hot water heater. (Courtesy of Rheem Water Heater Division, City Investing Co.)

FIGURE 3.3 Cross section of electric hot water heater. The dip tube, made from polysulfone plastic, allows water entering from the top of the tank to be directed to the bottom, or coldest part of the water heater. (Courtesy of Rheem Water Heater Division, City Investing Co.)

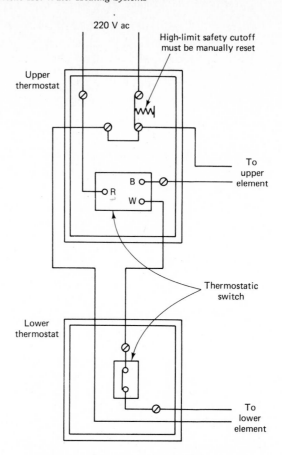

FIGURE 3.4 Wiring of electric hot water heater thermostats. Note that the insulating link disconnects the lower element when the upper element is on, preventing excessive current draw.

The use of electric hot water heaters offers the homeowner both advantages and disadvantages. Perhaps the least understood aspect of electric water heaters pertains to overall system efficiency. Since the conversion of electrical power to heat in a resistance heating element such as a hot water heating element is approximately 95% efficient, the efficiency of the electric hot water heater as an independent system is in the neighborhood of 90%. This figure can be misleading, however, since it represents only the efficiency of the water heater alone. The cost of operating the water heater also incorporates the electrical generating costs of the utility itself and the resulting inefficiencies inherent within this supply system. The thermal efficiency of most modern electric utility generators is in the neighborhood of 40%. This means that the generators can capture only 40% of the available power in the fuel; 60% of the power is lost up the smokestacks and cooling towers. Additional energy is lost in the transmission lines and in step-down transformers scattered throughout the utility grid network. Therefore, the price of delivered electrical energy to the consumer includes heat losses in the generator plant, line losses in the grid system, and the fuel adjustment and amortization costs of capital equipment and plant maintenance. All these cost factors combine to make electricity the most expensive and least efficient method of hot water heating almost everywhere in the United States. The overall end-use efficiency of most electric hot water heaters is in the range 25 to 30%. This figure is a combination of utility generating efficiency, transmission efficiency, and hot water heater design efficiency.

Electric hot water heaters are, however, the least expensive type of water heater to purchase and install. No auxiliary fuel sources, storage tanks, or fuel deliveries need to be arranged. Electric hot water heaters are ideal where hot water demand is light, or where fuels such as natural gas, propane, or fuel oil are unavailable. Summer and weekend residences often fit into this category. Electric water heaters require virtually no maintenance beyond occasional replacement of anode rods and draining tank sediment periodically. Indeed, electric water heaters make very desirable backup systems for solar installations, and the cost-effectiveness in this type of pairing is excellent.

Gas-Fired Heaters

Gas-fired water heaters burn the fuel in a fire chamber located under the hot water tank and within the confines of the outer heater jacket (Figure 3.5). Most gas-fired water heaters can

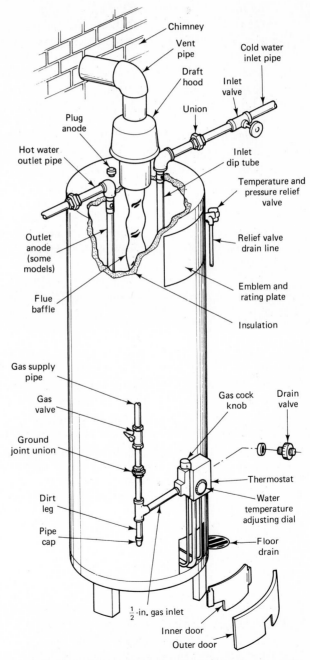

FIGURE 3.5 Gas-fired hot water heater. Centrally located atmospheric burner distributes heat through central baffle to water within the tank. (Courtesy of A.O. Smith Corp., Consumer Products Division.)

be set up to burn either natural gas or liquefied petroleum (LP) gas. When set up to burn LP gas, smaller gas orifices are used and minor modifications in the gas controls are made to accommodate the higher heat content of LP versus natural gas. (One cubic foot of natural gas contains approximately 1000 Btu; 1 ft^3 of LP gas contains approximately 2200 Btu.) A centrally located flue and baffle arrangement carries the combustion by-products into a chimney pipe located near the water heater. A typical gas control is illustrated in Figure 3.6. Located so that its flame comes in contact with a thermocouple, the pilot light burns constantly during normal operation. The thermocouple is made of dissimilar metals which, when heated, produce a small electrical current, which in turn operates a main solenoid valve. The solenoid valve opens and closes the gas line and is the most important safety feature of the gas feed system. Should the pilot light go out for any reason, the flow of electricity from the thermocouple ceases, closing the solenoid valve and stopping the flow of gas to the unit. The gas will not flow until the pilot light has been lighted again. Such a safety device is mandatory and prevents what could be a lethal buildup of gas should the pilot light go out. Operation of the gas water heater is thermostatically controlled. The temperature-sensing element in the gas thermostat contracts when there is a drop in water temperature, opening the main valve to the gas burner (Figure 3.7). Gas then flows to the main burner, where it is ignited by the pilot light. A high-limit safety cutoff (Figure 3.8) automatically closes the main gas solenoid should the water temperature exceed the preset maximum limit, generally between 180 and 200°F.

Gas-fired hot water heaters are relatively efficient devices (operating within the range of 60 to 70% overall system efficiency). Furthermore, due to the clean-burning nature of both natural and LP gas, these heaters require little maintenance. Periodically cleaning the combustion area to prevent a buildup of dust, draining sediment from the bottom of the tank, and occasional anode rod replacement are all that is required. When it comes to cost efficiency, however, a distinction exists between the operating costs of natural gas and LP heaters. LP gas is significantly more expensive than natural gas; in addition, LP heaters necessitate periodic deliveries of the fuel and suitable storage tanks. Deliveries and storage tanks are not required with natural gas heaters, simplifying their operation. In conclusion, long-term operating costs for LP gas heaters can be significantly higher than for their natu-

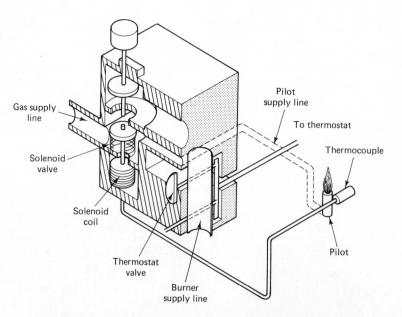

FIGURE 3.6 Valve control unit for typical gas-fired water heater. The main gas solenoid remains open as long as the pilot light remains burning. The thermostat valve opens and closes in response to water temperature acting on the heat-sensing element.

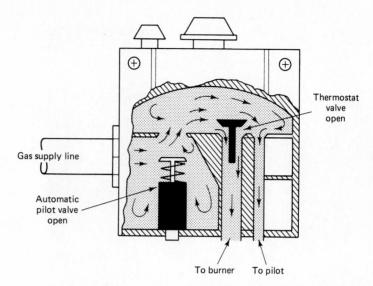

FIGURE 3.7 Gas water heater thermostat valve. As the water cools, the temperature-sensing element contracts, pulling open the gas valve through the action of connecting linkage.

ral gas counterparts. Both are, however, less expensive than electric water heaters. It should be noted that while the operation of pilot lights wastes energy in such appliances as central heating furnaces and boilers, pilot lights in gas heaters contribute to heating the water at the bottom of the tank, and therefore do not represent a significant waste of energy within the system.

Oil-Fired Heaters

Oil-fired domestic hot water heating systems are among the most efficient for either residential or commercial use. A modern oil-fired water heater is pictured in Figure 3.9. High-speed flame-retention oil burners heat water much faster than do most other fuels. The recovery

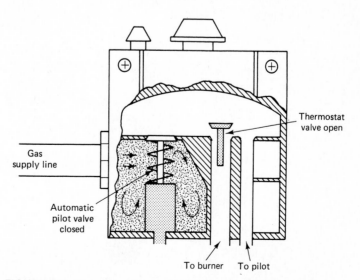

FIGURE 3.8 Gas water heater high limit. Whenever the water inside the heater exceeds the preset high limit, the thermostat breaks the electrical circuit to the thermocouple, closing the main gas line, preventing the water in the tank from overheating.

FIGURE 3.9 Oil-fired hot water heater. (Courtesy of Ford Products Corp.)

rate of a 30-gallon oil-fired water heater with a modern burner is in excess of 100 gallons of water per hour, heated 100°F above its incoming temperature. Modern flame-retention burners are much more efficient in their use of fuel oil than their slow-speed predecessors and represent a substantial updating of oil burner technology. In fact, these units can offer an additional 15 to 20% in combustion efficiency together with a significant reduction in fuel consumption. A typical flame-retention burner is pictured in Figure 3.10.

When there is a drop in water temperature, the thermostat closes the electrical circuit to the oil burner control mechanism that triggers a number of simultaneous operations: The transformer supplies high voltage across the tip of the oil burner electrodes, creating a spark in front of the oil nozzle assembly; the fuel pump provides a fine spray of fuel oil through the nozzle assembly and electrode spark; an air blower supplies high-velocity air for combustion of the fuel. These actions result in a high-temperature flame which spreads throughout the combustion chamber. Heat is directed through the baffled flue to the water within the storage tank. The modern oil burner employs a flame-sensing switch called a cadmium sulfide photocell (cad cell) as a safety device. If the cad cell senses that the burner has failed to ignite, it will automatically shut off the burner mechanism within 15 to 30 seconds to pre-

FIGURE 3.10 High-speed flame-retention oil burner. (Courtesy of R.W. Beckett Corp.)

vent a buildup of unburned fuel oil. This switch must be manually reset to start the oil burner. When the operation of the burner has satisfied the water temperature requirements within the tank, it automatically turns off.

Due to the high flame temperature characteristics of modern oil burners (flame temperatures often exceed 1800°F) the oil-fired hot water heater operates more efficiently and can deliver heated water at a lower cost than that for other types of water heaters. However, oil-fired water heaters are more complicated than gas or electric units in their design and construction characteristics and are therefore more expensive.

High initial investment notwithstanding, oil-fired water heaters offer their owners substantial savings over the life of the unit, generally between 15 and 20 years. These units require annual maintenance, which in most cases must be performed by licensed technicians. Venting requirements call for separate chimney flues to ensure safety. Other maintenance requirements for these units are the same as for other conventional water heaters, including occasional draining of tank sediment and anode rod replacement.

Corrosion Characteristics of Hot Water Heating Tanks

Since most domestic hot water heating tanks are constructed of steel, a variety of means are employed to prevent tank corrosion and ensure long service life.

Glass-Lined

One technique is the use of a glass lining bonded to the interior of the tank. This lining seals the inside surface of the water tank, preventing oxidation (rust). Sacrificial anode rods, made from magnesium, are inserted into the top of the tank to help prevent corrosion. The corrosion process, similar to the electrolytic corrosion that consumes the zinc container of a dry cell battery, will consume the anode rod first, thereby protecting the interior tank surface. Thus over a period of time the anode rod is consumed and must be replaced. Properly maintained, glass-lined tanks will last between 10 and 15 years with few problems.

Stone-Lined

The second method used by tank manufacturers to protect internal surfaces is to coat the interior of the tank with a concrete lining. These units are sometimes referred to as stone-lined water heaters. Figure 3.11 illustrates the method of applying concrete to the tank components. Stone-lined water heaters are very durable and have a longer service life than do glass-lined tanks, and depend on an entirely different method of corrosion protection.

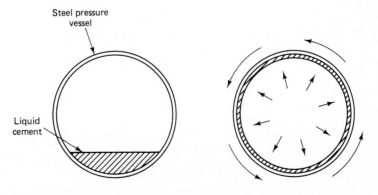

FIGURE 3.11 Centrifugal force caused by rotation of the steel pressure vessel is used to provide a uniform cement inner lining for the water heater.

The cement lining, approximately ½-in. thick when it is applied, is water absorbent. When the hot water heater is initially filled, water penetrates this concrete lining and comes into contact with the interior steel tank wall. This same water remains in contact with the interior tank wall for the entire life of the water heater. After some initial rusting, this charge of water is soon depleted of all oxygen and becomes an inert layer of water between the concrete lining and interior wall of the tank. Positive water pressure within the tank as well as the capillary action of the concrete lining prevents any exchange of the inert water with the domestic hot water within the water heater. No sacrificial anode rod is employed with stone-lined tanks.

Internal heat exchanger coils

Stone-lined solar storage tanks can employ internal finned copper heat exchangers. This component arrangement is illustrated in Figure 3.12. Internal heat exchangers of this type cannot be used with glass-lined water storage tanks, since so large an amount of dissimilar metal accelerates electrolytic corrosion and defeats the intended purpose of the magnesium anode rod. Internal heat exchangers are highly efficient. Coupled with the long-life characteristics of the stone-lined tanks themselves, many solar systems employ stone-lined water storage tanks with associated internal heat exchangers as standard equipment with their systems. These coils are generally fabricated from either heavy-walled copper or stainless steel tubing (Figure 3.13). The coils are available in either single- or double-walled configurations. Single-walled coils are used in any application employing a nontoxic antifreeze solution, such as food-grade propylene glycol. Double-walled coil configurations must be used, however, when a toxic antifreeze solution is used in the solar system and are sometimes made mandatory by local building codes. The double-walled construction prevents contamination of the potable water by the antifreeze should a leak develop within the heat exchanger tubing.

The coil is inserted into the solar storage tank during the initial phases of tank construction. The coils must be properly braced inside the tank to prevent metal fatigue and sagging which would impede coil performance. Also, the coils must be raised a sufficient height off the bottom of the tank to prevent sediment buildup from covering the surface of the heat exchanger, thereby impairing its efficiency.

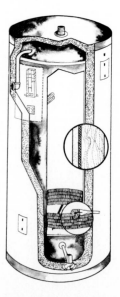

FIGURE 3.12   Cross section of solar storage tank. The finned copper heat exchanger is located at the bottom of the tank for maximum heat transfer. (Courtesy of Vaughan Corp.)

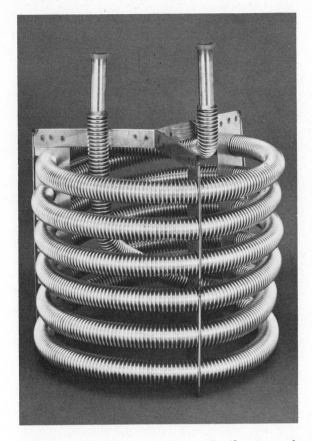

FIGURE 3.13 Heat exchanger coil. (Courtesy of Hydro-Flex Corp.)

Insulation Characteristics of Hot Water Heaters

The insulation characteristics of modern hot water heaters can be as important to the purchaser as the corrosion-protection features incorporated into the tank design. The amount of insulation can often dictate operating costs over the life of the unit. Modern hot water heaters are generally manufactured to specifications set forth by the American Society of Heating, Refrigerating, and Air-Conditioning Engineers, Inc. (ASHRAE). ASHRAE specifications spell out the permissible limits of heat loss in domestic hot water heaters, in addition to general specifications regarding overall tank design.* Water heaters manufactured prior to the adoption of these specifications are in most cases poorly insulated. For this reason a relatively large portion of their operating costs can be attributed to heat losses that occur through the tank walls and outer protective jacket. Also, hot water heaters are generally installed in unheated, or poorly heated, areas of a building, a factor that accelerates heat loss. Methods of bringing poorly insulated water heaters up to acceptable standards of heat loss are discussed in greater detail in Chapter 10.

There are two methods presently utilized to insulate hot water heaters: the use of a high-density fiberglass insulation and the use of injected urethane foam insulation. Water heaters insulated with fiberglass approach R values of 11. The fiberglass is wrapped around the tank shell prior to installation of the outer jacket. Tanks insulated with urethane foam insulation approach R values of 17. During the manufacture of the tank, foam insulation is

---

*ASHRAE Standard 90-75: *Energy Conservation in New Building Design*, is available from ASHRAE, 1791 Tullie Circle NE, Atlanta, GA 30329. This work covers a wide variety of energy requirements and design factors for new structures and is periodically updated.

injected between the inner and outer tank shell and jacket. Due to the relatively complex nature of manufacturing the foam-insulated models, they are more expensive than their glass-insulated counterparts. However, reduced life-cycle operating costs easily justify this additional expenditure.

General Installation Procedures

Satisfactory performance from any product depends on proper installation and maintenance of the unit. In fact, proper tank installation is just as important to maintaining a long service life as are the materials and manufacturing techniques employed.

Water storage tanks should not be placed directly on a floor that is subject to high moisture conditions. To do so accelerates corrosion of the bottom of the tank surfaces and can lead to premature tank failure. Since most water heaters are designed to withstand corrosion from the inside out and not the other way around, strips of pressure-treated lumber or similar material should be placed on the floor beneath the tank to prevent direct contact of the heater jacket with the floor surface. Also, sufficient headroom should be available for anode rod replacement. Other critical clearances, especially in regard to gas- and oil-fired heating units, should be followed as listed in the manufacturer's instructions or local building codes.

Most important, manufacturers' recommendations should be followed as closely as possible. The installation of water heaters are governed by state and local building codes that will specify critical clearances, pipe diameters, and general plumbing procedures that must be adhered to. A maintenance schedule should be set up and adhered to. Following these simple procedures, most water heaters will perform satisfactorily for long periods of time.

SOLID-FUEL WATER HEATERS

The use of solid-fuel water heating systems has increased together with the rise in the price of conventional fuels. Solid-fuel heating systems in current production are far different from their early twentieth-century prototypes. Modern heating technology has had a great impact on the solid-fuel industry, and the result has been a wide variety of products whose quality and efficiency approach those of conventional liquid- and gas-fueled domestic hot water systems.

Solid-fuel systems are used to produce domestic hot water in one of three ways: through the use of a separate, stand-alone solid-fuel water heater; with a corrosion-resistant water bottle mounted inside a solid-fuel radiant stove; or with a tankless coil inserted into a solid-fuel boiler.

The stand-alone water heater (Figure 3.14) is similar in function to liquid- and gas-fueled heaters in that it substitutes wood, coal, or pelletized fuel for conventional fossil fuels.

A thermostat, sensitive to water temperature, controls the amount of combustion air entering the fuel chamber (Figure 3.15). As the water temperature drops, the bimetallic temperature-sensing element rotates the lower arm of the thermostat upward, causing the air-draft door to open wider, allowing more combustion air into the firebox. Thus the fire burns hotter, raising the temperature of the water within the unit. As the water temperature increases, the thermostat automatically lowers the air-draft door, limiting combustion air and regulating interior water temperature.

A small-capacity hot water bottle, when inserted into a solid-fuel radiant stove, can easily supply 80 gallons of hot water at approximately 130°F per day. This type of installation is pictured in Figure 3.16. The water bottle within the solid-fuel heater, piped to the standard domestic hot water heater, absorbs heat from within the combustion chamber of the stove. When the water inside the bottle reaches the desired temperature, a surface-mounted temperature-sensing aquastat automatically turns on the circulating pump. The pump circulates the water from the domestic hot water heater, through the water bottle

FIGURE 3.14 Solid-fuel hot water heater. (Courtesy of Agua Heater Corp.)

within the stove, and back to the hot water heater. Water heated by the stove is constantly being added to the domestic hot water system. During the months when the solid fuel appliance is in use, it is possible to eliminate totally the use of conventional fuels for heating domestic hot water. It should be noted that this type of installation requires that the combustion chamber of the solid-fuel heater be of sufficient size to enable mounting the water bottle without restricting the interior of the combustion area. Both the water bottle and circulating pump should be manufactured from noncorrosive materials to ensure long service life. In addition, all electrical wiring within the vicinity of the solid-fuel appliance should be shielded and approved for high-temperature installations with adequate overcurrent and corrosion protection and installed in strict adherence with local electrical codes. Failure to follow safe installation procedures can have tragic consequences!

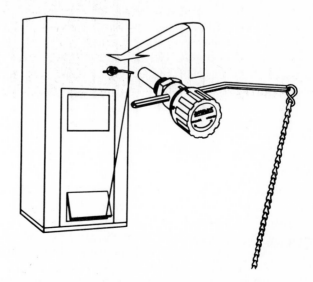

FIGURE 3.15 Thermostatic draft regulator. (Courtesy of Ammark Corp.)

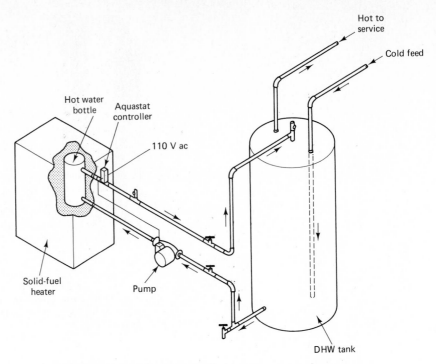

FIGURE 3.16 Domestic hot water bottle in solid-fuel heater.

The dramatic increase in the use of solid-fuel heating equipment is due largely to the economic advantages of solid fuels versus fossil-fuel-fired heating systems. The economic factors that affect solid-fuel heating systems include price and availability of wood and coal, type of wood and coal available, duration of the heating season, and price of solid-fuel appliances and associated costs of installation. In general, people who own their own woodlots, or have access to inexpensive fuelwood, or businesses which produce combustible solid fuels as a by-product of the manufacturing processes achieve significant savings by operating solid-fuel heating systems instead of, or in conjunction with, conventional heating systems. The seasonal use of solid-fuel heating systems requires a significant amount of manual labor as well as sufficient storage area for the fuel. Also, maintenance of solid-fuel systems is more time consuming than for oil, gas, or electric units. However, in most cases this maintenance is performed by the homeowner. The combustion characteristics of solid-fuel appliances mandate that they be vented into chimneys of sufficient strength and quality to withstand the corrosive effects of the effluents produced when they are burning. Also, most solid fuel chimneys are subject to occasional chimney fires and should be constructed from materials that can withstand temperatures that may reach as high as 2500°F. In most areas of the United States where solid-fuel heating systems have become popular, local building codes specify all solid-fuel appliance and chimney installation procedures. In addition, many states have building codes that deal with these appliances.

SUMMER/WINTER DOMESTIC HOT WATER SYSTEMS (TANKLESS COILS)

Many homes with central hot water or steam heating boilers are equipped with tankless coils to produce domestic hot water. Tankless coils are long coils of copper tubing wound into various configurations, depending on the design of the specific boiler (Figure 3.17). When the coils are immersed in the hot water inside the central heating boiler, heat is transferred from the surrounding boiler water through the coil to any cold water within. Hot water from the coil is then piped to the various plumbing fixtures in the house. The temperature of the water as it exits the tankless coil is dependent on the internal boiler wa-

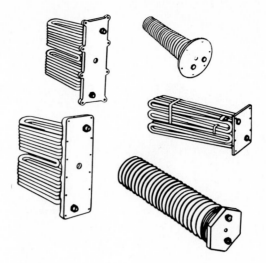

FIGURE 3.17 Tankless coil assembly. Tightly wound to fit into the confines of a boiler, such coils are usually 15 to 70 ft in length. (Courtesy of Amtrol, Inc.)

ter temperature. Assuming that the tankless coil is made from ½-in. copper tubing and is between 15 and 20 ft in length, hot water between 120 and 140°F will be produced when the internal water temperature of the boiler is between 170 and 190°F.

One obvious advantage of such a system is that a separate domestic hot water heater is not required. The major disadvantage of this type of system is that it is necessary to operate the heating boiler even during the summer months when space heat is not required. A common misconception exists that tankless coils provide "free" domestic hot water during the heating season, when, in fact, the production of domestic hot water represents a significant consumption of fuel. For example, a family of four during the winter months requires on the average over 22 gallons of fuel oil per month for domestic hot water production (assumed efficiency for a tankless coil is 55%). During the nonheating season, tankless coils turn out to be major consumers of fuel. Since domestic hot water is heated indirectly by the water in the boiler, rather than directly as in an oil- or gas-fired water heater, the efficiency of the coil system is greatly reduced. This nonheating season coil efficiency is between 15 and 20%. For example, during the nonheating season, the same family of four requires over 80 gallons of fuel oil for domestic hot water production each month (assumed coil efficiency of 15%). This loss of system efficiency has assumed greater importance as the cost of fuels has risen so dramatically, to the point that in certain parts of the United States the use of tankless coils installed with central heating boilers is discouraged in an effort to make more efficient use of available fossil fuels.

Aqua Booster Systems

One adaptation of the tankless coil system which helps to overcome low operating efficiencies during the nonheating season is the use of an aqua coil, or aqua booster, illustrated in Figure 3.18. This setup is a modification of a standard tankless coil and utilizes a separate storage tank with an internal heat exchanger for domestic hot water heating. Such a system offers greater flexibility than the common tankless coil in that it provides upward of 40 gallons of water storage. Tankless coil capacity, if limited to the water in the coil, is generally under 3 gallons. Also, the buildup of lime and other mineral deposits, a common problem with tankless coils, is eliminated with the separate booster tank, since the only water introduced into the heat exchanger coil is oxygen-free water present in the baseboard heating sys-

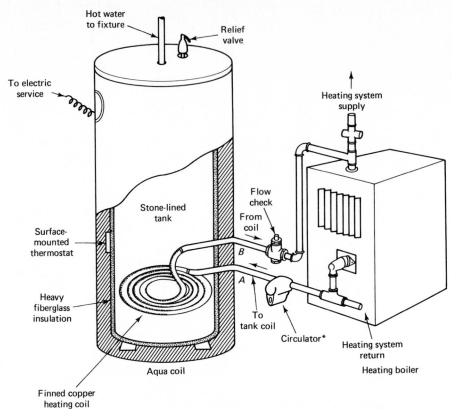

FIGURE 3.18 Aqua coil installation. Larger domestic hot water storage capacity increases the efficiency of the boiler during the nonheating season. (Courtesy of Ford Products Corp.)

tem. Unless the booster tank is equipped with internal resistance heating elements, however, the necessity of operating the boiler during the nonheating season can still account for considerable consumption of fuel, although it will be less than for standard tankless coil systems.

4

Solar
Domestic Hot Water
System Design

Solar domestic hot water and process hot water systems are generally the most cost-effective types of solar heating systems. The term *process hot water* refers to the hot water required for industrial and commercial applications. The cost-effectiveness of such systems is due to several factors: simplicity of design, low cost, year-round operation, eligibility for federal and state tax credits, and good collector performance. As discussed in Chapters 1 and 2, lower operating temperatures result in higher collector efficiency and therefore a higher return on investment. The possibility of expanding a domestic or process hot water system to provide additional output, such as space heating, should be considered in the initial design and installation of the system. In this chapter we treat the many types of solar hot water systems with emphasis on their respective designs and applications.

Solar-heated water is not a new concept. Many areas of the United States and other parts of the world as well were using solar systems to provide domestic hot water during the early twentieth century. Thousands of these systems have been providing inexpensive and reliable hot water for their owners for many years. However, many of the design concepts discussed in this chapter are new. The demand for solar systems that are workable in freezing climates, and advances made in the area of electronic controls, have hastened the development of the concepts we will now discuss.

REVERSE RETURN PIPING

Basic to the design of any solar system in the concept known as *reverse return piping*. Solar collectors should always be plumbed in a reverse return fashion. In other words, the cold feed and hot return should be at opposite ends of a diagonal line drawn through the collector array (Figure 4.1).

Water and antifreeze tend to flow the path of least resistance, the shortest path, for example. By plumbing the collectors in this fashion the heat transfer fluid will flow the same distance no matter which collector it flows through. Therefore, the flow will be balanced through all the collectors. The cold fluid pipe should be run the longest distance whenever possible to minimize heat loss.

In some cases, when there is more than one collector array, flow can not be adequately balanced by the plumbing configuration alone. In this instance, ball valves or gate valves may be necessary to assist in balancing the collector-loop flow.

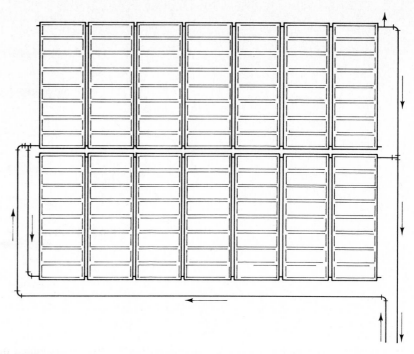

FIGURE 4.1 Reverse return piping.

MAJOR SYSTEM COMPONENTS

In spite of the vast number of solar system manufacturers in the United States, most major and peripheral components are similar from one manufacturer to another. Before entering into a discussion of the various types of system configurations available, a discussion of the major components will help to clarify the basic operating principles that these systems have in common.

Circulators

The circulator components that come into contact with potable water must be constructed from stainless steel or brass, illustrated in Figure 4.2. These metals will prevent corrosion of the pump caused by oxidation (rust). Fresh oxygen is constantly introduced into any open-loop system when potable water is used as the transfer fluid. A steel or cast-iron circulator will quickly rust and fail under such conditions. Therefore, stainless steel circulators are used in open-loop systems. Cast-iron circulators are used in closed-loop system. The power consumption of the circulators used in these applications is very low. Circulators chosen for these systems should have only enough horsepower to provide for the proper circulation of fluid within the system necessary for efficient operation of the solar collectors. Excess horsepower in a circulator wastes electricity and should be avoided. The normal horsepower requirement for a one- to four-collector pressurized domestic hot water system ranges from $1/35$ to $1/20$ hp. Power consumption on these units will be about 75 and 85 watts, respectively. These low-wattage circulators can be used because the fluid within the system does not have to be lifted up into the collector array by the circulator. The building's water pressure or pressure within a closed fluid loop keeps the collectors and piping full at all times. The circulator is required only to move the fluid in a loop between the collector(s) and storage. Circulator sizing and friction losses between fluids and piping should be kept to a minimum and are functions of proper pipe sizing and system design.

Check Valves

The check valve is used to prevent the loss of heat from solar storage during sunless periods. Use of this valve prevents the system from thermosiphoning in a reverse direction by allowing the collector fluid to move in one direction only. A check valve utilized in solar

FIGURE 4.2 Low-head stainless steel circulator.
(Courtesy of Grundfos, Corp.)

applications is illustrated in Figure 4.3. It is installed in the solar system to allow the collector-loop fluid to flow only in the same direction as the circulator is designed to move the fluid. The fluid should always enter the bottom of the collector array and exit the top. Thus the forced circulation in the system matches the natural convection tendency of the fluid in a sloped solar collector exposed to sunlight. During the evening or on cloudy days, the solar collector is almost always cooler than the water in the storage tank. This cold fluid, according to the laws of natural convection, sinks lower in the system. Without a check valve, the hot fluid would rise from the storage tank to the top of the collector, be cooled within the collector, and sink back from the bottom of the collector to the bottom of the storage tank. Heat loss from a solar system by natural thermosiphon action is illustrated in Figure 4.4.

Automatic Air Vents

The use of an automatic air vent allows air to be eliminated from the collector loop, illustrated in Figure 4.5. The vent is installed at the highest point of the system so that air

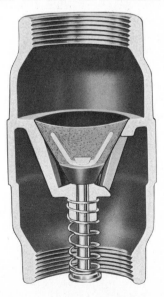

FIGURE 4.3 Check valve. (Courtesy of Strataflo Products, Inc.)

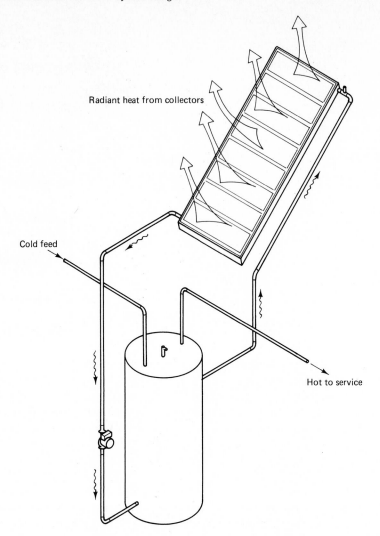

Radiant heat from collectors

Cold feed

Hot to service

FIGURE 4.4   Heat loss at night by thermosiphon.

cannot become trapped within the fluid loop, thus preventing fluid circulation. The air vent also acts as a vacuum breaker, allowing air to reenter the collector loop when the system is drained. The design and construction of an automatic air vent must take into account the extreme conditions under which it must operate: temperature extremes, ultraviolet radiation, and constant exposure to oxygenated fluid must not impede the efficiency of its operation. Since the air vent is exposed to the atmosphere on a year-round basis, it must be constructed of noncorrosive materials.

Differential Controllers

The differential solar controller is known as the "brains" of the solar system. The function of the controller is to energize the circulator automatically when and only when the solar collectors are warmer than the storage medium. A differential solar controller is illustrated in Figure 4.6.

The controller compares electrical resistance of the thermistor temperature sensor located within the solar collector to that of the thermistor located at the bottom of the storage tank. When the collector sensor becomes 5 to 20°F warmer than the storage sensor, the controller provides electricity to the circulating pump. If the collector cannot maintain a temperature at least 3 to 8°F warmer than storage, the controller will automatically turn off the

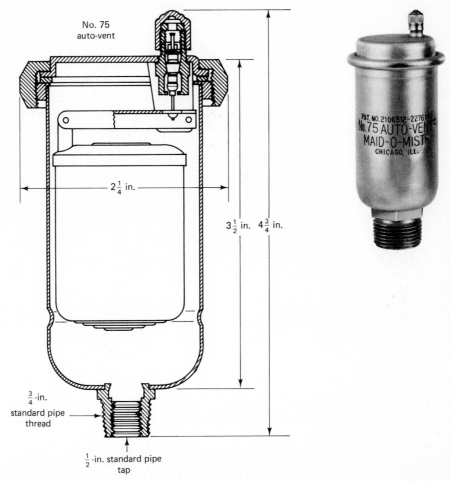

FIGURE 4.5 Automatic air vent. (Courtesy of Maid-O-Mist Division, Sunroc Company.)

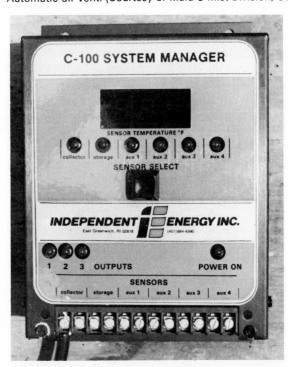

FIGURE 4.6 Differential temperature controller. (Courtesy of Independent Energy, Inc.)

circulator when the collectors are still a few degrees warmer than storage. This operating mode allows for heat loss from the return piping to storage. If circulation stopped when the two sensors registered equal temperatures, heat being lost from the return piping would result in a net heat loss from the system. Controllers are designed to start circulation only when the collector array is significantly warmer (5 to 20°F) than storage. This prevents short cycling during periods of marginal solar insolation.

During the past few years, important advances have been made in the design of differential solar controllers. During the infancy of the controller industry, the solar designer or system installer had to estimate the best transfer fluid flow rate for a specific collector loop. A flow restricting valve was then set to obtain the specified fluid flow rate. Determining the optimum flow rate for any solar system is always difficult due to the varying solar insolation on the collector array throughout the day. A recent improvement in controller design has eliminated the problem of setting fluid flow rate manually and has also resulted in significant increases in overall system performance. The variable or proportional-speed controller automatically adjusts the speed of the circulating pump based on the temperature difference between the solar collectors and the storage tank temperature. During operation, stronger solar insolation will cause a higher temperature differential between collector and storage, and the rate of circulation in the collector fluid loop will automatically be increased.

Temperature Sensors

The temperature sensors used for the differential comparisons made by the controllers are electrical devices called *thermistors*. A thermistor is a device that changes its electrical resistance value as the temperature changes. Common resistance values for thermistors used in conjunction with solar systems are 10,000 and 3000 ohms at room temperature.* These sensors increase in resistance with a drop in temperature and decrease in resistance with a temperature rise. Different controller manufacturers design their units for use with 10,000-Ω (10-kΩ) or 3-kΩ sensors. Some controllers also utilize a 30-kΩ sensor as a standard temperature sensor. The solar designer and installer must take care to use the proper sensors with the differential controller chosen for the specific installation. Ten-kilohm, 3-kΩ, and 30-kΩ sensors cannot be interchanged. Sensors must be matched in pairs to ensure proper system operation. It is most important that all sensors be well insulated from all but the surface that is to be monitored. The collector sensor should be installed within the solar collector, securely fastened to the underside of the absorber plate, sandwiched between the absorber and the collector insulation. The storage tank sensor should be the immersion type, enabling accurate temperature monitoring.

An upper-limit sensor is used to prevent a solar storage tank from overheating. Upper-limit sensors are not thermistors but rather thermostatic switches that open with a temperature rise at either 150°F, 165°F, or 190°F. Controller manufacturers generally design their units so that when a storage tank reaches the specified upper-limit temperature, thus opening the upper-limit sensor, this broken circuit will stop the circulator and all fluid flow within the collector array. This control logic ignores the possible problems of stagnating the fluid in the solar collectors.

A different, more versatile approach to solving the upper-limit-temperature problem is to wire the upper-limit sensor in series with the storage tank sensor. This wiring arrangement simulates very cold storage temperatures when the storage tank has reached the upper-limit setting. The controller regards a broken circuit in this type of wiring system (infinite resistance) as a very cold storage tank temperature. Thus as long as the collector sensor reads above 32°F, circulation within the fluid loop will occur. Circulation in this manner will continue even during sunless periods until the water storage has cooled back down to a temperature below the upper-limit setting. This method has been used successfully for years on

---

*An ohm (Ω) is a unit of measurement describing the ability of a device to resist the flow of electrical current. Table 5.3 lists the electrical resistance values of 10- and 3-kilohm (kΩ) sensors.

hundreds of solar systems from Maine to the Virgin Islands. The vexing problems of collector stagnation and fluid boiling are eliminated in this wiring arrangement. Even if the water storage reaches the upper-limit setting early in the day, the system temperatures will not rise very much higher since the efficiency of the system drops off significantly at elevated temperatures. The collectors will be reradiating about the same amount of energy into the atmosphere as they receive in insolation. Net heat gain in this operating mode is negligible.

BASIC SYSTEM CLASSIFICATIONS

Direct and Indirect Thermosiphon Systems

A thermosiphon solar system is the simplest of system designs. It relies on natural convection to transfer energy from the collector array to the storage tank. Natural convection precludes the need of forced circulation of fluid. A typical thermosiphon system is illustrated in Figure 4.7. During operation, as the fluid within the collector is warmed by the sun, it becomes less dense and rises. Thermosiphon systems require the storage tank to be placed at an elevation at or above that of the collector array so that the warmed fluid from the collectors will rise into the storage tank. Operating on a system of natural convection, there is no circulating pump, controller, or temperature sensors used in the thermosiphon system. In fact, the solar insolation performs the role of the controller; the more intense the sunlight, the faster the heated fluid will thermosiphon. Care must be taken that nothing in the system

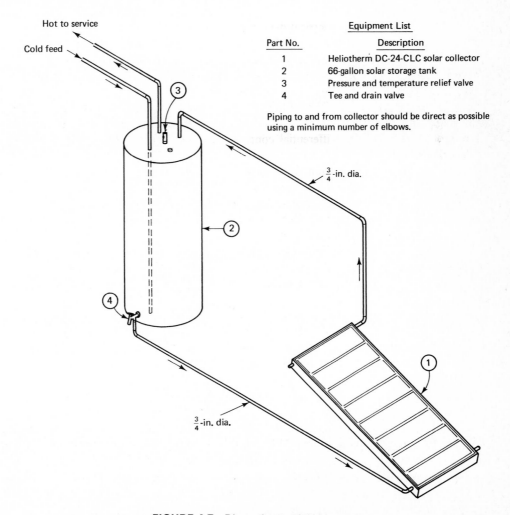

FIGURE 4.7 Direct thermosiphon system.

design or installation works to restrict fluid flow. The forces of natural convection are relatively weak and can be negated to a large degree by the friction losses caused by pipes, valves, and other plumbing fittings. All pipe runs in the system should be as short as possible and pipe diameter must be chosen so as not to cause excessive flow restriction. Generally, the collectors are placed as close as possible to the water storage, and the pipe diameters are one size larger than would normally be used for equivalent flow rates in forced convection systems. Most flat-plate collectors require between 0.5 and 1.0 gallon per minute of fluid flow to operate efficiently. In thermosiphon systems, ½-in.-internal-diameter copper pipe is generally used for a one-collector system if the pipe run is short. Three-quarter-inch-internal-diameter copper pipe is used for a system containing two to six collectors. The pipes must be sloped upward toward the tank at all points. Failure to do so will cause a "heat trap" which either reduces or stops the natural convection. The collectors themselves must also be installed on an upward slope of at least a 15° angle from horizontal to assure proper flow. This is true even on the equator, where it would otherwise seem natural to position the collectors perfectly flat for maximum yearly insolation.

There are two types of thermosiphon systems. The first and most common is the direct or open-loop type, illustrated in Figure 4.7. In the open-loop design, the storage water itself

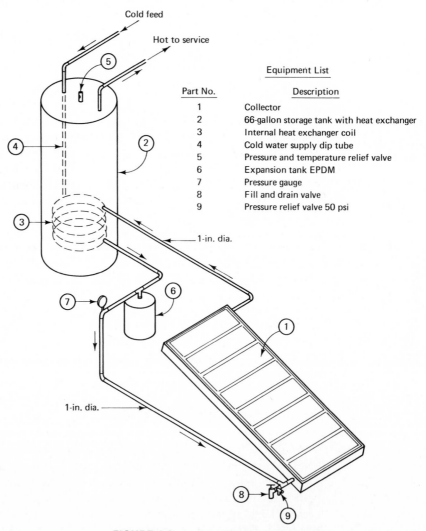

FIGURE 4.8   Indirect thermosiphon system.

flows through the solar collector. This type of system is very popular in areas where the temperature never falls below freezing but cannot be used in environments where subfreezing temperatures occur. The second type of thermosiphon system is the closed-loop or indirect type. This system is illustrated in Figure 4.8. The indirect type of thermosiphon system is designed for installation in cold climates and utilizes an antifreeze fluid or refrigerant to transfer heat from the collector array to water storage through the use of a heat exchanger. Figure 4.8 illustrates the use of an internal heat exchanger located within the solar storage tank. The closed-loop thermosiphon system requires an antifreeze fluid, a heat exchange coil, and an expansion tank (the design and purpose of which are described later in this chapter). This additional equipment and the yearly maintenance it requires makes the indirect thermosiphon system a bit more expensive than the direct system. Because of the heat exchanger, which requires the collector array to operate at somewhat higher temperatures to facilitate the transfer of heat to the storage water, the indirect system tends to be less efficient than the direct system.

An obvious problem with any thermosiphon system is the necessity of locating the storage tank above the collector. The weight of the water-filled tank is sometimes too great for the building's ceiling or roof structure. It is interesting to point out, however, that many parts of the world which are ideally suited for thermosiphon solar systems are also prone to a loss of water pressure due to the reliance on electric pumps to draw water from wells and cisterns. Electric outages are commonplace and are often scheduled. Thermosiphon systems offer the secondary benefit in these situations of supplying hot water in the event of power outages, an inherent advantage of the system which is sometimes overlooked.

Nonfreezing Environment System

When a thermosiphon system cannot be used in a warm climate, a nonfreezing solar system is installed, illustrated in Figure 4.9. This system features an open-loop system with the potable water from the storage tank circulating directly through the solar collectors. The collector(s) is elevated above the storage tank; therefore, forced convection is required for heat transfer from the collector(s) to the tank. This system utilizes a circulator, as well as the controller and associated temperature sensors. The sensors control the circulator on–off sequence, depending on available sunlight.

There are a number of reasons that a nonfreeze type of solar system might be used in place of a thermosiphon system. In cases where the roof structure might not support the weight of the storage, or in which the roof storage tank would be unsightly, a nonfreezing system would be called for. It is almost always easier for the installer to place the storage tank in the basement or on the ground floor. Whatever the reason for choosing it, the nonfreezing system offers simplicity of design and maintenance. If properly designed and installed, performance will be excellent. The additional components of a nonfreeze system compared to a thermosiphon system include a circulator, a check valve, an automatic air vent, a differential temperature controller, two thermistor temperature sensors, and an upper-limit sensor.

Recirculate Freeze-Protection System

A recirculate freeze system is simply a nonfreezing environment solar system installed with a pair of frost sensors that are wired in series with the tank storage and upper-limit sensors. This arrangement is depicted in Figure 4.10. Frost sensors are thermostatic switches which open when the temperature falls below 38°F. This switching action forces the controller to turn on the circulator in a manner similar to the operation of upper-limit sensors. The constant circulation of water from storage through the collector(s) prevents the collectors and

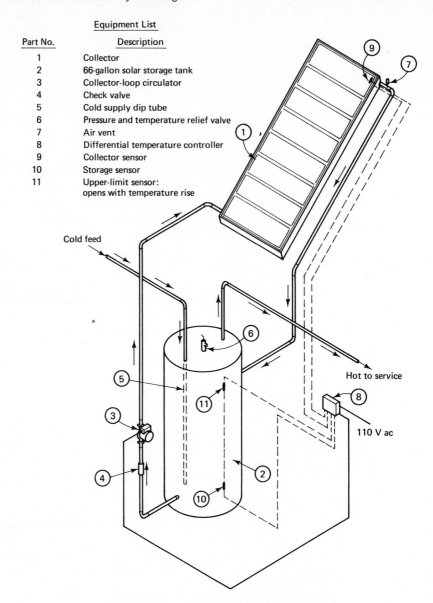

Equipment List	
Part No.	Description
1	Collector
2	66-gallon solar storage tank
3	Collector-loop circulator
4	Check valve
5	Cold supply dip tube
6	Pressure and temperature relief valve
7	Air vent
8	Differential temperature controller
9	Collector sensor
10	Storage sensor
11	Upper-limit sensor: opens with temperature rise

FIGURE 4.9   Nonfreezing environment system.

piping from freezing and subsequent rupturing. In this setup, heat from storage is purposely sacrificed to the environment in order to save the system from permanent damage.

Such freeze-protection systems are practical only in climates that rarely experience frost conditions. The use of the system in northern climates would be extremely wasteful in terms of collected solar energy and could, in fact, result in negative energy flows.

Two frost sensors are used in the freeze-protection circuit because the collector absorber plates can sometimes be 5 to 10°F colder than the ambient air. The absorber actually radiates heat to the night sky faster than heat from the environment can be replaced by conduction to the absorber plate. Therefore, a frost sensor must be attached to the bottom area of the absorber for accurate temperature sensing. The other frost sensor must be located outside the collector, where it is exposed to ambient air. There are many times when cold ambient air can freeze the water in the solar collector piping loop, although marginal solar insolation will maintain the absorber plate above 38°F but lower than the temperature necessary to start fluid circulation by means of the differential controller.

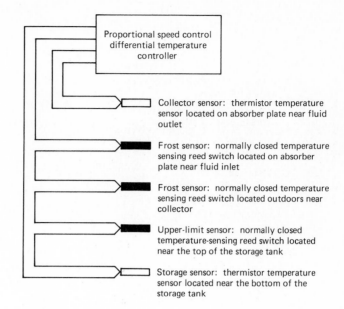

FIGURE 4.10 Wiring schematic for recirculate freeze-protection system.

Drain-Down System

Drain-down systems are open-loop systems that circulate potable water directly from solar storage tanks to the collectors (Figure 4.11). The water in the system is under street pressure [normally between 30 and 60 pounds per square inch (psi)]. Since the system is pressurized, the circulator that is used to move the water from storage to the collector array and back to storage can be very small. During sunless periods, when the collector or ambient temperature falls below 38°F, the water in the collectors and collector-loop piping automatically drains out of the system. In order to empty the collectors and collector-loop piping and prevent freeze damage, it is necessary that the water pressure in the solar storage tank be contained and isolated from the collector-loop pipes and the solar collectors. This is accomplished by the use of solenoid valves, heat-actuated valves, or check valves. Solenoid drain-down valves are illustrated in Figure 4.11.

The control logic of a drain-down system must be designed so that it is "fail-safe" in the event of a power failure. Thus, if electrical power is lost, the system will automatically drain down; otherwise, permanent damage could result. When the system drains down, a relatively small amount of water, generally between 2 and 4 gallons, is lost from the collectors and associated piping. Solar collectors usually contain approximately ½ gallon of water, while ½-in. copper pipe contains 1 gallon for every 92 ft of run. Three-quarter-inch copper pipe contains 1 gallon for every 40 ft of run.

It is important to note that many of the drain-down controllers on the market are designed to permit the use of only one, rather than two, frost sensors. It is emphasized that two frost sensors are mandatory for safe and efficient drain-down operation. When selecting controllers for this purpose, the two-sensor wiring option should be kept in mind.

Another desirable feature in the differential controller is an option referred to as *frost override*, which allows collector-loop circulation even when ambient temperature is below 38°F. Without frost override the system forfeits much available collection time during the winter months, reducing its contribution of winter domestic hot water.

Drain-down systems operate at very high efficiency because neither a heat exchanger nor antifreeze fluid is required. Installation requirements for drain-down systems are very rigid: all piping must be sloped at least ¼ in. per foot to assure proper drainage. Collectors

EQUIPMENT LIST

Part No.	Description
1	Solar Storage Tank
2	Collector
3	Pressure & Temperature Relief Valve
4	Circulator
5	Isolation Flange
6	Normally Closed Solenoid Valve
7	Drain Down Solenoid Valve
8	N.C. Backflow Solenoid Valve
9	Air Vent
10	Drain Down Controller
11	Collector Plate Sensor
12	Storage Sensor
13	Collector Frost Sensor
14	Ambient Frost Sensor

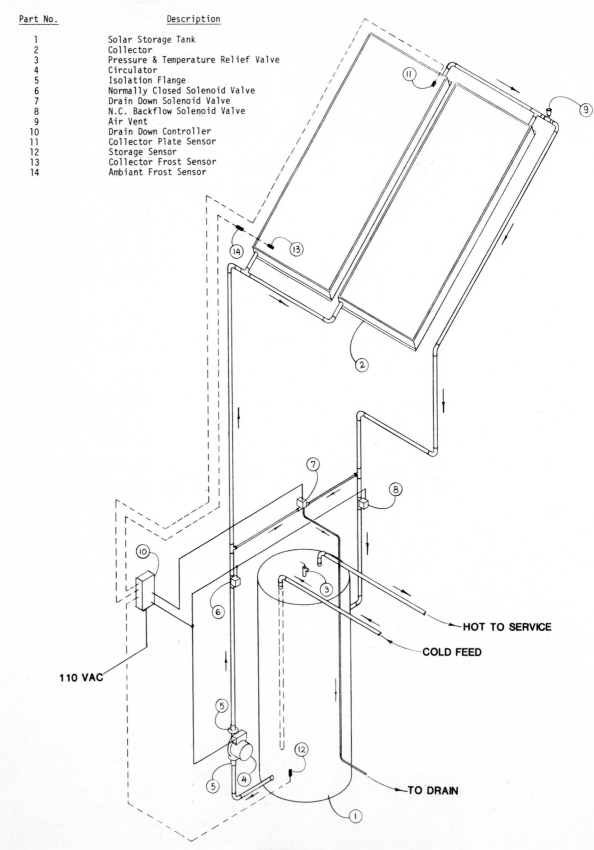

FIGURE 4.11 Drain-down system.

with internal header pipes should not be used in drain-down systems because they will freeze and rupture, due to their inability to drain properly. Very reliable and highly efficient drain-down systems are available that will perform satisfactorily for many years with little or no maintenance required.

Drain-Back System

A drain-back system is quite different from a drain-down system, although the two terms are often confused. A drain-back system uses a heat exchanger and therefore an indirect system of heat transfer, whereas in a drain-down system direct heat transfer is used. In a drain-back system the fluid in the collector loop is water and is not pressurized. During sunless periods, the collector-loop fluid is housed in the bottom portion of the collector-loop piping, which often contains a small tank that acts as a fluid reservoir. A drain-back system is illustrated in Figure 4.12. When the differential controller energizes the circulator, the water must be pumped with enough force to lift it up through the collectors. When the circulator is deenergized, the water drains back to its original location at the bottom of the system. In this system, water is never drained to waste as it is in a drain-down system. Assuming no leaks or evaporation losses, the same fluid may be used for the life of the system. An expansion tank is not required in drain-back systems because the fluid in the collector loop is not pressurized.

Drain-back systems require the most powerful circulators of all the domestic hot water and process systems, and as a result cost the most to operate. Such circulators should be carefully sized. If the circulator in the drain-back loop is too large, it will waste energy and therefore increase the operating costs of the system. If it is too small, it will not provide proper circulation, increasing the operating temperatures of the collectors and dropping the efficiency of the system. All piping in the drain-back arrangement must be sloped at least ¼ in. per foot to assure proper drainage. Collectors with internal header pipes should not be used in drain-back systems due to inadequate provision for proper drainage of the collector. Although some installers prefer drain-back systems due to their simplicity of design, they present a genuine challenge for the solar designer: Potential freezing remains a major concern, whereas system performance is no better than any other well-designed system utilizing a standard heat exchanger.

Closed-Loop System

The most common type of solar domestic hot water system installed in the United States is the closed-loop system, designed specifically for protection of the system in freezing climates. The collector loop is pressurized with an antifreeze/water solution, and by forced circulation this fluid transfers heat from the collectors to storage water through the use of a heat exchanger. This heat exchanger may be located within the solar storage tank, or it may be mounted externally. When using an external heat exchanger, an additional circulator is required to circulate storage water through the heat exchanger. A typical closed-loop system is illustrated in Figure 4.13.

With recent improvements in system components, such as increased tank insulation, large heat exchanger surfaces, variable-speed controllers, and prepackaged closed-loop assemblies, these systems now have an overall efficiency approaching that of direct transfer systems. There is very little risk of freezing or component failure with a properly designed and installed closed-loop system. Containing few moving parts, when properly designed and installed, the closed-loop system will operate efficiently and cost-effectively for an indefinite period of time. The only maintenance required is a semiannual inspection of the antifreeze chemistry and a check of the pressure gauge occasionally to ensure against leaks or fluid loss resulting from stagnating conditions within the closed loop.

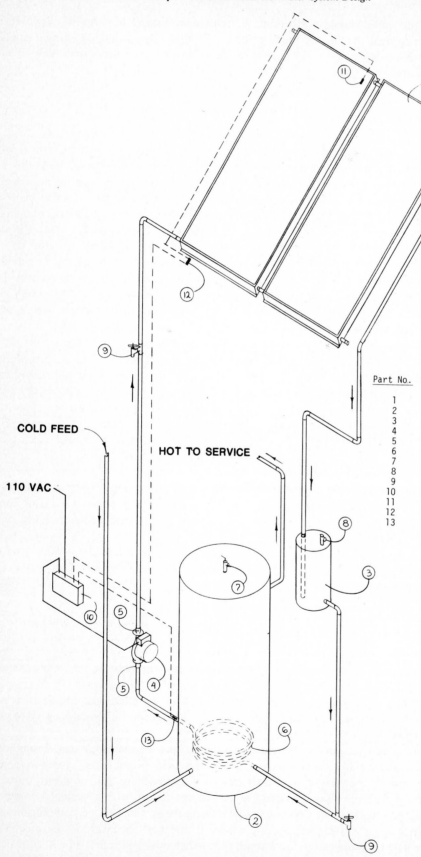

COLD FEED

110 VAC

HOT TO SERVICE

EQUIPMENT LIST

Part No.	Description
1	CEC-24 Collector
2	Solar Storage Tank
3	Reservoir Tank
4	High Head Circulator
5	Isolation Flange
6	Internal Heat Exchanger
7	Pressure & Temperature Relief Valve
8	Pressure Relief Valve
9	Fill & Drain Valve
10	Differential Controller
11	Collector Sensor
12	Extreme Frost Sensor
13	Storage Sensor

FIGURE 4.12   Drain-back system.

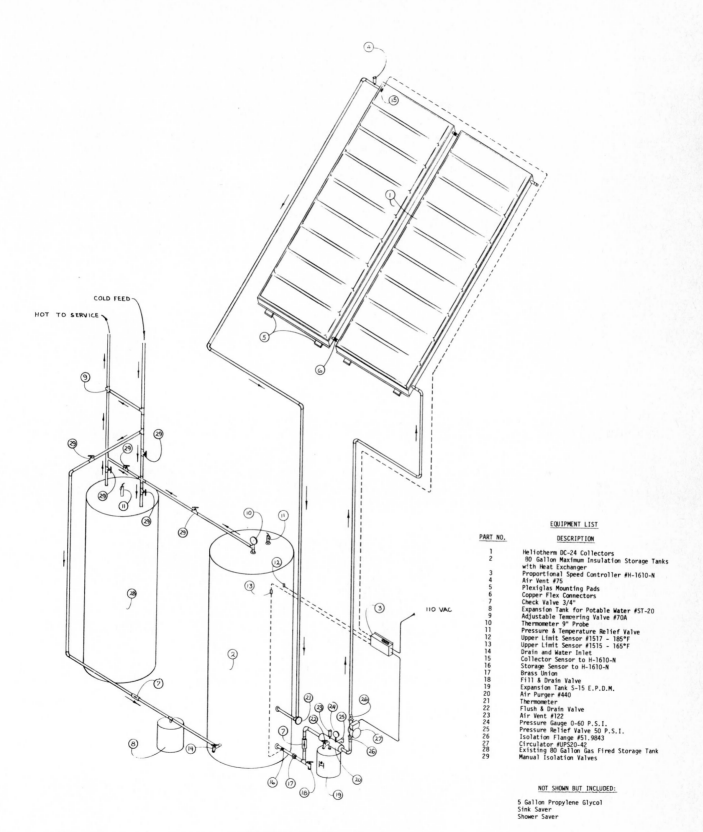

EQUIPMENT LIST

PART NO.	DESCRIPTION
1	Heliotherm DC-24 Collectors
2	80 Gallon Maximum Insulation Storage Tanks with Heat Exchanger
3	Proportional Speed Controller #H-1610-N
4	Air Vent #75
5	Plexiglas Mounting Pads
6	Copper Flex Connectors
7	Check Valve 3/4"
8	Expansion Tank for Potable Water #ST-20
9	Adjustable Tempering Valve #70A
10	Thermometer 9" Probe
11	Pressure & Temperature Relief Valve
12	Upper Limit Sensor #1517 - 185°F
13	Upper Limit Sensor #1515 - 165°F
14	Drain and Water Inlet
15	Collector Sensor to H-1610-N
16	Storage Sensor to H-1610-N
17	Brass Union
18	Fill & Drain Valve
19	Expansion Tank S-15 E.P.D.M.
20	Air Purger #440
21	Thermometer
22	Flush & Drain Valve
23	Air Vent #122
24	Pressure Gauge 0-60 P.S.I.
25	Pressure Relief Valve 50 P.S.I.
26	Isolation Flange #51.9843
27	Circulator #UPS20-42
28	Existing 80 Gallon Gas Fired Storage Tank
29	Manual Isolation Valves

NOT SHOWN BUT INCLUDED:

5 Gallon Propylene Glycol
Sink Saver
Shower Saver

FIGURE 4.13 Closed-loop system.

**SYSTEM
PERIPHERAL
COMPONENTS**

Heat Exchangers

The internal storage tank heat exchanger in the smaller closed-loop system is normally located in the bottom of the stone-lined storage tank. While the collector-loop fluid moves by forced convection throughout the system, transfer of heat between the heat exchanger and surrounding storage water takes place by natural convection.

The heat exchangers utilized in solar applications are most often constructed from either type L or type K rated copper tubing.* Stainless steel alloys are also available for solar heat exchangers if corrosive fluids will be utilized in the system. The raised fins on the surface of the heat exchanger increase available surface area exposed to the water for heat transfer. After the heat exchangers have been wound into a coil, they are inserted into the bottom of the storage tank and brazed into place (Figure 4.14). Proper mounting of the heat exchanger within the tank is critical to efficient coil performance. The exchanger must be high enough off the bottom of the tank to prevent sediment from covering the coils, yet low enough to permit adequate convective currents to keep a fresh supply of cold water circulating over the coils. The actual square footage of surface area of the heat exchanger will depend on the manufacturer, but the most common sizes are 10, 15, 20, and 40 ft². In order to have a heat exchanger with 40 ft² of surface area, nearly 80 linear feet of raised-fin ¾-in. copper tubing is employed. The dependence on natural convection and conduction for heat transfer within the tank dictates the relatively large surface area of the heat exchanger. Other types of heat exchangers, such as shell-and-tube heat exchangers that employ forced convection of both the collector fluid and storage water, require less surface area than natural convection heat exchangers. However, the cost of such external heat exchangers, plus the additional circulator and plumbing required, often makes them less cost-effective for use in small systems. The internal heat exchanger is generally the best option for residential closed-loop systems.

The size of the heat exchanger to be used must be determined on the basis of the energy output expected from the collector array. The size, number, and efficiency of the collectors, as well as the amount of solar insolation available at a given geographical location, must be considered as well. Methods used to determine the proper size of the collector array for a particular application are discussed in Chapter 8.

FIGURE 4.14 Placement of heat exchanger in stone-lined water heater. (Courtesy of Ford Products Corp.)

---

*Copper pipes are rated as K, L, and M. M pipe is soft copper for light-duty applications. L pipe is hard-copper pipe for applications up to 125 psi, and is generally used in residential plumbing systems. K pipe is industrial-strength pipe suitable for high-temperature high-pressure applications.

Once the proper square footage of collector surface area is determined, the size of the heat exchanger required can be established. In general, 0.30 to 0.45 ft² of heat exchanger surface area is required for each square foot of collector aperture area when the heat exchanger is of the internal tank type. When an external shell-and-tube type of heat exchanger is employed, 0.10 to 0.15 ft² of heat exchanger per square foot of collector aperture area will suffice.

Pressure Gauges and Expansion Tanks

The pressure gauge is used to determine if there is a sufficient quantity of fluid within the collector loop. If the loop is not totally filled, the circulator will attempt to lift the fluid rather than merely circulate it through the system, causing fresh oxygen continually to enter the air vent at the top of the collector array. This leads to oxidation of the closed-loop components. The circulators used in these systems are not designed to lift fluid, but rather to move the fluid within the loop. A typical pressure gauge utilized in solar applications is illustrated in Figure 4.15.

Pressure gauges utilized in solar systems generally read in terms of pounds per square inch (psi). The pressure of the collector-loop fluid is caused by the vertical height of the system and the compression of air within the closed-loop expansion tank, part 19 in Figure 4.13. As the collector loop is filled, the fluid level increases in elevation. This increase in elevation will cause a proportional increase in pressure at the pressure gauge. Each 1-ft increase in elevation above the pressure gauge will cause a 0.433-psi pressure increase. Each 1-psi increase is equal to a 2.31-ft increase in fluid elevation. If the number of vertical feet between the pressure gauge and the air vent at the top of the system is known, the desirable pressurization of the system can be calculated to fill the entire collector loop properly by multiplying the height by 0.433.

For fluids with a specific gravity of approximately 1.0:

$$1 \text{ psi} = 2.31 \text{ ft of head}$$

$$1 \text{ ft of head} = 0.433 \text{ psi} \tag{4.1}$$

It is good practice to overfill the collector loop by about 5 psi. The installer forces additional fluid into the collector loop with a force pump. The injection of the fluid into the collector loop compresses the air contained within the expansion tank. The air within the ex-

FIGURE 4.15 Pressure gauge. (Courtesy of Marsh Instrument Co., Unit of General Signal Corp.)

pansion tank is separated from the collector-loop fluid by an EPDM rubber diaphragm. The diaphragm allows the expansion tank to be installed within the system in any position and prevents the air cushion from being dissolved into the fluid within the loop and being eliminated by the action of the automatic air vent. The extra fluid injected into the system in this manner is held in reserve. If there are small pockets of air trapped within the collector-loop piping which are eliminated, or as small quantities of fluid are removed from the loop for periodic testing, the reserve fluid within the expansion tank will automatically be pushed into the system to maintain proper system pressure and fluid level. The diaphragm type of expansion tank is illustrated in cross section in Figure 4.16.

An air purger is a very helpful device for solar designers to incorporate into the collector loop. It helps to eliminate air trapped within the fluid and is a very inexpensive item which will pay for itself during the initial system flushing and pressurization. A typical air purger is illustrated in Figure 4.17.

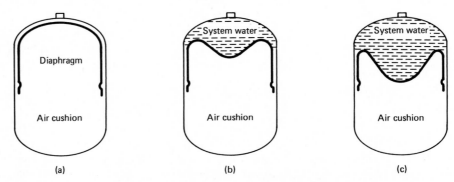

FIGURE 4.16 Expansion tank operation. (a) When the system is first filled with cold water, the charge pressure keeps the diaphragm flush against the tank's inner lining. (b) As the system comes up to temperature, the expanded water is received by the expansion tank. (c) As the water temperature reaches its maximum, the diaphragm flexes against the air cushion to allow for the increased water expansion. (Courtesy of Amtrol, Inc.)

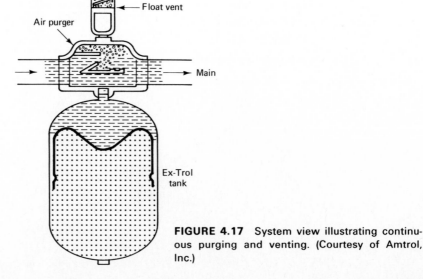

FIGURE 4.17 System view illustrating continuous purging and venting. (Courtesy of Amtrol, Inc.)

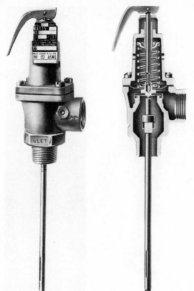

FIGURE 4.18 Pressure relief valve. (Courtesy of Watts Regulator Co.)

Pressure Regulation

A pressure relief valve must be included within the collector loop. This valve prevents the buildup of potentially dangerous pressures within the loop and automatically vents these pressures when they reach a preset limit. The maximum working pressure of some of the components within the closed loop is 75 psi. Exceeding this pressure will damage these components and can cause a rupture within the closed loop. A collector loop with the high point of the system located 50 ft above the pressure gauge requires a fluid pressure of 27 psi (50 ft × 0.433 + 5). Therefore, a pressure relief valve which is designed to discharge at 50 psi is a good choice for almost all solar systems. A typical pressure relief valve for this purpose is illustrated in Figure 4.18.

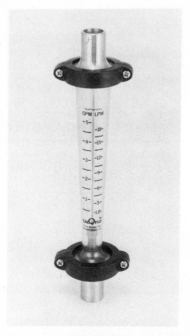

FIGURE 4.19 Flow meter. (Courtesy of Blue White Industries.)

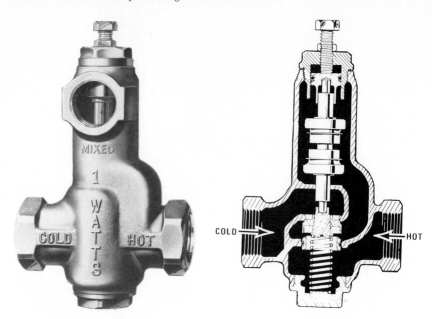

FIGURE 4.20 Hot water mixing valve. (Courtesy of Watts Regulator Company.)

Flow Meters

A flow meter is an instrument which gives a visual indication of the amount of fluid moving within a pipe. Such devices are useful in a solar system to ensure proper operation of both the circulator and the differential controller (Figure 4.19).

Mixing Valves

A mixing valve is a device used to deliver water at a predetermined temperature into the hot water service lines. These valves are installed on the output side of the hot water tank or tankless coil. Both the cold and hot water feed lines enter the valve. The user selects the water temperature by simply adjusting a dial. A heat-sensitive regulator, usually consisting of a coiled bimetallic spring or similar sensing element, automatically regulates the amount of hot and cold water entering the valve to deliver water at the desired temperature level (Figure 4.20).

Table 4.1   PROPERTIES OF SOLAR SYSTEM FLUIDS

Fluid	Specific Heat (Btu/lb-°F)	Freezing[a] Point (°F)	Boiling Point (°F) at 30 psi	Density (ft³)	Viscosity	Heat Conduction
Water	1.0	32	220	62	Low	Good
Propylene glycol and water (50/50)[b]	0.88	−28	240	63	Low	Good
Synthetic oil	0.58	NA[a]	NA	52	High	Poor
Silicone oil	0.36	NA[a]	NA	57	High	Poor

[a]NA, not applicable.

[b]Ethylene glycol and water mixtures have properties very similar to propylene glycol and water but are highly toxic and therefore in limited use for solar application.

Heat Transfer Fluids

There are a number of different fluids available for use in the solar collector loop. The fluid should be chosen based on its cost, operating temperature range, specific heat, viscosity, surface tension, toxicity, longevity, and heat transfer ability. The properties for various collector fluids are listed in Table 4.1.

PREPACKAGED CLOSED-LOOP MODULE

A recent system improvement by solar system manufacturers is the prepackaged closed-loop assembly. As the popularity of domestic hot water systems increased, system manufacturers often found that the installers were responsible for incorrectly designing and installing the closed-loop assembly components. This caused premature system failure and/or poor system efficiency. To remedy this situation, many system manufacturers began to factory assemble, inspect, and test closed-loop assemblies to help ensure proper installation and operation. Also, prepackaging helped to ensure that all components were properly matched and engineered for the specific applications in which they were to be used. An example of a prepackaged closed-loop module is illustrated in Figure 4.21.

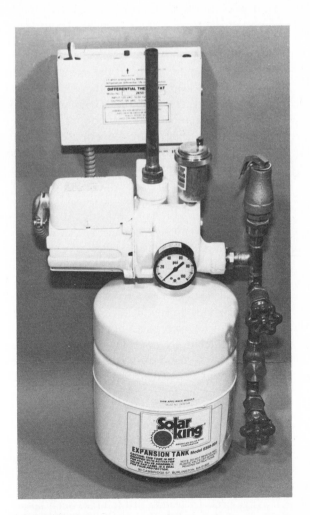

FIGURE 4.21 Factory-assembled closed-loop module.
(Courtesy of American Solar King.)

COMMERCIAL AND INDUSTRIAL (PROCESS) HOT WATER SYSTEMS

Industrial solar energy applications are generally referred to as *process* applications. Process loads are normally the result of some stage of manufacturing or product treatment, and the uses for warm and hot fluids in industry are numerous.

Industrial and commercial process hot water systems are simple in design and have proven to be very cost-effective. Large system design duplicates that of the smaller residential domestic hot water systems in most respects. These larger systems can be of any type, chosen on the basis of climatic conditions. However, most large process systems in the United States are closed-loop antifreeze systems. Generally, only small modifications of the design of the system are required to change from a closed-loop to other types of collector-loop arrangements.

Some of the best industrial solar applications occur in situations where water must be heated from street temperature continually. These situations occur in cleaning operations and food processing. For example, the poultry processing industry is an excellent example of this type of application. Many thousands of gallons of well and river water are needed for two major operations; some water is used to remove feathers from the animals and is heated from temperatures as low as 32°F up to 126°F; other water is simply warmed from ground temperature up to 70°F so that the hands of the workers will be comfortable. These are excellent solar applications because of the low requested water temperatures being used and the need to heat fresh cold water daily. On the other hand, a situation where water is simply maintained at 126°F rather than being freshly heated on a continual basis would not allow the collectors to operate efficiently since their operating temperatures would be fairly high. However, applications where very high temperatures are requested can be good solar applications as long as fresh groundwater or well water is heated daily. In this manner, the solar system is used for preheating and therefore used efficiently.

In addition to low- and medium-temperature requirements, the pattern of hot water usage is also important to the operating efficiency of the solar system. The best potential situations for solar energy are those in which hot water usage coincides with the availability of solar energy. For example, a commercial photographic laboratory is an excellent solar application because a constant flow of warm water is required for the film processing baths. Also, these facilities are often in use 5 to 7 days per week. In the 7-day-use situation, a solar system can be designed with no storage, since the collector array can be sized so that the resulting solar contribution never exceeds the use requirement. Normally, however, a small amount of solar storage is incorporated into the system in order to handle short periods of reduced system demand.

Process hot water systems will vary in size and can feature from one or two collectors to several hundred in the array. As with any solar system, the primary design feature is to ensure that the collectors operate as coolly as possible at all times. This requires adequate and equal fluid circulation throughout all the collectors. Piping, valving, circulation, and heat exchangers must all be properly sized to minimize friction losses and pressure drop. Friction losses caused by pipes of various lengths and diameters and assorted fittings are discussed in greater detail in Chapter 5. Properly designed and installed, process systems will provide many years of cost-effective service.

A relatively small seven-collector process system is depicted in Figure 4.22. As can be seen, this system is primarily an enlarged residential domestic hot water system. The solar storage tanks, part 17, are plumbed in parallel to decrease friction losses and pressure drop resulting from high water demands placed on the system. Two 120-gallon storage tanks are used rather than one larger tank because pressure-rated storage tanks with internal heat exchangers specifically designed for solar system use are not manufactured in tank sizes larger than 120-gallon capacities in most situations. An external heat exchanger and an additional circulator can be used with one 240-gallon tank, but the cost will be higher. The use of larger tanks also presents a number of other problems. In many applications, tanks must have American Society of Mechanical Engineers (ASME) certification. ASME approval requires high-pressure testing of the tank. Such large tanks must be galvanized or otherwise specially lined to prevent internal tank corrosion. Galvanized steel tanks should not be used to store

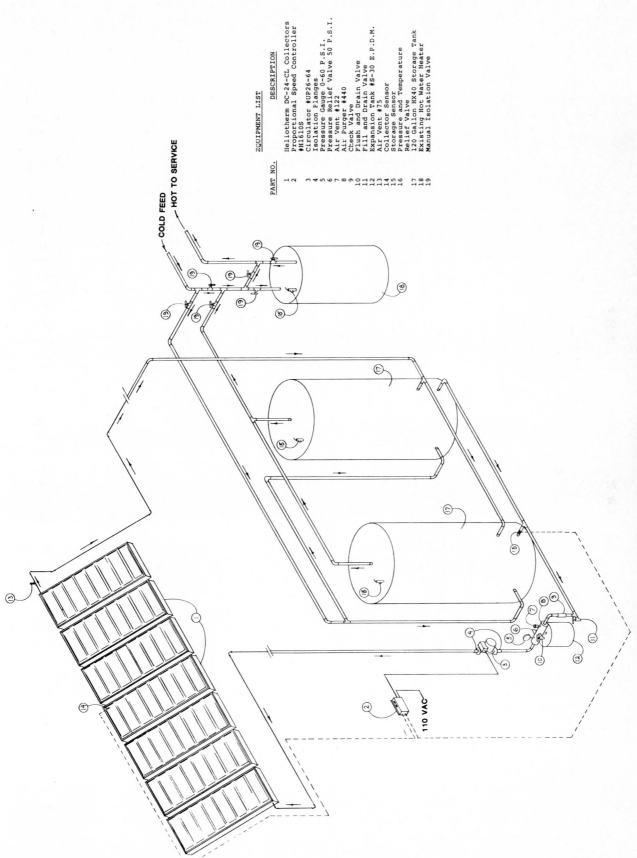

EQUIPMENT LIST

PART NO.	DESCRIPTION
1	Heliotherm DC-24-CL Collectors
2	Proportional Speed Controller #H1610S
3	Circulator #UP26-64
4	Isolation Flanges
5	Pressure Gauge 0-60 P.S.I.
6	Pressure Relief Valve 50 P.S.I.
7	Air Vent #122
8	Air Purger #440
9	Check Valve
10	Flush and Drain Valve
11	Fill and Drain Valve
12	Expansion Tank #S-30 E.P.D.M.
13	Air Vent #75
14	Collector Sensor
15	Storage Sensor
16	Pressure and Temperature Relief Valve
17	120 Gallon HX40 Storage Tank
18	Existing Hot Water Heater
19	Manual Isolation Valve

COLD FEED

HOT TO SERVICE

110 VAC

FIGURE 4.22 Seven-collector domestic hot water system. (Courtesy of Heliotherm.)

water at temperatures over 165°F because they are subjected to accelerated corrosion at these elevated temperatures. Larger tanks must also be insulated after they are installed. These considerations increase the cost of the single-high-capacity-tank concept for larger systems. The use of smaller 120-gallon tanks with internal heat exchangers are generally more economical in these instances. There are two factors, however, that work against the use of small solar storage tanks: greater heat loss and additional plumbing required for their installation. As the system becomes larger, the cost advantage of the smaller-tank concept begins to diminish. The additional heat loss and plumbing requirements will reach the point of making small tank usage economically impractical. The point at which a single large tank and external shell-and-tube heat exchanger becomes the best design option is reached when the water storage in the system reaches between 500 and 700 gallons.

Regardless of how many tanks are used, each should have its own pressure and temperature relief valve located at the top of the unit. Standard relief valves are designed to open between 125 and 150 psi and/or 210°F. In some instances, especially when very large tanks are used, a vacuum breaker should also be used in conjunction with the pressure/temperature relief valve. Storage tanks are designed only to withstand internal water pressure, not external pressure that would result should a vacuum develop within the tank. The incorporation of the vacuum breaker allows atmospheric pressure to enter the tank, preventing damage. A typical vacuum breaker valve suitable for solar applications is illustrated in Figure 4.23.

The double-bypass arrangement of the plumbing circuitry, illustrated in the right-hand portion of Figures 4.24 and 4.25, is ideal for the solar and auxiliary hot water interface. This double-bypass plumbing arrangement allows the complete isolation and bypass of either the solar or the backup hot water system. This permits servicing of either system without disturbing the flow of hot water to service. Care must be taken, however, when isolating the solar storage tanks. If the tanks are pressurized and isolated, and allowed to be heated by the collector array at this time, the storage water will become less dense, expand, and possibly cause dripping of the pressure/temperature relief valves. To help prevent this occurrence, potable water expansion tanks should be plumbed into the system to absorb this expansion. This arrangement is illustrated in Figure 4.25. This tank, specifically designed to operate under higher working pressures than closed collector-loop expansion tanks, is illustrated as part 14. A check valve, part 15, is often installed in the cold water feed pipe to prevent hot water expansion from filling the cold water feed pipe line with solar heated water. A 0- to 150-psi pressure gauge, part 20, should be installed whenever potable water expansion tanks and check valves are employed within the system. The use of this pressure gauge will alert the owner to an undersized expansion tank, or to a loss of its air charge.

Large collector arrays must be divided into subarrays of no more than 10 or 12 collectors. The reason for this is that when too many collectors are plumbed together in a row, the flow rates through them tend to be unequal. The collectors located in the center of a

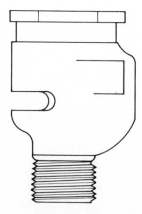

FIGURE 4.23 Vacuum breaker valve.

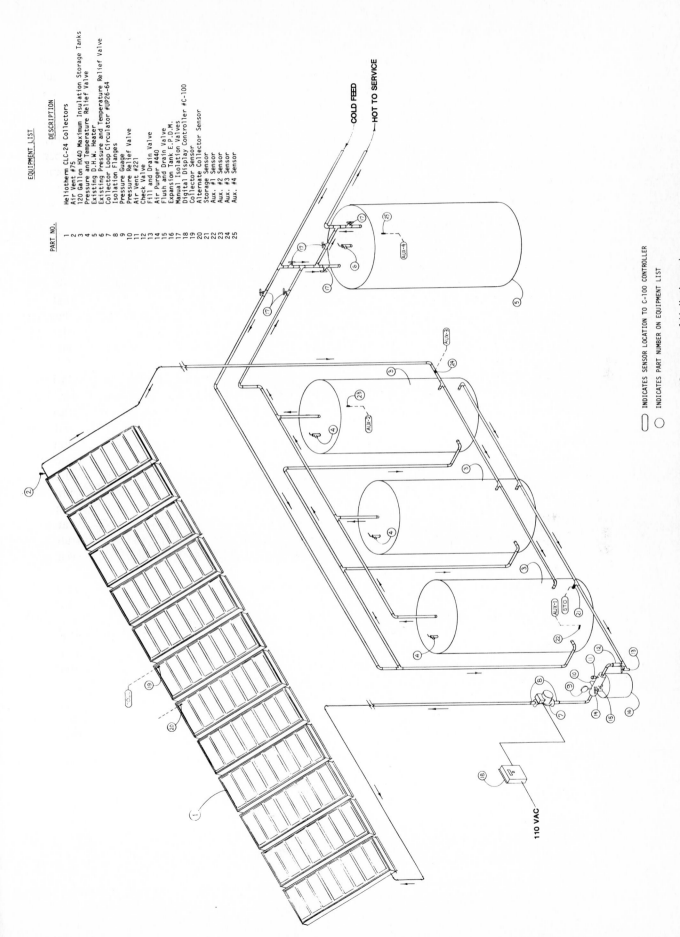

EQUIPMENT LIST

PART NO.	DESCRIPTION
1	Heliotherm CLC-24 Collectors
2	Air Vent #75
3	120 Gallon HX40 Maximum Insulation Storage Tanks
4	Pressure and Temperature Relief Valve
5	Existing D.H.W. Heater
6	Existing Pressure and Temperature Relief Valve
7	Collector Loop Circulator #UP26-64
8	Isolation Flanges
9	Pressure Guage
10	Pressure Relief Valve
11	Air Vent #221
12	Check Valve
13	Fill and Drain Valve
14	Air Purger #440
15	Flush and Drain Valve
16	Expansion Tank E.P.D.M.
17	Manual Isolation Valves
18	Digital Display Controller #C-100
19	Collector Sensor
20	Alternate Collector Sensor
21	Storage Sensor
22	Aux. #1 Sensor
23	Aux. #2 Sensor
24	Aux. #3 Sensor
25	Aux. #4 Sensor

COLD FEED

HOT TO SERVICE

110 VAC

☐ INDICATES SENSOR LOCATION TO C-100 CONTROLLER

○ INDICATES PART NUMBER ON EQUIPMENT LIST

FIGURE 4.24 Twelve-collector domestic hot water system. (Courtesy of Heliotherm.)

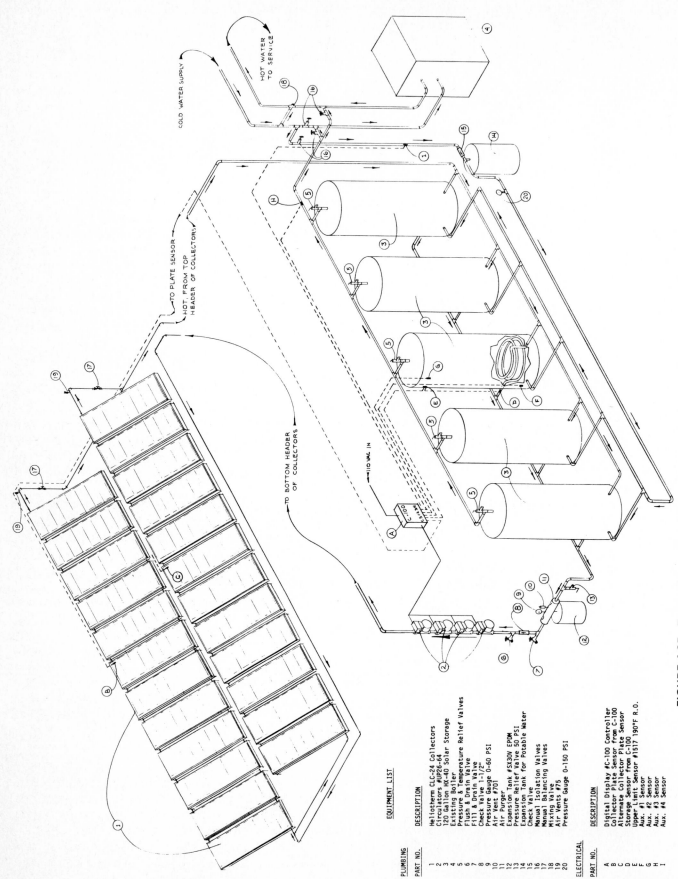

FIGURE 4.25 Twenty-four-collector domestic hot water system. (Courtesy of Heliotherm.)

COLD WATER SUPPLY

HOT WATER TO SERVICE

TO PLATE SENSOR

HOT FROM TOP HEADER OF COLLECTORS

TO BOTTOM HEADER OF COLLECTORS

110 VAC IN

EQUIPMENT LIST

PLUMBING

PART NO.	DESCRIPTION
1	Heliotherm CLC-24 Collectors
2	Circulators #UP26-64
3	120 Gallon HX-40 Solar Storage
4	Existing Boiler
5	Pressure & Temperature Relief Valves
6	Flush & Drain Valve
7	Fill & Drain Valve
8	Check Valve 1-1/2"
9	Pressure Gauge 0-60 PSI
10	Air Purger
11	Air Vent #701
12	Expansion Tank #SX30V EPDM
13	Pressure Relief Valve 50 PSI
14	Expansion Tank for Potable Water
15	Check Valve
16	Manual Isolation Valves
17	Manual Balancing Valves
18	Mixing Valve
19	Air Vents #75
20	Pressure Gauge 0-150 PSI

ELECTRICAL

PART NO.	DESCRIPTION
A	Digital Display #C-100 Controller
B	Collector Plate Sensor from C-100
C	Alternate Collector Plate Sensor
D	Storage Sensor from C-100
E	Upper Limit Sensor #1517 190°F R.O.
F	Aux. #1 Sensor
G	Aux. #2 Sensor
H	Aux. #3 Sensor
I	Aux. #4 Sensor

long array will receive less fluid than those on the ends. In order for the solar designer to provide equal flow rates, and therefore equal cooling for every collector, the total number of collectors should be divided into equal parts if possible. These subarrays should then be separately balanced with ball valves when the system is first started up. It is good design procedure to install separate temperature sensors located in a center collector of each subarray to facilitate flow balancing. The valves are manipulated until the temperature sensors located in each central collector give the same temperature value when the system is in full operation, ensuring proper flow rates.

Proportional-speed controllers should be used whenever possible to eliminate the need to set the collector-loop fluid flow rate. Controllers that have the capability of digital temperature display of various points throughout the system are very helpful in monitoring system performance. This controller feature is especially useful in performing the subarray balancing discussed. It should be kept in mind that the output power rating of any differential controller is limited. For starting large circulators, a heavy-duty motor starter relay, illustrated in Figure 4.26, part D, should be installed. This will prevent the controller from being overloaded by the heavy starting current draw of circulators utilized in large systems. Controllers often require a minimum output load of 10 to 15 watts for proper operation, so care must be taken to provide at least that much load on the output side of the controller at all times. A heat-generating resistor may sometimes be required to be wired in parallel with the motor starter relay coil.

The motor starter relay depicted in Figure 4.26 is one of those used in larger commercial domestic hot water systems, which often include a design feature known as a *recirculation loop*. Recirculation loops are sometimes employed in larger domestic hot water systems because of the distance between the hot water fixtures and the hot water supply tank. This distance often requires the hot water tap to be turned on, allowing the water to run for several minutes until the hot water reaches the fixtures. To prevent water from being wasted in this manner, use of the recirculation loop features instant hot water at any location within the system because it is constantly circulated from the most distant hot water service location in the building back to the hot water heater. A great deal of energy can often be expended for the convenience of instant hot water in this arrangement. The circulation is accomplished with a relatively small circulator, part 5, and an aquastat, part B, to control the flow within the recirculation loop. If the aquastat senses that the temperature of the water within the hot water pipe has dropped too low (generally 110 to 120°F), the circulator is automatically switched on, bringing the water temperature in the hot water pipe up to a preset limit (generally between 130 and 140°F). Due to the generally wasteful way in which energy is employed in standard recirculation loops, an interesting and energy-saving option which can be designed into any solar hot water system employing circulation is to allow the solar storage tank to provide all or some of the heat required for the recirculation loop when it is available. This can be done by using one additional aquastat, part C, and two solenoid valves, part 16. The additional aquastat is wired from the existing circulator as illustrated in the wiring diagram of Figure 4.26. This aquastat senses the hottest part of the solar storage tank. The solar aquastat should be set to divert the flow of the recirculation loop through the solar storage tank when this tank exceeds the temperature setting of the existing recirculation aquastat, part B. In this way the solar system will keep all the hot water pipes hot within the system, and thus the collectors will also operate more efficiently. All the hot water piping in this arrangement should be well insulated.

DESIGN FOR THE APPLICATION

Before the design of any solar system is undertaken, the requirements of the specific application must always be carefully considered. For example, liquid flat-plate solar collectors have an effective operating range of up to 180°F, but they operate most efficiently at much lower temperatures. This must be considered in the initial stages of any solar design. Using flat-plate collectors, for example, to help maintain an electrically heated storage tank at 160°F by trying to solar-heat that tank directly is not good solar design. The operating tempera-

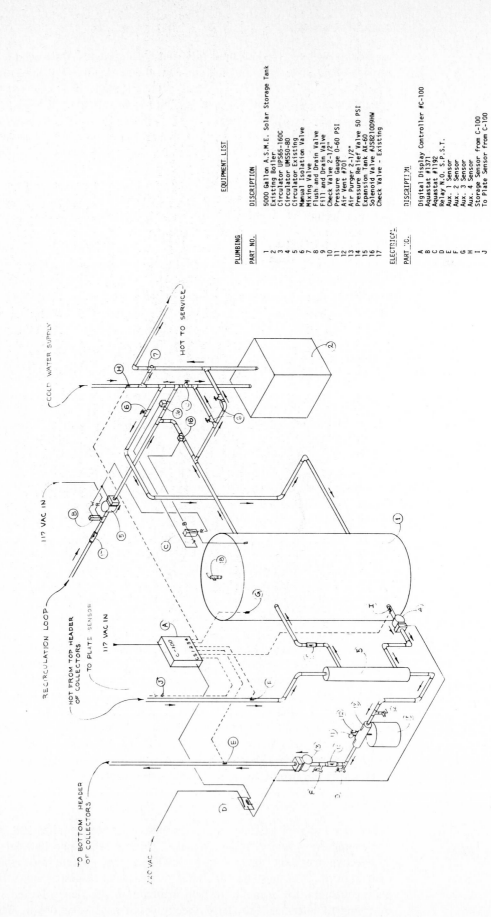

FIGURE 4.26 Commercial/industrial solar domestic hot water system.

EQUIPMENT LIST

PLUMBING

PART NO.	DISCRIPTION
1	5000 Gallon A.S.M.E. Solar Storage Tank
2	Existing Boiler
3	Circulator UPS65-160C
4	Circulator UMS50-80
5	Circulator Existing
6	Manual Isolation Valve
7	Mixing Valve
8	Flush and Drain Valve
9	Fill and Drain Valve
10	Check Valve 2-1/2"
11	Pressure Gauge 0-60 PSI
12	Air Vent #701
13	Air Purger 2-1/2"
14	Pressure Relief Valve 50 PSI
15	Expansion Tank AX-60
16	Solenoid Valve #JS821009HW
17	Check Valve - Existing

ELECTRICAL

PART NO.	DISCRIPTION
A	Digital Display Controller #C-100
B	Aquastat #1371
C	Aquastat #1192
D	Relay N.O. S.P.S.T.
E	Aux. 1 Sensor
F	Aux. 2 Sensor
G	Aux. 3 Sensor
H	Aux. 4 Sensor
I	Storage Sensor from C-100
J	To Plate Sensor from C-100

tures of the collectors will be high, reducing their operating efficiency. Preheating fresh cold water before it enters the electrically heated tank is good solar design, however, because the collectors will operate at lower temperatures, thus increasing their efficiency.

Many of the decisions made by the designer and installer must be based on information about the existing auxiliary hot water heating system. A site inspection that covers everything from collector and solar storage location to the interface with the auxiliary system is prerequisite to good solar design. All the critical components of the system, such as circulators, air purgers, storage tanks, and heat exchangers, must be chosen on the basis of characteristics needed for a specific installation.

System Sizing: Preliminary Design

Before a solar system can be designed, information must be obtained concerning the intended use and the desired energy contribution of the system. Potentially poor solar energy applications should be eliminated from consideration during the early stages of project development. A cost-effective flat-plate collector solar system is one that will operate within a temperature range 35 to 150°F and be in use all year long. For example, a solar domestic hot water system for a weekend vacation home, or a flat-plate collector system used to maintain industrial process water at 180°F, are poor solar applications. In the first example, a weekend home limits full potential use of the solar system, while the industrial process system forces the collectors to operate inefficiently at high temperatures. The proper questions must be asked during the initial design phases of a project to gain information necessary for correct system design.

Prospective solar system purchasers are rarely aware of the size or price of the system required to provide desired energy contributions. Too often, solar energy is looked upon as the ultimate solution to a person's energy problems. Disappointment soon follows when it is learned that the proposed system will deliver only between 50 and 70% of the required energy contributions, requires a relatively high installation price, and offers a 5- to 10-year payback period. Those desperate for relief from high fuel costs should be made aware that high energy loads require large and expensive solar systems. A short cut sizing and pricing method, if used properly, can prevent wasted time and engineering skills by first qualifying the prospective system.

The six basic steps for thorough preliminary design are:

1. Determine energy load.
2. Determine requested solar contribution.
3. Estimate required number of solar collectors.
4. Observe and qualify site characteristics.
5. Estimate total system cost.
6. Make final modifications in system design.

A preliminary design analysis for a domestic hot water system is, in general, relatively simple to undertake. The first step in the procedure is to determine the hot water load. This is accomplished using one or more of the following three methods.

Method 1: Use of national averages for per capita hot water consumption. The most widely used figure for consumption in residential applications is 20 gallons per person per day. For apartments, the figure drops to 18 gallons. Office buildings and nonprocess applications use 1 to 2 gallons per person per day.

Method 2: Calculations of hot water loads from fuel consumption data. Fuel consumption data can be used to calculate the hot water load by converting fuel consumed to a Btu value, multiplying this figure by the overall efficiency of the hot water heating system, and converting the resulting Btu figure to gallons of hot water. The final step in this method takes into account changes in incoming water temperature due to seasonal variations. City water supplies, for example, often operate at close to freezing temperatures in the winter and can warm up to 80°F in the summer.

Method 3: Measurement of hot water directly with a water meter. This method need not be drawn out over a long period of time. Fortunately, hot water consumption remains relatively constant from month to month in most situations.

Practical applications of these methods allow relatively quick system sizing.

Example 4.1

An apartment building has a separate oil-fired hot water heater. During the month of July, when the central heating boiler is shut off for servicing, the incoming water temperature was measured at 75°F, the set temperature was 130°F, and the oil consumption was 1000 gallons of No. 2 fuel oil. Heat losses from the recirculation loop and standby tank losses reduce the overall efficiency of the water heating system to 50%. What is the average daily domestic hot water (DHW) consumption for the building?

Solution

1000 gallons oil $\times$ 140,000 Btu/gal $\times$ 0.5 efficiency = 70 million Btu

$$\frac{70 \text{ million Btu}}{8.33 \times (130°F - 75°F) \times 31 \text{ days}} = 4929 \text{ gallons DHW per day}$$

Once the hot water load per day has been calculated, it is divided by the maximum number of gallons of hot water produced daily by one collector. The resulting figure yields the number of collectors required for a cost-effective system.

If the number of collectors necessary for a given solar system thermal performance can be quickly estimated, various site characteristics such as available roof area, orientation and inclination, and the available area for solar storage can be worked into initial pricing calculations to yield approximate installed system costs. In some cases the need for collector racks or specialized storage facilities was previously misunderstood or overlooked entirely. The cost for these items can result in a negative purchasing decision. Since site characteristics have a significant impact on the installed price of most solar systems, they can limit the size of the system in order to minimize total capital outlays. Lack of adequate space for collectors or limited area for solar storage might act to reduce the overall size of the system. While overall energy contribution from a smaller solar system decreases as opposed to a larger system, the actual energy delivered per collector will increase. For example, a 60-collector system for a public school's domestic hot water might supply 50% of the yearly requirements, while a 30-collector system in the same situation will supply 34% of the annual hot water load. The smaller system in this instance delivers more energy per collector since they will be operating more efficiently at lower temperatures. The 34% contribution system may never deliver solar heated water to the building at the requested temperature of 120°F, but will nonetheless pay for itself more quickly than will the larger system.

Thus initial discussions concerning system pricing and sizing should pinpoint the expectations of delivered system energy. At this time, the magnitude of the heating load and the approximate temperatures of collector operation can be used to estimate the number of collectors needed for the system.

Example 4.2

Using the hot water requirement calculated in Example 4.1, determine the number of collectors for a preliminary design analysis, based on a manufacturer's specifications that their collector is known to heat 37 gallons of water from 75°F to 130°F on a warm sunny day.

Solution

$$\frac{5000 \text{ gallons of water per day required}}{37 \text{ gallons per collector}} = 135 \text{ collectors}$$

The value used for "gallons per collector" will vary for each collector and will be dependent on climatic and site conditions as well as the solar system application. This figure must be ascertained either by close observation and monitoring of operating systems, or by computer simulation. Although this value is approximate in nature, it is nonetheless very helpful in preliminary design and pricing calculations. It is important to emphasize that a solar system should not be oversized, and surplus production of hot water should be avoided. It is for this reason that a maximum value of "gallons of water heated per collector" is used. As the production of hot water per collector decreases in poor weather, the yearly contribution of the solar system to the energy load will decrease as well. Thus conservative estimates should be used to maximize accuracy of all preliminary design calculations.

Once an approximate number of solar collectors has been established, the installation site must be examined carefully. The following factors can greatly affect the size and price of the resulting system: location of the solar collectors relative to the storage tanks, the use of roof racks or ground mounts that must be custom designed and fabricated, and available area for solar storage tanks.

Once all these factors have been considered, the simplest method for determining approximate installed cost of the system is to establish a cost per collector and multiply this figure by the number of collectors in the array. Installed costs per collector will vary from one location to another and are influenced by the degree of difficulty of the particular installation. Current values for this figure vary from $1500 to $2200 per collector. If, for example, $1750 per collector is used, the 135-collector system calculated in Example 4.2 will cost approximately $235,250 to install. It should be kept in mind that this figure does not include federal, state, or local tax credits. Nor does it take into account depreciation. Final investment calculations must be made in consultation with an accountant for an accurate representation of final system costs. Computer analyses can be programmed to accept tax credit and depreciation inputs and are discussed in Chapter 8.

TESTING AND CERTIFICATION OF SOLAR DOMESTIC HOT WATER SYSTEMS

The standard procedure for testing and certifying the output of complete solar domestic hot water systems is called ASHRAE 95-1981. This procedure differs from the ASHRAE 93-77 test in that the complete system, including backup auxiliary tanks and electrical consumption, is carefully monitored in a controlled laboratory environment rather than simply predicting collector performance. In the 95-1981 procedure, the tests yield a series of Q numbers to describe how a system operates. The appropriate Q numbers are as follows:

Q_{NET} Considered to be the most important value in the series, Q_{NET} is a measure of the solar energy, in Btu/day, that is delivered by the domestic hot water system being tested. Q_{NET} takes into account all standby energy losses as well as auxiliary energy inputs to the system to deliver a rated amount of hot water.

Q_{DEL} Total amount of daily energy delivered by the solar domestic hot water system, including all energy supplied by either gas or electric for backup water heating.

Q_{LOSS} Standby losses of the backup water heater.

Q_{AUX} Backup energy consumed by the solar system to deliver the required volume of domestic hot water in the testing procedure. If the solar system being tested has no backup heating element or if use of the element was discontinued during the test procedure, no Q_{AUX} figure appears on the certification label of the solar system.

Q_{RES} Reserve capacity of the solar domestic hot water system. The higher this figure, the better able the solar system is to handle variations in domestic hot water loads.

Q_{CAP} In solar system providing backup, this figure is used to measure the energy storage capacity of the auxiliary tank.

Q_{PAR} Q_{PAR} (PAR for "parasitic") represents the amount of energy consumed by circulating pumps, differential controllers, and other devices utilized in the solar system during normal operation.

Thus

$$Q_{\text{NET}} = Q_{\text{DEL}} + Q_{\text{LOSS}} - Q_{\text{AUX}} - Q_{\text{PAR}}$$

As with any testing or certification procedure, these figures are most helpful in comparing the operation of one system with another.

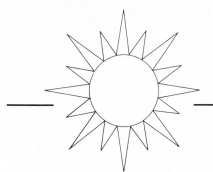

5

Installing a Solar Domestic Hot Water System

Because the widespread use of solar domestic hot water heating systems is a relatively recent phenomenon, many people think that the installation of such a system is a complicated affair. However, with recent manufacturer improvements that focus on prepackaged modules, and simplification of the major solar system components, many domestic hot water systems can be installed by the purchaser. Proper system installation is mandatory if the solar system is to work efficiently and cost-effectively. To this end, a step-by-step guide to the various aspects of a typical domestic hot water installation is presented. Such an installation is usually accomplished in the following seven steps:

1. Researching local building and plumbing codes; obtaining the required permits
2. Installing and interfacing the solar hot water storage tank with the existing domestic hot water supply system
3. Siting, mounting, and plumbing the solar collector array
4. Installation of the peripheral components
5. Electrical connections
6. System flushing and testing
7. Operating techniques and troubleshooting

CODE REQUIREMENTS

Most localities in the United States are governed by local building codes. These codes detail the specific requirements and procedures to be followed during the course of residential and commercial building construction, including the installation of all plumbing, heating, ventilating, and electrical subsystems. In the absence of local town building codes, state codes cover these aspects of the building trades. Some localities may require that a building permit be issued before a solar energy system can be installed on a residential or commercial building. Other localities have no such requirement. To determine what procedures and permits need be obtained prior to an installation, a meeting should be requested with the local building inspector to outline the proposed installation and determine what permits are required. In addition, this meeting will also help determine if there are any special factors that may affect basic system design. For example, some codes may specify the methods of mounting and securing the solar collectors to the roof framing members. Others may specify the type, size, and insulation characteristics of all copper tubing to be used within the solar system's interconnection to the existing domestic hot water system.

In addition to the installation procedures to be followed, the system purchaser is urged to research the area of tax incentives that are often associated with solar system installations. Aside from federal tax incentives, many areas of the country offer property assessment immunity as well as local and state tax incentives to purchasers of solar energy systems. Since these incentives vary widely from one area to another, professional tax consultants can be most helpful in this regard.

INSTALLING AND INTERFACING THE HOT WATER STORAGE TANK

Although the size of the storage tank and heat exchanger will vary with the number of occupants in the home and the size of the solar collector array, the installation procedures for most storage tanks are similar.

To minimize heat loss from the solar system, it is preferable to locate the solar storage tank as close as possible to the existing domestic hot water heater in the home. To help determine if sufficient space is available to install the solar storage tank Figure 5.1 gives the basic dimensions of a typical solar storage tank together with the location of the various plumbing connections.

It is good installation procedure to prevent the bottom of the solar storage tank from coming in direct contact with any flooring surface. Figure 5.2 illustrates the use of pressure-treated dimensional lumber to elevate the solar storage tank, preventing corrosion damage to the exterior tank shell. Also, this type of installation procedure allows air to circulate below the base of the tank, preventing condensation from taking place that often causes dry rot on wooden floors.

Pipe Fastening Techniques

Installation of a solar system requires knowledge of two common methods of joining pipes: threaded connections and sweat soldered connections. A brief description of each follows to aid the installer with these procedures.

Threaded Connections

All threaded connections should be well "doped" before they are inserted into the appropriate fittings. One successful procedure is to apply several windings of Teflon tape to the male fitting. After the tape has been applied, a layer of Teflon pipe dope is brushed over the tape. The two final threads of the fitting should be kept free of both tape and pipe dope to prevent contamination of the female fittings. This procedure will virtually eliminate the possibility of leaking connections. Teflon pipe dope is impervious to the action of glycol/water solutions commonly used in the closed loop of the collector array and is therefore recommended in solar applications.

Soldered Connections

The majority of pipe fastening techniques used in the installation of a solar domestic hot water system involve the technique referred to as *sweat soldering.* Solder is an alloy composed of varying amounts of lead, tin, and antimony. When heated, the solder melts, filling all the spaces in a typical tube-and-socket pipe fitting. As the solder cools, it hardens into a durable, watertight joint that will last for many years.

Most soldering applications use 50/50 or 60/40 solder (these numbers refer to the percentage of tin and lead in the solder, by weight, respectively). For solar applications it is recommended that 95/5 solder (tin, antimony) be used. Due to excessively high temperatures that can develop within the solar system, especially within the vicinity of the solar col-

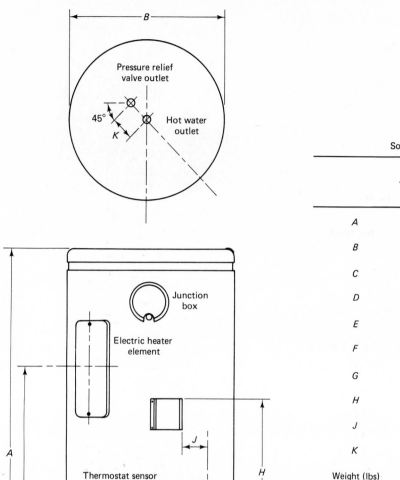

	Solar Tank Dimensions (in.)*		
	Tank Dimensions		
	65 gal	80 gal	120 gal
A	$55\frac{1}{4}$	$69\frac{1}{2}$	68
B	24	24	28
C	38	48	49
D	$17\frac{5}{8}$	$17\frac{5}{8}$	$17\frac{5}{8}$
E	$6\frac{1}{2}$	$6\frac{1}{2}$	$6\frac{1}{2}$
F	15	15	15
G	9	9	9
H	20	20	20
J	$4\frac{1}{2}$	$4\frac{1}{2}$	$4\frac{1}{2}$
K	$3\frac{1}{2}$	$3\frac{1}{2}$	$3\frac{1}{2}$
Weight (lbs)	330	390	500

*All tappings are $\frac{3}{4}$ ips.

FIGURE 5.1 Typical solar storage tank dimensions. (Courtesy of Ford Products Corp.)

lectors, the use of 95/5 solder will eliminate the possibility of ruptured solder joints due to high operating temperatures and pressures.

The following step-by-step guide outlines the basic techniques necessary for successful sweat soldering results.

1. *Cut and ream the copper tubing*: Using a wheel-type tubing cutter, cut the measured end of the copper tubing square. Most tubing cutters contain a reamer which is built into the body of the cutter. Use this reamer to remove burrs from the inside of the copper tube. If no reamer is available, a round file can be substituted for this purpose. If no tubing cutter is available, a hacksaw can be used to cut the pipe. If a vice is to be used to hold the pipe while cutting, make sure that a pipe vice is used rather than a

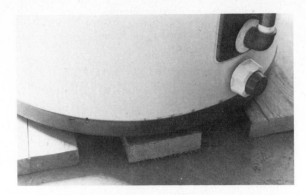

FIGURE 5.2   Elevation of solar storage tank.

flat-jaw vice to eliminate the possibility of flattening out the ends of the soft-copper tubing.

2. *Clean all fittings*: Both the end of the tubing and the socket end of the fitting should be well cleaned. Cleaning involves the removal of copper oxide from the surface of the pipe and fittings. Failure to remove this oxide completely will result in a "cold" joint, one that will continually leak and eventually rupture from system pressure. High-grade aluminum oxide sandpaper, 180 grit or finer, should be used to clean the end of the tubing. The sandpaper should be wrapped around the tubing and rotated until the end of the tubing is bright copper and no oxide appears. To clean the inside of socket-type fittings, wire brushes are sold that are used in conjunction with specific sizes of pipe, and are available ranging in size from ½ to 1 in.

3. *Application of flux*: Soldering flux should be applied liberally to both the tubing end and the socket end of all fittings. Special flux brushes are sold for this purpose. It is not recommended that fingers be used to apply the flux since it is highly irritating if it remains on the skin for short periods of time. After the flux has been applied, the tube and socket ends of the joint should be pushed together and twisted several times to distribute the flux evenly.

4. *Application of heat*: Before any soldering can take place, the joint must be sufficiently heated to allow the solder to melt and run into the joint. Heat should be applied uniformly around the joint. It should be noted that 95/5 solder requires more heat than a standard propane torch can always deliver. Acetylene torches are therefore highly recommended for this purpose. Solder always flows to the hottest part of the joint; therefore, after the flux begins to fry, move the flame directly to the base of the fitting.

5. *Application of solder*: After the flux has begun to fry, touch the end of the solder to the joint occasionally to determine when the joint has reached the melting and flow point of the solder. This will become apparent when the solder begins to melt on contact with the tubing. When this occurs, feed the solder into the joint and withdraw the flame. Capillary action will draw the solder into the joint. On larger pipe sizes, it is good practice to feed the solder into the joint around the circumference of the pipe while keeping the flame on the joint to ensure adequate temperatures.

6. *Finishing procedures*: After all the solder has been applied, take a damp cloth and wipe off the joint. This action removes excess solder from the joint and gives it a clean, professional appearance.

When the installer is not familiar with these basic soldering techniques, several pieces of scrap pipe and excess fittings should be cut, cleaned, fluxed, and used for practice. One good practice procedure is to solder several joints completely, then heat them up and take them apart using insulated pliers or welding gloves. Inspect the two ends of each joint for evenness of solder flow and absence of cold spots on either the socket or tube end of the fitting. Continue practicing until all the joints come out coated uniformly with solder on both the tube and sockets. These practice sessions should also include soldering pipe to gate valves and other assorted plumbing fittings. Soldering tubing to valves and other peripheral components involves more heat than is required for standard elbows and couplings. After the installer has mastered these basic techniques, the installation can proceed.

The pressure temperature relief valve should be installed in its proper location on the storage tank and plumbed to an existing drain or waste pipe if possible. If no drain is available for this purpose, the waste line from the valve should be plumbed to within 6 in. of floor level. It should be emphasized that no shutoff valve should be installed in this waste line, nor should the diameter of the line be reduced anywhere along its length.

Valving Arrangements

Standard Bypass

Solar domestic hot water tanks are generally installed with a bypass valving arrangement as illustrated in Figure 5.3. In this setup, the incoming cold water is diverted to the solar storage tank, where it is preheated. This preheated water is then fed to the incoming line of the existing hot water heater. Thus the solar system functions as a cold water preheating system for the existing domestic hot water heater. In addition to the bypass arrangement shown, provision should be made to feed the hot water from the solar storage tank directly to the hot water service lines, bypassing the existing domestic hot water heater.

Tankless Coil Modifications

Although the hot water service bypass is an optional valving arrangement when a solar system is installed in conjunction with an existing stand-alone domestic hot water heater, it is highly recommended when interfacing the solar storage tank with a tankless coil in a hot

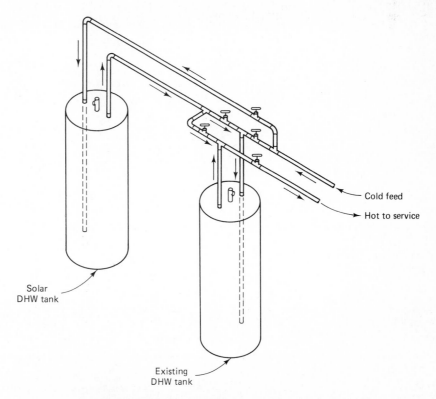

FIGURE 5.3 Bypass valving arrangement.

water boiler. Utilizing the bypass arrangement, the solar-heated water can either bypass the tankless coil and go directly to the hot water service lines or be routed to travel first through the tankless coil. The option to bypass the coil is often taken during the nonheating season when the boiler is shut down. It might be preferable at times, however, to run the solar-heated water through the tankless coil before going to the hot water service lines even though the boiler is not operating. Running the solar-heated water through the tankless coil during the summer will serve to keep the internal boiler water warm, reducing the temperature of the solar heated water in the process. At first glance this type of arrangement might seem somewhat counterproductive. After all, why heat the boiler water during the summer, which results in cooler water temperatures from the solar system? However, when the boiler is located in a damp basement or lower level of the home, this type of arrangement will prevent condensation and resulting corrosion from affecting the boiler during the humid summer months. During the winter when the boiler is in operation, the system owner has the option of either bypassing the coil or running preheated water from the solar system through the coil first. Running preheated water through the tankless coil will relieve much of the stress on the boiler for domestic hot water production. Since the solar-heated water enters the tankless coil in a warmed state, the fuel consumption necessary for domestic hot water production will be reduced accordingly. A typical tankless coil bypass arrangement is illustrated in Figure 5.4.

Installation of Pipe Insulation

After all plumbing connections have been completed, the tank should be pressurized and all joints checked for leaks. Pipe insulation should be installed only after the watertight integrity of all joints has been established. In heated areas of the home, a minimum of ½-in.-thick-wall closed-cell rubber insulation or equivalent should be used. If pipe runs in the system are extremely long, or they pass through unheated basement areas, heavier insulation should be used. Pipe insulation comes in a variety of lengths and materials and is available in most plumbing and heating supply houses.

Figure 5.5 lists the *R* values and heat loss in piping runs associated with the use of closed-cell rubber insulation. Although the closed-cell structure of the material reduces mois-

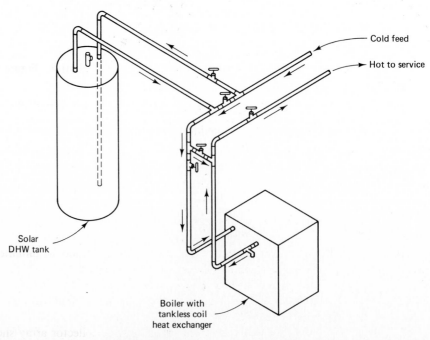

Cold feed

Hot to service

Solar
DHW tank

Boiler with
tankless coil
heat exchanger

FIGURE 5.4 Bypass valving of tankless coil.

HEAT LOSS

INSULATED PIPES

Btu/hr. ft. - length °F

INSULATION WALL THICKNESS

PIPE SIZE	1/4in.	3/8 in.	1/2 in.	3/4 in.	1 in.
3/8″ O. D.	0.117	0.105	0.0945	0.0822	
1/2″ O. D.	0.140	0.123	0.110	0.094	
5/8″ O. D.	0.160	0.141	0.125	0.106	0.0934
3/4″ O. D.	0.185	0.158	0.139	0.117	0.112
1/2″ IPS	0.204	0.175	0.153	0.128	0.130
3/4″ IPS	0.234	0.199	0.173	0.143	0.147
1″ IPS	0.289	0.243	0.209	0.170	0.164
1 1/4″ IPS	0.337	0.281	0.239	0.193	0.180
1 1/2″ IPS	0.377	0.312	0.265	0.212	0.196
2″ IPS	0.457	0.375	0.316	0.250	0.212
2 1/2″ IPS	0.540	0.441	0.369	0.290	0.228
3″ IPS	0.665	0.539	0.449	0.349	0.244
4″ IPS	0.811	0.654	0.542	0.417	

In order to determine heat loss in a particular application, the appropriate figure from the above table must be multiplied by the temperature difference between the pipe and ambient air. For example: with 1 in. IPS pipe at 190°F, 1/2 in. wall insulation and room temperature of 70°F, then 190 − 70 = 120 × 0.209 = 25.08 Btu/hr. ft. · length. In a 20 ft. length the total loss would be 20 × 25.08 = 501.6 Btu/hr., etc.

R FACTORS

TUBING INSULATION

INSULATION I. D. (min.) or PIPE O. D.	3/8″ WALL R	1/2″ WALL R	3/4″ WALL R	1″ WALL R
3/8″	2.2	3.2	5.4	
1/2″ (1/4″ IPS)	2.0	2.9	4.9	
5/8″ (3/8″ IPS)	1.9	2.8	4.6	6.7
3/4″	1.9	2.6	4.4	6.1
7/8″ (1/2″ IPS)	1.8	2.6	4.2	5.7
1 1/8″ (3/4″ IPS)	1.7	2.4	4.0	5.4
1 3/8″ (1″ IPS)	1.6	2.3	3.8	5.2
1 5/8″ (1 1/4″ IPS)	1.6	2.2	3.6	5.0
1 7/8″ (1 1/2″ IPS)	1.6	2.2	3.5	4.9
2 1/8″	1.5	2.2	3.5	4.8
2 3/8″ (2″ IPS)	1.5	2.1	3.4	4.7
2 5/8″	1.5	2.1	3.3	4.6
2 7/8″ (2 1/2″ IPS)	1.5	2.1	3.3	4.5
3 1/8″	1.5	2.0	3.2	
3 5/8″ (3″ IPS)	1.5	2.0	3.2	
4 1/8″ (3 1/2″ IPS)	1.5	2.0	3.1	
4 1/2″ (4″ IPS)	1.5	2.0	3.1	
5 1/2″ (5″ IPS)	1.4	1.9	3.0	
6 5/8″ (6″ IPS)		1.9	3.0	
8 5/8″ (8″ IPS)		1.9	2.9	

Note: These "R" factors were calculated from the thermal conductivity and equivalent (flat sheet) wall thickness in each case.

FIGURE 5.5 *R* values and heat loss of closed-cell rubber pipe insulation. (Courtesy of Rub tex Corp.)

ture absorption, this type of insulation is recommended only for indoor applications. When small sections must be exposed to the outdoors, a high-quality latex paint or waterproof vinyl tape must be wrapped around or applied to the insulation. This coating will protect the rubber from ultraviolet deterioration that would occur otherwise. The use of tapes or paints that are neither ultraviolet nor water resistant should be avoided. Closed-cell rubber insulation can be easily cut and mitered for a neat, professional appearance. It must be securely sealed along the seams after being installed on the pipe. Special adhesives and sealants are available from insulation manufacturers for this purpose. The installer should take care to insulate all sections of pipe and not leave small, uninsulated sections that will result in inordinately large heat losses when the system reaches high operating temperatures.

In addition to closed-cell rubber insulation, add-on foam-type insulation with rigid polyvinyl chloride (PVC) outer shells is available. In this type of insulating system, the pipe runs are insulated after all piping has been pressure-tested. The foam insulation slips over the pipe. The protective outer sleeve of PVC is then slid over the foam and sealed. In addition, prefabricated 45° and 90° elbows and tees are available. This type of insulation offers a higher *R* value than that of rubber insulation and is rated for outdoor use. However, its cost is significantly higher than the rubber insulation. Figure 5.6 illustrates the use of slide on special fittings.

Support for all pipe runs should be provided at intermediate intervals, preferably every 4 to 6 ft. The use of suitable brackets and/or strapping material will prevent the pipes from sagging, which can induce stresses that can lead to both pipe and joint failure. Various types of straps and brackets are available for this purpose. After the solar storage tank has been installed and its proper operation verified, the collector array is installed.

SITING, MOUNTING, AND PLUMBING THE COLLECTOR ARRAY

Orientation of Collectors

For best operation, the solar collectors should be mounted facing as close to true south as possible. Significant deviation from true south can diminish system performance. The allowable parameters for east or west deviation from true south fall within 15 to 20°. Figure 5.7 highlights how solar system performance diminishes as the collector orientation deviates from true south.

It should be remembered that the collector array should face true south, not magnetic south. Since magnetic compasses do not point to true north, but to the magnetic north pole,

FIGURE 5.6 Wraparound foam insulation with rigid outer PVC jacket. (Courtesy of Insultek, Inc.)

the compass reading must be corrected to determine true south based on geographic location. An isogonic chart (Figure 5.8) can be used to determine true north and true south from the magnetic compass readings.

For proper collector orientation, when the compass reads east of true north, the solar collectors should be faced east of magnetic south; conversely, when the compass reads west of true north, the solar collectors should be oriented west of magnetic south.

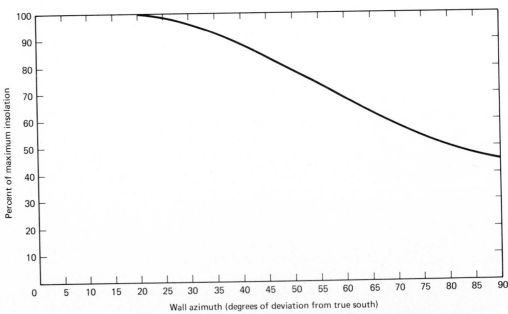

FIGURE 5.7 Approximate loss in solar insolation based on deviation from true south.

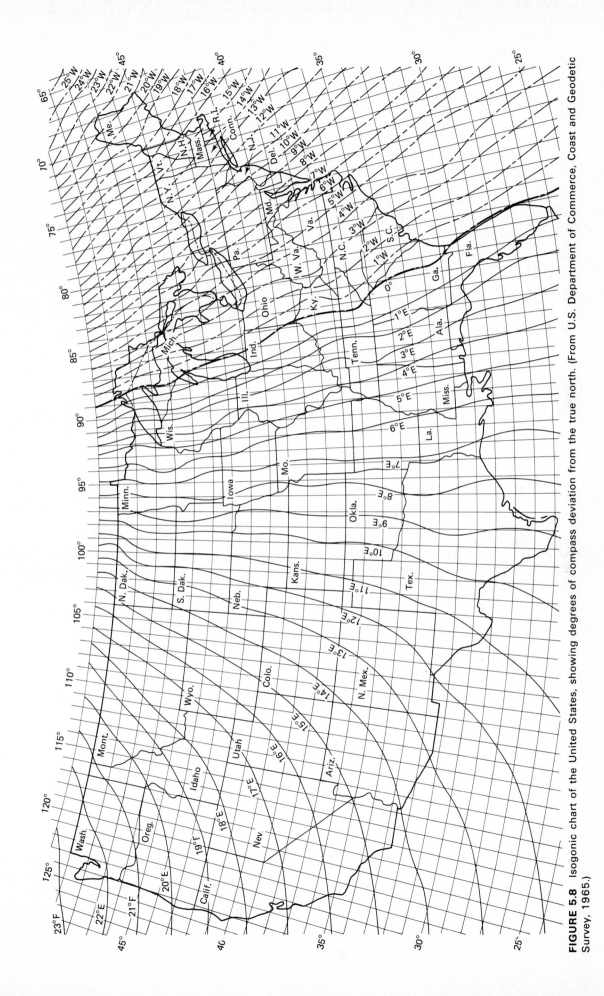

FIGURE 5.8 Isogonic chart of the United States, showing degrees of compass deviation from the true north. (From U.S. Department of Commerce, Coast and Geodetic Survey, 1965.)

Collector Slope

After the proper orientation of the collector array has been determined, the proper slope or angle of inclination of the collector array must be determined. Collector slope is largely determined by the latitude of the installation site. Since the latitude of the site determines the angle of the sun in the sky, hence the angle at which the sunlight will strike the collector array, a slope angle must be chosen that will give good performance throughout the year. Although it might seem advantageous to arrange the collector array so that the angle of inclination can be changed during the year, this is both impractical and generally unnecessary. Figure 5.9 illustrates the north latitude locations of various parts of the United States. Once the north latitude of the installation site is known, an angle of inclination can be chosen which will maximize system performance on an annual basis.

Just as there is a degree of flexibility in orienting solar collectors east or west of true south, there are a range of slope angles within which the solar collectors can be properly installed. Figure 5.10 illustrates available insolation for various collector inclination angles based on the north latitude of the installation site for a typical domestic hot water system. Note from Figure 5.10 that although optimum system performance is obtained when the collector slope is between − 5 and − 10° of latitude, little overall loss in system performance results as long as the deviation from north latitude remains within ± 20°. It should be noted, however, that the negative effects of less than optimum orientation and inclination are synergistic. A rack may be required to optimize one if the other is marginal. In areas that experience heavy winter snows, the collector array should be angled at a minimum of 45° to aid in the shedding of snow from the collector surfaces. It is mandatory that the collector not be shaded during the day for more than brief periods of time. Almost 90% of the energy collected by a solar system occurs between 9 A.M. and 3 P.M. solar time. Observation of the specific location where the collectors are to be installed should be done during these hours to determine if shading will be a significant problem. In heavily wooded building lots, the installer might have to make a choice between removing shade trees or forgoing the solar installation if it appears that shading will be a significant problem.

One successful method to aid in the determination of site suitability involves the use of specialized reflection instrumentation (Figure 5.11). Site analysis instruments are available in a variety of designs and sophistication. Most of these units reflect the daily path of the sun over the collector area and illuminate all pertinent obstructions. Readings from these instruments are available on an hourly basis by the day, month, and year. Thus an accurate prediction can be made as to the annual energy available to the collector array based on daily radiation figures.

FIGURE 5.9   North latitude locations of the continental United States.

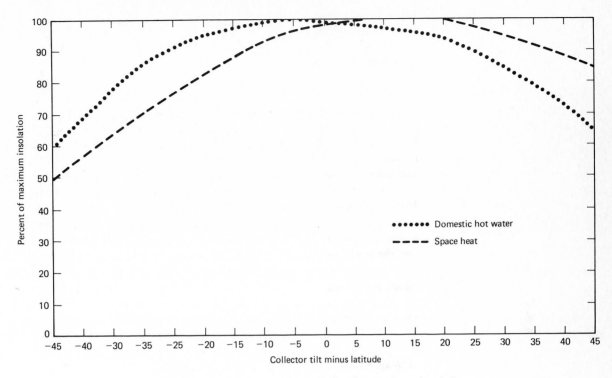

FIGURE 5.10 Effect of collector inclination angle on insulation.

Once the slope of the array has been determined, installation of the collectors can begin. The following section describes the installation of solar collectors on a variety of surfaces, from ground mounts to flat-roof installations.

Mounting the Collector Array

Although most solar systems are similar in appearance and major components, each installation is as unique and different as the homes and buildings they become a part of. What follows is a discussion of the mounting procedures encountered by the vast majority of system installers. These procedures involve the following five types of collector installations:

FIGURE 5.11 Reflection instrumentation for solar site analysis. (Courtesy of Ensar, Inc.)

1.   Pitched-roof installation, no rack required
2.   Pitched-roof installation, rack required
3.   Flat-roof installation, rack required
4.   Vertical wall mount, rack required
5.   Yard mount, rack required

Pitched-Roof Installation: No Rack Required

When the pitch of the roof angle falls within the allowable orientation and inclination angles for proper system performance, the collectors can be installed directly on the roof without the use of any racking arrangement. A typical flush-roof-mount installation is pictured in Figure 5.12. The collectors are bolted to the roof through holes in the mounting flanges or by the use of auxiliary mounting clips. Provision should be made to raise the collectors approximately 1 in. off the roof surface. This procedure will ensure that there is adequate air circulating beneath the collectors and sufficient area for water runoff to prevent the roofing material from dry rotting. If the collectors are raised too high off the roof surface, debris will build up beneath the collector array; therefore, spacing should be kept to within 1 in.

The mounting bolts from the collector tie into 2 in. $\times$ 4 in. stringer studs that are positioned at right angles below the roof trusses or rafters. This procedure ensures adequate wind load resistance for the collectors. A minimum thickness of $\frac{3}{8}$-in. cadmium- or zinc-coated threaded rod should be employed. The collectors may be lag-bolted directly to the roof rafters if the spacing of the rafters and collector mounting holes is such that each collector can be secured to at least two supporting roof members. Sealant should be placed around all penetrations to prevent leaks (Figure 5.13). A good grade of plastic roofing cement should be used for this purpose. Liberal amounts of cement should be applied around

FIGURE 5.12   Solar collectors mounted directly onto roof surface.

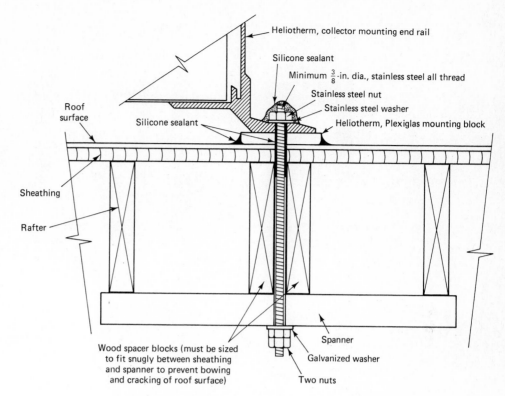

FIGURE 5.13 Sealant pitched around roof penetrations.

the rod and bolt penetrations and worked into the hole. A small bead of sealant should be built up around this joint to help water runoff.

Pipe penetrations must be weatherproofed differently from penetrations made by fastening devices such as threaded rods. Since the pipes in the collector loop are subject to wide variations in temperature during normal collector cycling, they must be flashed with materials that not only withstand these temperature extremes but are corrosion resistant as well. A typical flashing used for this purpose is illustrated in Figure 5.14. To form a weather-tight seal, the flashing is made with a large base which slips under the roof shingles. The collar of the flashing through which the piping passes forms a tight compression fit around all pipe penetrations, effectively preventing moisture and condensation from penetrating through the roof.

FIGURE 5.14 Weatherproof pipe flashing.
(Courtesy of Oatey Corp.)

FIGURE 5.15 Racking to increase collector slope angle. (Courtesy of Sunracks, Division of Sunsearch, Inc.)

Pitched-Roof Installation; Rack Required

Pitched roofs often require the use of a rack in order to increase the collector slope angle for optimum system performance. This type of racking arrangement is pictured in Figure 5.15. The installer has the option of either purchasing a factory-made rack, or site-building the rack. Many collector manufacturers sell racks which are adjustable to a wide variety of roof angles, and are easy to install. To determine the existing slope angle of the roof, two methods are commonly employed. The first method involves the use of a bubble indicator (Figure 5.16). By placing this indicator on the surface of the roof, the device will automati-

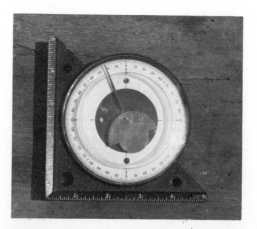

FIGURE 5.16 Solar inclinometer. (Courtesy of Solar City, Inc.)

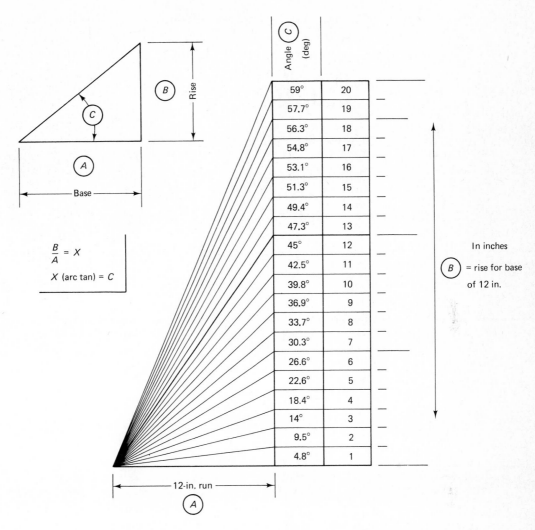

FIGURE 5.17   Roof angles. (Courtesy of Solar City, Inc.)

cally read out the pitch angle of the roof. This procedure is quick and reliable and requires access to the roof surface. The second method involves the use of a chart, illustrated in Figure 5.17, which is used to determine the roof angle based on the framing dimensions. Once the angle of the roof has been determined, the dimensions of the racking members can be determined, as illustrated in Table 5.1.

Let's examine the installation procedures for both factory-assembled and site-built collector racking systems.

Factory-assembled collector racks

A completed factory-assembled rack is illustrated in Figure 5.18.

In Figure 5.19, note the use of threaded rod to hold the bracket in position. Two- by four-inch stringers can be used in place of the steel plates to secure the brackets to the roof rafters. This type of factory design is flexible since the top of the rack is easily adjustable, permitting its use on a variety of pitched roofs.

A rack of this type can withstand winds in excess of 100 miles per hour. This type of rack is recommended where the installer has either had little experience in the design and construction of site-built collector racks, or simply wishes to purchase rather than fabricate the rack.

Table 5.1 MOUNTING DIMENSIONS[a]

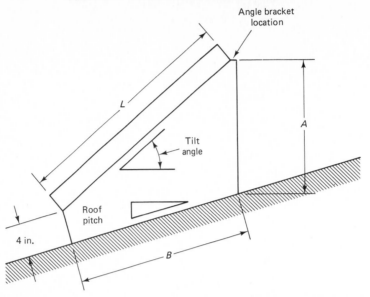

Collector Length, L (in.)	Collector Tilt Angle (deg)	Roof Pitch												
		Flat		3/12		4/12		5/12		6/12		9/12		12/12	
		A	B	A	B	A	B	A	B	A	B	A	B	A	B
77-in. collector, vertical	30	43	67	26	68	20	69	15	71	9	73	—	—	—	—
	40	54	59	39	60	34	61	29	63	24	64	10	71	—	—
	50	63	50	51	50	47	51	42	52	38	54	26	60	14	67
	60	71	39	61	39	58	40	55	40	52	42	42	46	32	52
93-in. collector, vertical	30	50	80	31	82	23	84	18	86	10	89	—	—	—	—
	40	64	72	46	73	40	74	35	76	28	78	12	87	—	—
	50	76	60	61	61	55	62	51	63	45	66	32	72	17	81
	60	84	46	73	47	69	48	66	49	61	50	57	56	40	62
35-in. collector, horizontal	30	22	31	14	31	12	31	7	32	7	32	—	—	—	—
	40	27	27	20	27	18	27	14	29	14	29	8	31	—	—
	50	31	23	26	22	24	23	20	24	20	24	15	26	10	28
	60	35	18	30	17	29	17	26	18	26	18	22	19	19	21

[a]A, angle bracket location (inches); B, roof length (inches).

SOURCE: Revere Solar and Architectural Products.

FIGURE 5.18 Factory-assembled collector rack. (Courtesy of Sunracks, Division of Sunsearch, Inc.)

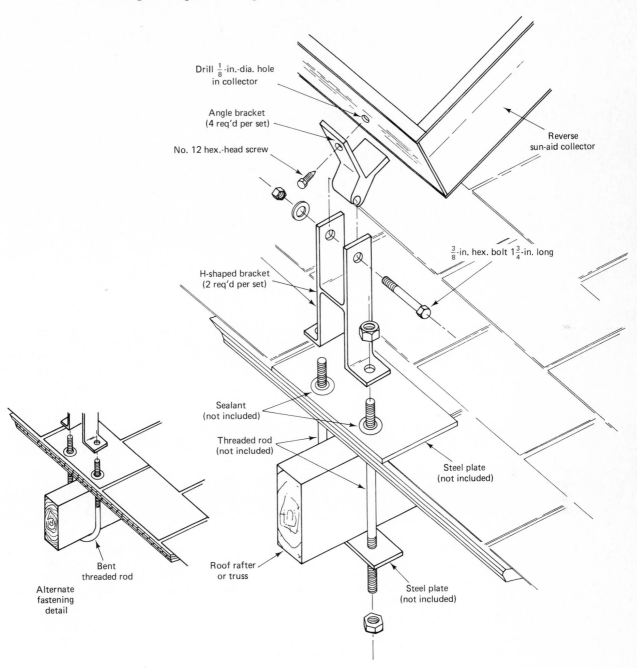

Drill $\frac{1}{8}$-in.-dia. hole in collector

Angle bracket (4 req'd per set)

No. 12 hex.-head screw

Reverse sun-aid collector

H-shaped bracket (2 req'd per set)

$\frac{3}{8}$-in. hex. bolt $1\frac{3}{4}$-in. long

Sealant (not included)

Threaded rod (not included)

Steel plate (not included)

Bent threaded rod

Alternate fastening detail

Roof rafter or truss

Steel plate (not included)

FIGURE 5.19 Typical rack mounting assembly. (Courtesy of Revere Solar and Architectural Products.)

Site-built collector racks

The use of site-built custom-designed solar collector racks offer several advantages over their factory-manufactured counterparts. First, site-built racks are considerably less expensive than the manufactured versions. Second, site-built racks can be custom-engineered to fit unique installation circumstances, a design feature not often found in factory units. Properly engineered and constructed, site-built racks will perform as well, if not better, than the manufactured versions. In Figure 5.20 the basic arrangement of a site-built rack is illustrated. A top and a bottom plate are required to stabilize the collector array. Struts are used to raise the top end of the collectors to the required slope angle.

Note from the detail of Figure 5.21 that the solar collectors rest on the bottom plate of the rack. Both the top and bottom plates in this rack are fabricated from pressure-treated

FIGURE 5.20   Site-built collector rack.

FIGURE 5.21   Bottom plate rack detail.

4 in. × 4 in. dimensional lumber. The angle iron used for the struts is 2 in. × 2 in. × ³/₁₆ in. galvanized steel or aluminum. It is most important that all rack components be made from corrosion-resistant material. All fasteners should be either stainless steel or zinc- and cadmium-plated automotive-grade quality. Note that the threaded rod penetrates the bottom piece of angle iron, the 4 in. × 4 in. lumber, the roof sheathing, and terminates through a 2 in. × 4 in. stringer running perpendicular to the roof rafters. Installed in this manner, the collector array offers excellent wind resistance and long-term durability. All roof penetrations should be adequately sealed, as detailed previously. The top of all threaded rods should be sealed with either roofing cement or silicone rubber caulk to prevent water penetration. If the collector rack is placed midway down a long roof, water channels should be cut into the top and bottom 4 in. × 4 in. plates to facilitate water runoff. If galvanized steel or aluminum is not available, regular steel can be used for the rack assembly if it is properly rustproofed prior to installation. The procedure for rustproofing begins by removing all surface scale from the metal with a stiff wire brush. After ensuring that all the surface oxidation has been removed, apply two coats of a good-quality primer, and finish with two coats of rust-resistant paint. Although this type of surface preparation will necessitate "touching up" the rack every 4 or 5 years with rustproof paint, it will provide excellent corrosion protection for the life of the system. An alternative to the rustproofing procedure described is the use of special epoxy and urethane enamels to coat the steel. Although these types of coatings are more expensive than standard primer and rust-resistant paint, they are maintenance-free for extended periods of time.

The procedures and designs offered in this section represent only a few of the many available for on-site fabrication of collector racks. It is advised that before any site-built rack is installed, the basic design features of the rack be checked by a person knowledgeable in the area of mechanical or structural engineering to help ensure the validity of the design.

Flat-Roof Installation; Rack Required

When installing a solar system on a flat roof, the use of a rack is mandatory to achieve proper collector slope. Before attempting a flat-roof installation, the roof surface should be carefully inspected to determine the following:

1. Type of roof surface (hot asphalt, etc.)
2. Condition of the roof surface where the racks are to be installed
3. Condition of the roof surface adjacent to the collector array that may be affected during the installation of the solar collector racks (weak areas of the roof that might develop leaks, for example, due to the walking and stress caused by workers installing the system)

Once the condition of the roof has been determined to be satisfactory, the installation can proceed.

A typical flat-roof solar collector racking arrangement is illustrated in Figure 5.22. The rack struts in this type of arrangement should be fabricated from either aluminum or corrosion-protected steel angle iron. When installing this type of rack on a flat hot-tar roof, it is imperative that the surface of the roof be scraped flat, to remove all the stone and gravel surface where the racking members will rest on the roof surface. Failure to do this will cause the rack to rest on the surface gravel of the roof rather than the roof surface itself, making a water-tight seal around rack foot plates virtually impossible. Angle iron should be used as a stringer between steel roof trusses to ensure that the rack has adequate wind resistance and dimensional stability.

Although the illustrations in Figures 5.18 and 5.22 depict factory-assembled units, this type of rack can be fabricated on-site from dimensional angle iron, aluminum angle, and steel plate.

FIGURE 5.22   Flat-roof racking arrangement.

Vertical Wall Mount

When the roof surfaces on a building are undesirable for installing solar collectors, vertical wall surfaces can be used for this purpose. Undesirability of roof surfaces is usually due to inadequate strength of the roof framing members, improper roof orientation, or the necessity of keeping the roof area of the dwelling unobstructed. While vertical wall surfaces are perfectly adequate for installing solar collectors, modifications need to be made in the racking arrangement.

A vertical-wall-mount system incorporates features of both roof-pitch racks and flat-roof racks to space out the bottom of the collector array to yield a suitable inclination angle. A typical site-built wall mount is illustrated in Figure 5.23.

Sloped Wall Mount

An installation situation similar to a straight vertical wall is encountered when installing a collector array on a nearly vertical surface, such as found on sloped A-frame houses. This type of installation is pictured in Figure 5.24. Note that 2 in. $\times$ 6 in. pressure-treated stringers have been lag-bolted to the roof prior to the installation of the rack struts. All struts are lag-bolted to the stringers and securely fastened to the solar collectors, keeping the number of roof penetrations to a minimum. The stringers can either be directly bolted to the roof rafters, or threaded rod can be used to penetrate the roof surface completely and tie into a 2 in $\times$ 4 in. stringer that runs perpendicular to the roof rafters. Due to variations in roof pitch and wall construction, vertical wall mounting racks are almost always site-built.

Ground Mount; Rack Required

When installation of the solar collector array on the building structure is impractical, the collectors are often mounted on racks that are anchored directly into the ground. These racks are commonly referred to as ground mounts. A typical ground mount is pictured in

FIGURE 5.23 Vertical wall collector mounting.

Figure 5.25. Note from this illustration that a yard mount is basically a flat-roof rack mounted on concrete footings. In areas experiencing significant snow accumulation, the mount must be designed to prevent snow buildup on the surface of the collectors. The basic configuration of this rack, together with selection of structural materials and fasteners, is similar to flat-roof racks.

The major differences between ground-mounted and roof-mounted solar systems is the necessity of running the collector piping underground and through the foundation wall of the building when a ground mount is employed. All underground piping should be placed at least 1 ft below the bottom of the frost line to protect it from both ground heaving and low temperatures, which can lead to large heat losses from the pipes. The insulation materials for these installations require the use of waterproof coatings. One such type of pipe and insulation suitable for this application is pictured in Figure 5.26. If possible, the pipe runs from the collector array through the foundation wall of the house should be kept as straight as possible to allow for expansion. If the pipe runs must be angled underground, specially designed compensators must be employed to absorb linear expansion. To install the pipe, a trench in the ground is opened, and a bed of approximately 6 in. of crushed stone is placed at the bottom of the trench. The pipe is then placed on top of the crushed stone prior to backfilling. The stone offers a buffer zone for the pipe and will protect it from minor ground heaving and settling that occurs after the trench has been back-filled. All penetrations through the building foundation wall should be adequately waterproofed. Once the pipe has penetrated the foundation wall, closed-cell rubber insulation can be used to insulate the pipe run to the storage tank.

Collector Interconnections

Several methods are employed to join the individual solar collectors together within the array: hard-soldered flexible joints, hard-soldered solid couplings, and flexible high-temperature hose.

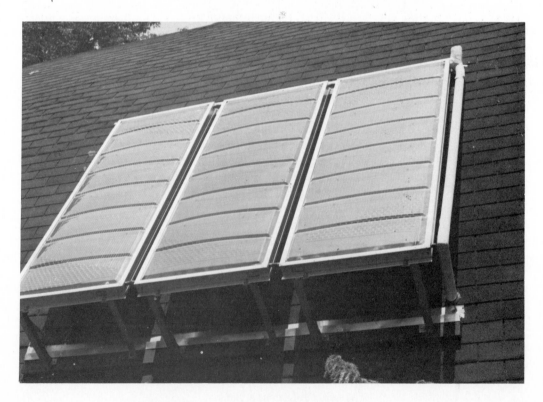

FIGURE 5.24   A-frame and sloped roof mounting.

FIGURE 5.25 Ground mount. The rack is securely mounted into concrete footings which extend below frost level.

FIGURE 5.26 Foam insulated vinyl-jacketed pipe suitable for underground installation. (Courtesy of Insultek.)

FIGURE 5.27   Flexible sweat-soldered manifold connection.

Flexible Joints

Hard-soldered flexible joints offer the greatest versatility of all methods available for joining solar collectors: the durability of a hard-soldered joint, the flexibility of the joint to allow for variations in absorber manifold location, and provision for expansion and contraction of the solar collectors. These features enable collector installation and operational cycling without inducing stress on the system from rigid components (Figure 5.27). The flexible connector is a bellows joint, available in either copper or stainless steel, depending on the application and fluid to be used. Although more expensive than most other methods used to join collector manifolds, the durability of the resulting joint is worth the additional expenditure. Only 95/5 solder should be used when making connections to the absorber manifolds.

Solid Couplings

Copper couplings (Figure 5.28) are standard plumbing fittings available at most plumbing supply establishments. It is preferable to purchase this coupling without internal stops so that it can be moved out of the way until the two adjacent manifolds are properly aligned. The coupling is then positioned so that it equally straddles both manifolds and is soldered into place.

FIGURE 5.28   Slip coupling manifold connection.

FIGURE 5.29 High-temperature silicone hose manifold connection.

High-Temperature Hoses

High-temperature hoses are sometimes used to connect adjacent solar collectors. Silicone is the most common type of material used in high-temperature hoses (Figure 5.29). The hose is slipped over both manifolds and secured with stainless steel hose clamps. Alignment of the manifolds is not critical since the hose is flexible and will adjust easily to offset manifolds. This type of installation is the easiest of all to perform, but offers the least amount of durability. The hose clamps used to secure the hose should be made from solid stainless steel banding to prevent the silicone from squeezing through the clamp at elevated temperatures. Avoid overtightening the clamps to prevent cutting through the hose.

All fittings should be carefully cleaned and prepared prior to installation. Leaks in the manifold connections are very difficult to repair after the collectors have been secured. Often, defective joints must be physically cut out of the array in order for repairs to be made. In addition, stress placed on the manifolds during collector installation and alignment should be kept to a minimum to avoid dislodging brazed connections within the collector between the manifold and absorber plate riser tubes.

INSTALLATION OF SYSTEM PERIPHERAL COMPONENTS

Closed-Loop Module

Peripheral components of a solar system include those items that are utilized in the solar collector loop to transfer heat from the collector array to the storage medium. In a typical closed-loop antifreeze system, these components include the circulating pump, expansion tank, and air-purger float valve assembly, check valves, fill and drain valves, flow meters, temperature gauges, and the intermediate piping. Proper design and installation of this closed-loop assembly is critical to the efficient functioning of the solar system. Due to this fact, many system manufacturers offer these components in one factory-assembled and factory-tested package, requiring simple insertion in the field. Figure 5.30 illustrates one type of factory-assembled closed-loop module requiring only two plumbing connections to complete the installation. One obvious advantage of a prepacked module is that all the components are compactly and conveniently arranged (Figure 5.31). Note the use of unions on either side of the air purger to facilitate removal of the circulator and air purger without physically cutting out these components from the plumbing array. A boiler drain valve located upstream of the check valve is used to fill and pressurize the system after installation is complete. The differential controller should be mounted as close as practical to the module to facilitate subsequent temperature sensor and circulator connections. All the installer need do is

FIGURE 5.30 Closed-loop module. (Courtesy of Heliotherm.)

plumb the module into the system. Preplumbed modules can save a lot of space and greatly enhance the appearance of an installation, as opposed to plumbing all the peripheral components into the system separately.

When glass-lined solar storage tanks are used instead of stone-lined storage tanks, external heat exchangers are employed in the closed-loop arrangement. A typical external heat exchanger arrangement available as a prepackaged module is illustrated in Figure 5.32. The external heat exchanger system requires the use of two circulating pumps. One circulator moves the water from the solar storage tank through the external heat exchanger. The other circulator moves the antifreeze fluid throughout the collector array and through the heat exchanger, then back to the collectors. In this arrangement, both circulators operate simultaneously.

Drain-Down Module

Drain-down domestic hot water systems are also available with prepackaged heat transfer modules. A typical unit for a drain-down application is illustrated in Figure 5.33.

It is often desirable to insert temperature gauging into the system in order to monitor its performance. Both manual and automatic recording devices are available for this purpose.

A Btu monitor (Figure 5.34) can record the number of Btu delivered by the collector array to the solar storage tanks. These devices incorporate memory capability, are sophisticated in nature, and offer the system owner a high degree of accuracy in determining overall system output and performance.

Manual temperature gauges (Figure 5.35) are available with either bimetallic or mercury temperature sensing. These gauges are installed with standard plumbing fittings for spot

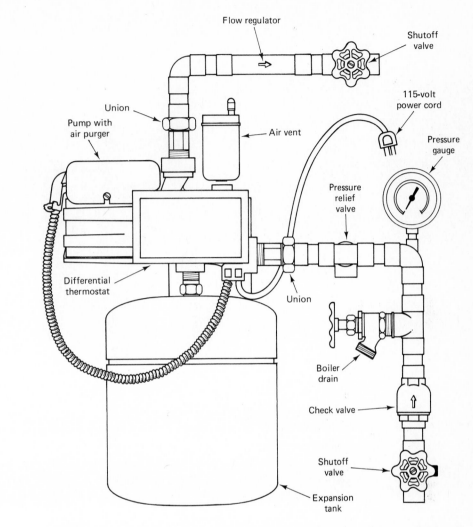

FIGURE 5.31 Components of closed-loop module. (Courtesy of Revere Solar and Architectural Products.)

FIGURE 5.32 External heat exchanger module. (Courtesy of Mor-Flo, Inc.)

FIGURE 5.33   Drain-down module. (Courtesy of Mor-Flo, Inc.)

temperature readings at any location within the system. They are generally used to measure the temperatures in the collector feed and return lines, solar storage tanks, and the hot and cold water feed lines. Although these gauges do not offer recording capability, their incorporation within the solar system greatly increases knowledge of its operational characteristics.

Pipe Sizing

The average solar domestic hot water system employs from two to four collectors. Since most collectors are equipped with either ¾- or 1-in. internal manifolds, it is recommended

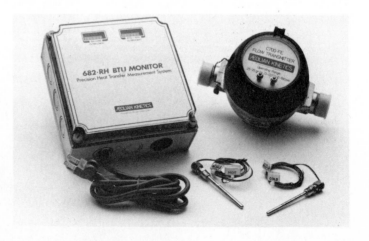

FIGURE 5.34   Btu meter. (Courtesy of Aeolian Kinetics.)

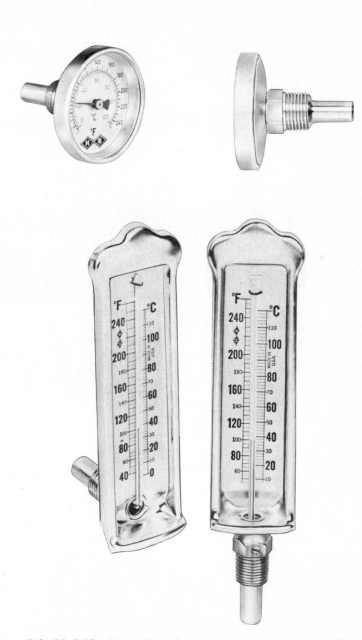

FIGURE 5.35 Bimetallic and mercury temperature gauges. (Courtesy of H-B Instrument Co.)

that a minimum of ¾-in. copper tubing be used in all piping runs. The use of ¾-in. pipe will also simplify the connection of peripheral components, most of which are equipped with ¾-in. threaded or sweat connections. Copper tubing is readily available in type L or type M ratings. Type L is high-pressure hard-copper tubing, and more expensive than its low pressure type M counterpart. Type M tubing is adequate for plumbing the solar collector loop, provided that the loop is a low-pressure system as in a closed-loop antifreeze system. When plumbing drain-down and recirculate freeze systems that employ high-pressure potable water in the collectors, only type L tubing should be used. All domestic hot water piping in the system should be done using type L tubing, since higher operating pressures are encountered in the service lines.

Pipe runs should be kept as short as possible. Table 5.2 indicates the maximum recommended pipe lengths between the solar storage tanks and the solar collector array for a given pipe size.

Collector Grounding

It is important to ensure that a good electrical ground exists between the collector array of copper tubing and the cold water feed pipe in the building water supply system. This ground path should be via the collector plumbing and should not be broken by the use of plastic pipe fittings, high-temperature hose, Teflon taped joints, and so on. If it is suspected that a good electrical ground is not inherent in the system, the collector array should be

Table 5.2   MAXIMUM PIPING DISTANCE (ft)[a,b]

Nominal Tube Size (in.)	Flow Rate (gpm)			
	1	1.5	2	2.5
½	200	125	40	—
¾	200	200	200	[c]

[a]*Supply plus return with standard flow rate.*

[b]*General notes on pipe sizing:*

1. Table 5.2 is based on a closed-loop system containing a solution of 50/50 water/propylene glycol. The use of other antifreeze solutions may necessitate larger pipe sizes than those indicated.
2. Unnecessarily large tubing sizes in the collector loop should be avoided, since this only increases the amount of fluid in the system. Increased fluid increases the response time of the system, lowering overall efficiency.
3. When installing a small domestic hot water system with an eye toward a possible expansion of the system in the future, a minimum of ¾-in. pipe is recommended.
4. For conditions other than those described above, the individual solar system manufacturer should be consulted.

[c]*Some solar systems are equipped with flow regulators to provide approximately ½ to 2 gallons per minute per collector automatically. Flow regulators are used when single-speed differential controllers are employed. When using variable-speed differential controllers and circulators, the flow rate in the system will range between ½ to 5 gallons per minute, depending upon the capacity of the circulator.*

connected to a properly installed ground rod using No. 8 copper ground wire. This ground rod should be installed as close to the collectors as possible.

The major electrical wiring to be done in the installation of a solar domestic hot water system involves the connection of the differential controller and temperature sensors, and the optional wiring of the electrical backup resistance heating element in the solar storage tank. Let's begin with the differential controller, the "brains" of the solar system.

Differential Controllers

Controllers are available in a wide variety of configurations with many optional features. Certain decisions on the part of the system purchaser need to be made prior to the selection of a differential controller. These decisions are:

1. *Is variable-flow-rate circulation required?* This option can increase system efficiency by as much as 15% by constantly matching the fluid flow rate in the solar system to the amount of incoming solar radiation.
2. *Does the system purchaser want digital display capability in the controller?* Digital display gives the system owner more knowledge as to exactly how the solar system is operating. While some controller manufacturers offer digital display as an integral part of the features of the controller, others offer it as an add-on unit to the basic controller.
3. *Does the controller offer variable limit controls?* Many solar controllers come with adjustable circuitry for freeze protection, recirculate freeze protection, and high-limit shut-down protection. A wide array of sensors for these applications is also available.
4. *What are the warranty options that come with the unit?* Some manufacturers offer a one-year replacement warranty. Others offer extended 5-year warranties. Consult the controller manufacturer for specifics.
5. *Does the solar system require more than one controlled output?* A simple domestic hot water system requires only one controlled output, the circulation pump. Larger solar systems involving space heating, hot tub, and solid-fuel appliance loops may require two or more outputs to be operated by the differential controller.

Sensors

Installation of the controller involves connecting the sensor wires to the appropriate controller terminals. A basic domestic hot water system employs two temperature sensors: One sensor is located either on the output of the storage tank heat exchanger, or on the coldest part of the solar storage tank; the other sensor is located on the hottest part of the collector array. This sensor should be factory installed on the underside of a collector absorber plate where the fluid exits the collector.

The sensors must be securely fastened directly to the surface to be monitored to operate properly. Storage tanks usually have mounting clips welded onto the tank shell to facilitate sensor installation. Stainless steel hose clamps can be used to attach the sensors directly to copper tubing. Where possible, sensors equipped with immersion-type fittings for direct insertion into water tanks or pipe fittings are recommended.

After the sensors have been installed, they must be adequately insulated to ensure that they will be unaffected by weather and ambient temperature conditions that might otherwise affect the validity of the readings. Methods of installing and insulating temperature sensors are illustrated in Figure 5.36.

Note from the illustration that the collector sensor has been installed on the external header or manifold piping exiting the solar collector. In Chapter 4 it was indicated that the

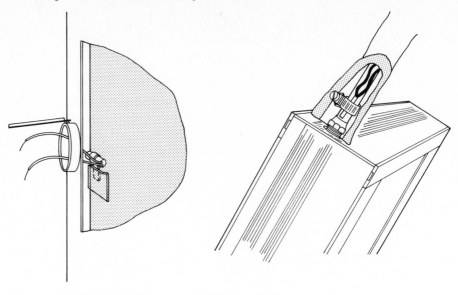

Placement of tank sensor   Placement of collector panel sensor

FIGURE 5.36 Placement of tank and collector temperature sensors.

ideal location for the solar collector sensor was directly on and under the collector absorber plate, completely surrounded by the internal collector insulation. This type of sensor installation is usually done at the factory during final collector assembly, and it is not recommended for field installation since it would necessitate disassembly of the collector components. In the absence of factory-mounted internal temperature sensors, Figure 5.36 illustrates the external mounting procedure which will yield satisfactory operating results. Sensor wiring should be done using two-conductor No. 18 or No. 20 gauge, stranded and spiraled wire. Polarity of the wiring need not be observed. All wiring within the sensor circuitry must be both waterproof and physically durable. The recommended procedure for accomplishing this is as follows:

Materials required:
    Crimp-type wire nuts (2)
    Crimping tool
    Silicone rubber or RTV caulking

Strip the wire leads and twist together the wires to be connected. Cut exposed strands to approximately ⅜-in. length. Fill the "cup" of the wire nut with silicone rubber or RTV caulking material. Using the crimping tool, crimp the wire nut to secure the connection. The connection should be checked to ensure that no bare wires are exposed. To provide strain relief for the connection, the wired assembly should be tied as illustrated in Figure 5.37.

It is good installation practice to check the accuracy of all temperature sensors when they are first installed. Since the sensors utilized in solar systems are thermistors that change electrical resistance based upon their temperature, their accuracy can be determined by measuring their electrical resistance in ohms and comparing this value to the ambient temperature of the sensor. The electrical resistance of standard 10-kΩ and 3-kΩ sensors are given in Table 5.3. Note that as the sensor increases in temperature, its internal electrical resistance decreases. Where tests indicate that sensors yield readings that are inaccurate by more than 3%, new sensors should be used.

Wiring the Storage Tank and Service Outlet

The service wiring to the solar hot water heater electric resistance elements should be a minimum of No. 10 gauge two-conductor wire with ground. A separate fused circuit should be

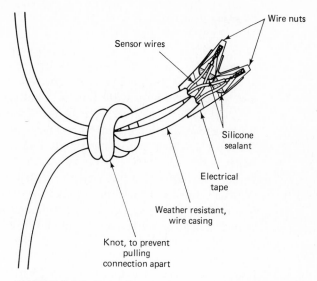

Wire nuts

Sensor wires

Silicone
sealant

Electrical
tape

Weather resistant,
wire casing

Knot, to prevent
pulling
connection apart

FIGURE 5.37 Preparing durable and weatherproof wire connections. *Note*: Use No. 22 gauge stranded and shielded cable.

provided for this line. All wiring must conform to the local electrical codes governing such installations. In addition, a separate branch circuit should be run for the solar system to power the controller and circulator pump, which will avoid possible overloads.

The controller circuitry should not be energized until the collector loop has been filled and tested. Failure to do this could result in overloading of the controller circuitry and overheating of the circulator, causing premature failure of both of these components.

The only remaining steps prior to system startup is to check the watertight integrity of the collector loop. System purging and testing procedures follow.

Table 5.3 ELECTRICAL RESISTANCE (kΩ) of 10-kΩ AND 3-kΩ SOLAR SENSORS

	Sensor			Sensor	
°F	10 kΩ	3 kΩ	°F	10 kΩ	3 kΩ
32	32.6	9.81	125	3.38	1.01
35	30.0	9.00	130	3.05	0.907
40	26.1	7.84	135	2.75	0.824
45	22.8	6.98	140	2.49	0.747
50	19.9	5.97	145	2.25	0.677
55	17.4	5.24	150	2.04	0.612
60	15.3	4.93	155	1.85	0.557
65	13.5	4.05	160	1.69	0.505
70	11.9	3.55	165	1.54	0.461
75	10.5	3.16	170	1.40	0.421
80	9.29	2.79	175	1.28	0.385
85	8.25	2.48	180	1.17	0.351
90	7.33	2.20	185	1.07	0.321
95	6.53	1.96	190	0.982	0.295
100	5.83	1.75	195	0.901	0.271
105	5.21	1.57	200	0.828	0.248
110	4.66	1.40	205	0.761	0.227
115	4.18	1.25	210	0.701	0.211
120	3.76	1.13	212	0.679	0.204

Flushing and Pressure Testing

Prior to filling and pressurizing the collector loop with antifreeze solution, it is necessary to purge all the dirt, solder, and debris left from the installation procedures. A flushing system as illustrated in Figure 5.38 is designed into most closed-loop modules, or can be plumbed into the collector loop at the time of installation.

To flush the system, water is introduced into boiler drain No. 2 by means of a double female-ended washing machine hose. One end of the hose is connected to boiler drain No. 2, the other end to a cold water supply line (the cold water inlet line on the solar hot water heater is convenient for this purpose). An alternative to a cold water supply line is to use a bucket of water and a force pump. With the shutoff valve closed (or check valve in some systems), water will travel up through the collectors and down the return piping through the heat exchanger in the solar storage tank and out of boiler drain No. 1. As a general rule, the system should be flushed with a minimum of 10 times the amount of fluid that the closed loop will contain. For example, a small closed-loop domestic hot water system usually contains between 3 and 4 gallons of fluid in the collector loop. The flushing procedure in this instance would involve approximately 25 to 40 gallons of water to ensure removal of all debris. If possible, warm water should be used to flush a system to facilitate removal of soldering flux. Care must be taken to prevent freezing of the water used to flush out a system if this procedure is being undertaken during the winter months.

To determine the amount of fluids contained in various domestic hot water systems, Table 5.4 gives the volumetric capacities for various sizes and lengths of pipe runs.

After the flushing has been completed, close boiler drain No. 1, open the shutoff valve, and allow the system pressure to rise to 45 psi, then close boiler drain No. 2. Allow the system to remain at this pressure for a period of at least 2 hours. All the pipe joints should be visually inspected at this time to determine if there are any leaks in the collector-loop piping. Should there be a leak, quickly open both boiler drains, drain all fluid from the system, repair the faulty connection, and repeat the pressurizing procedure. During this trial period, examine the pressure gauge in the closed-loop system frequently to determine if there are any leaks in the system which will be indicated as a loss in system pressure. Once it has been determined that the loop maintains pressure, all water should be drained out of the system in preparation for charging with antifreeze solution. (Note: Boiler drains should be installed at all low points of the system to facilitate complete draining of the loop.)

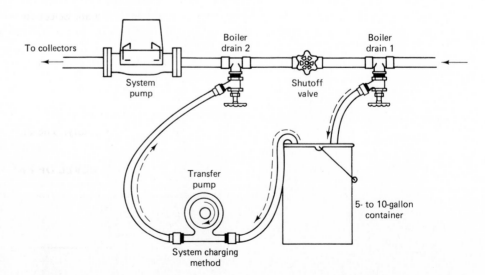

FIGURE 5.38 System flushing procedure. *Note*: This procedure should be done early in the day or late in the afternoon when the collector absorber plate is cool. Cold water introduced into a hot absorber plate can cause severe deformation due to thermal shock, reducing the life of the absorber plate.

Table 5.4 VOLUMETRIC CAPACITIES OF STEEL PIPE AND COPPER TUBE (GAL/FT)[a]

Pipe Size	1/2	3/4	1	1 1/4	1 1/2	2	2 1/2
Copper Tube	0.012	0.025	0.043	0.065	0.092	0.161	0.250
Steel Pipe	0.016	0.028	0.045	0.078	0.105	0.172	0.250
Pipe Size	3	4	5	6	8	10	12
Copper Tube	0.357	0.625	1.00	1.40	2.43	3.78	5.40
Steel Pipe	0.385	0.667	1.00	1.50	2.63	4.20	5.90

[a]*To use this table: (1) Enter the left-hand column at the desired type of pipe; (2) move to the right to the desired pipe size; (3) read the resulting gallons per foot of pipe.*
SOURCE: *Burnham Hydronics.*

Charging the Collector Loop

Although this charging procedure assumes the use of propylene glycol as the antifreeze solution, the use of other solutions will involve similar procedures. Antifreeze solutions should be chosen based on their heat transfer abilities as well as their toxicity. If the solar storage tank contains a single wall internal heat exchanger, only nontoxic food-grade antifreeze solutions can be used in the collector loop. When using toxic antifreeze solutions, the solar storage tank must be equipped with a double-wall vented heat exchanger to prevent contamination of the potable hot water by the antifreeze in the event of a failure of the heat exchanger. Local building codes sometimes specify both the type of antifreeze to be used and the specifications governing the type of internal heat exchanger permissible in the solar storage tank.

The antifreeze solution should be diluted with the proper amount of water, based on the volumetric capacity of the solar system and the required freeze-protection level. Table 5.5 indicates the freezing point of selected mixtures of propylene glycol and water.

The system can be filled and pressurized using one of two methods described below.

Method 1: The proper amount of propylene glycol/water solution is placed into a container. The fluid is then forced into the system using either a hand force pump or electrical transfer pump, as illustrated in Figure 5.38. Select the proper system pressure from the information provided in Chapter 4. On initial system charging, approximately 5 pounds of additional pressure should be added to the system, since the action of the air purger will reduce the system pressure as trapped air is released from solution. After the system has been pressurized, the differential controller should be energized, placing the solar system in the automatic operating mode.

Method 2: If electrical transfer pumps or force pumps are not available for charging the system, the procedure can be done manually. The air valve vacuum breaker on top of

Table 5.5 FREEZE LEVEL OF PROPYLENE GLYCOL/WATER MIXTURES

Propylene Glycol (Percent by Volume or Weight)	Freezing Point	
	°F	°C
30	+8	−13
40	−6	−21
50	−26	−32
60	−57	−50

the collector array is removed to fill the system manually. The system can be designed so that the vacuum breaker is left in place, with a copper "tee" and a plug installed adjacent to the air vent for manual filling (Figure 5.39). After the system has been completely filled, the air valve or filler plug should be replaced. The threaded male connections on these fittings should be wrapped with Teflon tape and liberally coated with Teflon pipe dope prior to installation. After completing this operation, a double-ended female washing machine hose equipped with a check valve to prevent backflow is used to introduce enough water into boiler drain No. 2 to bring the system pressure up to the desired level. Since extra water is being added to the system for final pressurization, it is recommended that the antifreeze/ water solution contain additional antifreeze to compensate for this final "watering down" of the solution. Once the operating pressure of the system has been reached, the boiler drain should be closed and the hose removed from the valve. The system can now be activated. The air valve caps located on the collector vacuum breaker air vent and the expansion tank air vent should be left open to facilitate air purging.

FIGURE 5.39 Fill valve for manual system charging.

SYSTEM OPERATING TECHNIQUES AND TROUBLESHOOTING

Once the system is in operation, the owner should become familiar with the techniques that will ensure proper and efficient operation of the solar system (see Table 5.6). As soon as the system has been energized, it must be determined if fluid circulation has been established within the collector loop. Assuming that the system has been charged or put into operation late in the day to eliminate thermal shock in the absorber plate, it will be necessary to turn the differential controller to manual operation in order to energize the circulator. After the circulator has been operating for several minutes, the installer should be able to hear the fluid circulating in the return piping as it approaches the heat exchanger. This noise is due to trapped air in the collector loop and will usually disappear after a few hours of operation. However, it is useful to use this method to ensure that circulation has been established and that no air locks exist to prevent the fluid from circulating.

Allow the controller to operate the system on automatic if the sun is out. If fluid circulation has been established, the return pipe will feel warm to the touch as the fluid returns from the collector array prior to entering the heat exchanger. When the system contains a digital display controller, readings can be taken of the collector temperature to determine if the heat transfer fluid is circulating through the collectors. Ordinarily, the collector fluid will cool down the solar collectors quickly, and this will be reflected on the digital temperature readout from the collectors.

Table 5.6 TROUBLESHOOTING PROCEDURES

Problem	Component at Fault	Cause	Suggested Solution
Insufficient volume of hot water production	Solar collectors	Improper orientation of collectors	Check direction with compass; reorient if necessary
		Improper inclination angle of collectors	Check inclination angle (see Chapter 5); remount if necessary
		Shading of collector array during the hours of 9 A.M.–3 P.M. solar time	Remove shading objects or relocate collector array
		Insufficient collector area versus system demand	Increase
		Unequal flow in collector array	Drain and purge system; install ball valves for balancing; check to ensure reverse return plumbing
	Differential temperature controller	Improper electrical connections	Check wiring schematic; redo improper connections; check all connections for tightness
		Sensors not securely fastened to tank or piping surfaces	Fasten securely; use of stainless steel clamps, heat conducting grease, and immersion units may be necessary
		Sensors not properly insulated	Insulate adequately
		Faulty sensors	Test resistance of sensors and replace faulty units
		Faulty differential controller	Follow manufacturer's checklist; replace unit if faulty
	Solar storage tank	Insufficient storage volume	Install second tank or increase collector array
		Improper gas/oil connections to backup unit	Check power/fuel source; check electric elements; repair or replace
		Excessive temperature losses in tank	Add insulation; relocate from cold areas to warmer ones
		Thermosiphon heat losses at night	Clean out or replace check valve
		Overall piping heat loss	Check for uninsulated or deteriorated insulation in pipe
	Mixing valve	Adjusted improperly	Check adjustment; set higher if necessary
		Valve supplies only cold water	Replace valve

Table 5.6   (*continued*)

Problem	Component at Fault	Cause	Suggested Solution
Potable water leaks	Collectors	Leak at manifold connections	Remove, clean; install new flexible connector
		Internal absorber leak	Field repair if possible or contact manufacturer
	Pressure/temperature relief valves	Set too low	Adjust pressure setting if possible; replace fuse in valve or replace valve
		Valve does not reseat after discharge	Replace
		Backflow preventer in cold water supply line	Install potable water expansion tank
Drop in close-loop pressure	Collectors	External leak	Remove old fittings and replace
		Internal leak	Field-repair or consult manufacturer
	Closed-loop relief valve	Releasing fluid	Replace
	Air vents	Drop in pressure due to purging	Repressurize to factory specifications
		Releasing fluid	Replace
	Expansion tank	Loss of pressure	Check for integrity of gasket; replace if defective or recharge; inspect air valve
	Solar storage tank	Leak in internal heat exchanger	Replace tank
	Control module	Leak in external heat exchanger	Replace exchanger
	Piping	Joints leaking	Repair
	Interruption of electrical service	Drop in pressure due to boiling	Reset circuit breaker; purge and recharge system
System operates with excessive noise	Piping	Trapped air	Purge system
		Defective air purger	Replace
		Vent caps tight	Loosen and purge system
		Pipe vibration	Install vibration pads where necessary
	Air vents	Omitted during installation	Install at high points
		Valve malfunction	Operate manually; replace if defective
No fluid flow in collector loop	Collectors	Air lock	Drain and purge system; check for tight caps on air vents and loosen if necessary
	Circulator	Undersized	Check head requirements; add second circulator in parallel or replace with larger unit
		Insufficient fluid in loop	Check system pressure and repressurize if necessary

Table 5.6 (*continued*)

Problem	Component at Fault	Cause	Suggested Solution
		Air lock	Check venting caps for looseness; purge system if necessary
		Locked impeller	Free up or replace circulator
		Pump installed backward	Reverse flow direction
		No fluid reaching circulator or moving up to collectors	Check isolation flanges and gate valves on fluid loop for proper position
		Low speed on pump	Increase pump speed to high; verify proper operation of proportional speed controller
	Check valve	Valve stuck in closed position	Manually purge system with transfer pump to free up; replace if needed
	Piping	Air locks	Check vents and caps; purge system
		Excessive pressure drop in pipe	Increase pipe size or add additional circulator
	Air vents	Improperly located	Install at system high points; install air scoop with purger assembly
		Vents malfunction	Replace
		In closed position	Open
	Isolation flanges and manual shut off valves		
	Flow meters and flow regulators	Blocked input or output fittings	Remove from system; clean and reinstall
		Improperly installed	Reverse connections if necessary
Performance level of system decreases as system ages	Collectors	Increased shading as trees grow	Top trees or remove them
		Dirt on collector glazing	Clean
		Deterioration of absorber plating	Contact collector manfacturer
	Insulation	Deterioration with age	Replace defective areas
	Storage tank	Sediment buildup around heat exchanger	Flush tank periodically
System does not sequence either on or off automatically	Differential thermostat controller	Improper location of temperature sensors	Relocate and refasten
		Faulty sensors	Replace
		Incorrect switch setting on controller	Check and reset
		Controller incorrectly wired	Check manufacturer's wiring diagram and replace faulty connections

Source: Adapted from Revere Sun-Pride System Installation Manual, *Revere Solar and Architectural Products, Rome, N.Y., 1979.*

After the system has been running for several hours, a small amount of antifreeze solution should be drawn from either boiler drain No. 1 or No. 2 to check the freeze-protection level of the solution. Most antifreeze manufacturers supply inexpensive test kits that allow the installer to check the level of freeze protection and corrosion inhibitors. The fill valve should be tagged to indicate the date and the freeze level of the initial charging.

The system pressure should be checked from time to time to ensure the integrity of the collector-loop piping. Gradual loss in system pressure indicates a small leak in the closed loop that must be repaired. Under ordinary operating conditions, system pressure will fluctuate as system temperatures change. As the collector fluid heats up, it expands, and system pressure will increase. Conversely, system pressure will decrease as the temperature of the collector-loop fluid drops.

Troubleshooting Charts

Malfunctions in a solar domestic hot water system may occasionally occur. To aid in diagnosing and correcting malfunctions, the troubleshooting chart shown in Table 5.6 has been included. If the chart does not address the specific problem encountered, the manufacturer should be consulted.

Now that the installation of solar domestic hot water systems has been examined in detail, it is time to move on to an analysis of conventional space heating system design. Chapter 6 details the design and operating characteristics of fossil-fuel and solid-fuel heating systems. An understanding of the operational aspects of these systems will then form a foundation for the study of solar space heating system design and interfacing techniques.

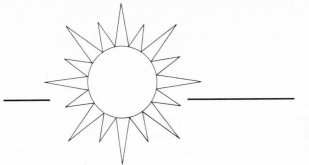

6

Conventional Space Heating Systems

Efficient home heating is the result of properly chosen heating system components engineered to work together. If these basic engineering conditions have been met, the heating system will provide even temperatures throughout the home regardless of the weather conditions. Although the application of solar technology for space heating is not new, the abundance of relatively cheap fossil fuels has restricted the use of solar energy for space heating until recently. Coupled with the dramatic rise in the price of fossil fuels has been an increased use of solar energy for providing space heating as well as domestic hot water.

To gain a full understanding of solar space heating technology, one must first understand the design of conventional fossil-fueled space heating systems which will interface directly with the solar system.

Regardless of the fuel utilized for combustion, all heating systems consist of the following basic components: the heating plant, the heating distribution system, system controls and safety devices, and auxiliary accessories such as humidifiers and air cleaners. The systems that are examined are classified into five categories: forced warm air heating, baseboard hot water heating, electric resistance heating, heat pumps, and radiant floor heating. The heat distribution network for these systems varies according to the fluid or material that is being heated. For example in the case of radiant heating, the floor or ceiling of the dwelling is the distribution system. System controls regulate the amount of heat produced and the temperature limits. Safety controls prevent the system from overheating and building up excessive temperatures and pressures that can damage both the heating system and building occupants. The ultimate purpose is to provide maximum year-round comfort with a high degree of safety at a minimum of cost. Basic to a discussion of any type of heating system are the sizing techniques for specific applications.

SIZING THE HEATING SYSTEM

The first decision to be made when installing any type of heating system is to determine the maximum amount of heat that the system will be called upon to produce. Heating systems are rated in Btu per hour and indicate both the gross fuel input to the heating appliance as well as the net energy delivered to the distribution system. All ratings are based on anticipated operating efficiencies of the burner mechanism. For accurate sizing of the heating system, the heat loss of the building must be calculated. These calculations involve a determination of the following: the number of heating degree-days per year for the specific location of the system, the minimum interior temperature to be maintained by the heating system, the insulation characteristics of the building, and the possibility of additions to the dwelling or heating system at some future date.

It can be stated as a generalization that the majority of home heating systems are oversized in terms of Btu capacity. For example, consider a house with a heat loss of 100,000 Btu per hour at a design temperature of −10°F. Many houses with these heat loss characteristics are found to have a central heating system with a Btu capacity of between 125,000 and 150,000. Oversizing the system in this manner causes it to cycle on and off frequently during operation, which wastes fuel and reduces overall efficiency. On the other hand, a smaller Btu capacity system will cycle less frequently, use less fuel during its operation, and attain higher operating efficiencies. Specific methods for calculating heat loss can be found in Chapter 7. With this in mind we move on to the study of the types of central heating systems encountered when installing a central solar heating system.

SYSTEM CLASSIFICATIONS

Forced Warm Air Heating

Forced warm air heating systems utilize a furnace to heat air which is forced throughout a series of distribution ducts by an electrically operated blower. A typical warm air heating system is illustrated in Figure 6.1. Warm air systems are commonly fueled with gas or oil, although electric heating elements are occasionally used. Heat pumps are sometimes used for warm air heating systems and are discussed later in this chapter. Note the arrangement of the components in a typical warm air furnace illustrated in Figure 6.2.

The fuel burner is located directly below a hot air heat exchanger and plenum assembly. Heat from the combustion of the fuel rises through the heat exchanger. When the heat

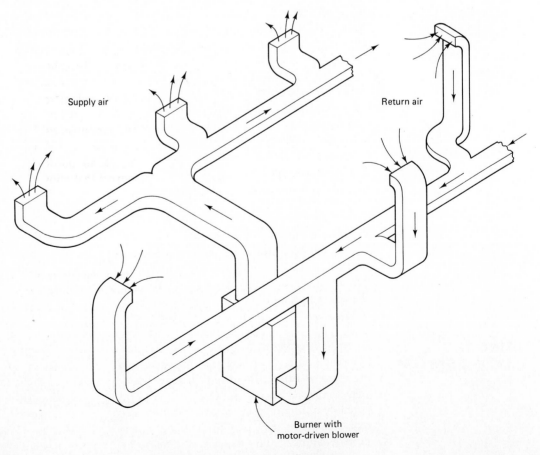

Supply air   Return air

Burner with
motor-driven blower

FIGURE 6.1 Warm air heating system.

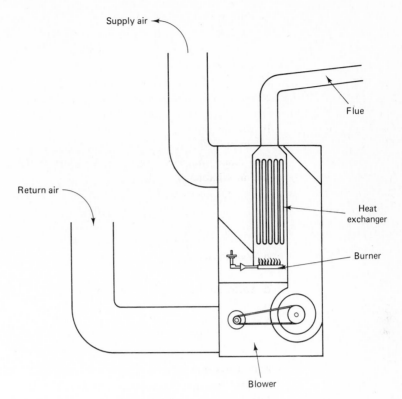

FIGURE 6.2 Warm air furnace.

exchanger reaches a preset temperature, usually 150°F, a motor-driven blower is activated. The blower pulls cold air from the rooms in the house through a series of cold air return ducts and channels it over the heat exchanger. The cold air is heated by conduction as it comes into contact with the large surface areas of the heat exchanger. The heat exchange that takes place in this operation causes air entering at room temperature of approximately 68°F to exit the heat exchanger at temperatures ranging between 100 and 160°F. This hot air now travels through a series of ducts to the various parts of the home. Most hot air ducts contain manual or automatic dampers that can be closed if desired. This feature is especially helpful if the homeowner wishes to heat only certain portions of the home. In addition, most hot air room registers contain adjustable louvers that allow for the control of both direction and volume of airflow (Figure 6.3).

Zoned Systems

In a zoned warm air system, multiple thermostats control the operation of electrically operated dampers located in each of the main hot air trunk lines that serve each heating zone (Figure 6.4). When, for example, thermostat A calls for heat, damper A is electrically opened. When damper A is fully opened, the burner mechanism is automatically activated to provide the necessary heating. If zone B is not calling for heat, the damper controlling zone B air ducts remains closed, preventing hot air from traveling through the zone B ductwork.

The sequence of operation for a hot air heating system begins when the thermostat calls for heat and the burner mechanism in the furnace is ignited. The heat from this combustion is confined in a chamber that is molded and shaped according to the flame characteristics of the fuel and burner mechanism being used. The heat from the combustion chamber rises into the heat exchanger located directly above. A heat-sensing electrical switch, called a fan-limit switch, is mounted on the exterior of the furnace, with its heat-sensing probe projecting into the warm air plenum. When the turn-on set point of the fan-limit switch has been reached, indicating that the furnace plenum is hot enough to supply

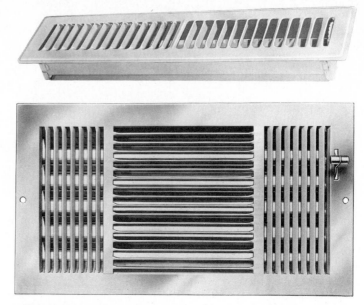

FIGURE 6.3   Manually adjustable air registers. (Courtesy of Hart and Cooley.)

warm air to the dwelling, the switch automatically turns on the circulating blower. Cold air is drawn through the return ducts located on the suction side of the blower. This cold air is then pushed over the heat exchanger, where it is warmed and circulates through the hot air ducts into the rooms of the building. Continuous action of the cold air flowing over the heat exchanger gradually drops the temperature of the plenum until it is no longer warm enough to heat the house effectively. When this low-limit set point has been reached, the blower is automatically turned off by the fan-limit switch until the temperature of the plenum and heat exchanger assembly rises again to the turn-on point. This action continues until the thermostat setting in the home has been satisfied and the burner mechanism shuts off. The blower will continue to run until the heat exchanger has cooled down to the low-limit setting on the fan-limit switch. Most hot air systems are set to activate the blower when the heat exchanger reaches approximately 150°F, and turn off the fan when the temperature drops below 100°F. Fan-limit switches also contain a high-limit safety cutoff point. Should

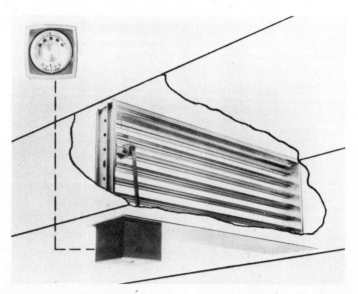

FIGURE 6.4   Electrically operated dampers. (Courtesy of Trolex Corp.)

the fan fail to operate, the limit switch will automatically shut down the burner mechanism when the plenum reaches the high-limit set point, usually between 180 and 200°F, to prevent overheating of the heat exchanger and associated components.

Electrostatic Air Cleaners and Power Humidifiers

The main advantages of hot air heating systems over other types is that interior spaces can be quickly heated. A house can be brought from a temperature of 35 to 40°F up to 70°F from a cold-start furnace within 15 to 20 minutes. Hot air systems lend themselves easily to the installation of auxiliary heating components such as power humidifiers and electrostatic air cleaners, enabling the homeowner to "customize" the air in specific heating applications. For example, electrostatic air cleaners are highly effective in filtering out dust, pollen, and suspended dirt particles from the air within a home. A typical electrostatic air cleaner is illustrated in Figure 6.5. These cleaners work by imparting dust and dirt with one type of electrical charge, and the walls of the filter elements of the air cleaner with the opposite electrical charge. The dirt particles are attracted to, and become enmeshed in, the filtration system of the air cleaner. Electrostatic air cleaners are especially helpful in eliminating allergy symptoms associated with pollen and dust.

Power humidifiers are accessories that are commonly found in central hot air heating systems. A typical power humidifier is pictured in Figure 6.6. During the winter months, the relative humidity in the air can drop to as low as 5 to 10%. Extended periods of low humidity can cause such diverse problems as skin irritations to cracks in wooden furniture. These problems are easily rectified by the installation of power humidifiers.

A further advantage claimed for forced warm air heating systems is that the room air is constantly mixed, which tends to minimize temperature differences between individual rooms and reduces the stratification layers of heated air between the floor and the ceiling. The annual maintenance involved with forced hot air systems consists of cleaning the burner assembly, heat exchanger, and flue passages.

Natural Convection Warm Air Heating

A modification of forced warm air heating is known as a natural convection warm air heating system. These types of heating systems are most often found either in older homes or those located in southern climates, where central heating plays a less important role in

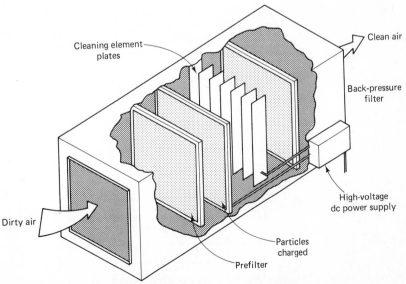

FIGURE 6.5 Electrostatic air cleaner.

FIGURE 6.6   Power humidifier. (Courtesy of Auto-Flo Co., Division Masco Corp.)

maintaining interior temperature levels. Burner operation in a gravity or natural convection warm air system is similar to that of its forced-air counterpart. A thermostat controls the ignition of the burner mechanism. Air is heated within a heat exchanger located within the furnace. This air rises into the room located directly above the furnace through a large floor grate. Cooler air in the home falls to the floor, where return air ducts channel it back to the furnace, continuing the heating cycle. Movement of the air is accomplished by natural convection. Gravity-fed systems are adequate for heating homes in areas where the heating

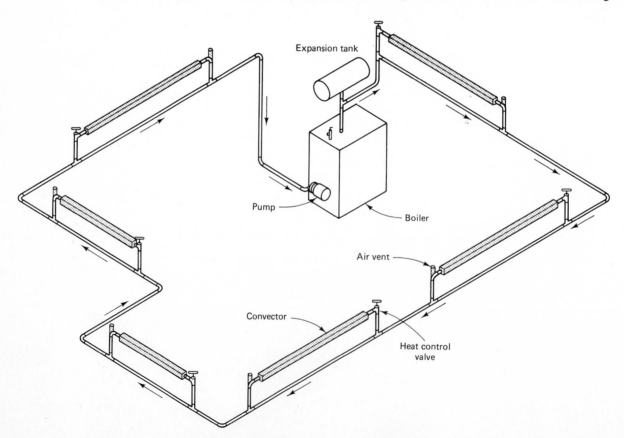

FIGURE 6.7   Hot water baseboard heating system.

loads are minimal; however, in northern climates forced convection heating systems are a necessity if even temperature levels are to be maintained.

Baseboard Hot Water (Hydronic) System

In a baseboard hot water heating system, sometimes referred to as a hydronic heating system, a boiler heats water which is then circulated through finned convection heating units located throughout the house. A small circulating pump is used to move the hot water through the finned baseboard units, and heat transfer takes place from the hot water to the interior of the room. The water finally returns to the boiler, where it is reheated. A typical hot water baseboard system is illustrated in Figure 6.7. There are several system variations in baseboard hot water heating.

Series Loop System

In the series loop type of heating system, the heated water travels through all the baseboard connections in sequence until it returns to the boiler located at the end of the circuit. This type of system is illustrated in Figure 6.8.

The series loop system is the least expensive type of baseboard hot water system to install, since there are no electrical zone controls or separate heating loops. A centrally located thermostat controls the entire heating circuit, and there is very little control of temperature from one room to the next. A modification of the series loop system is the supply loop design.

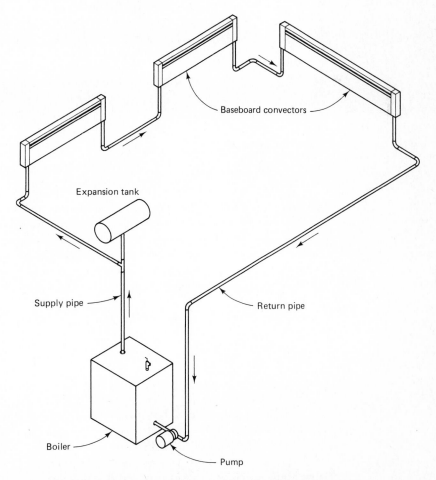

FIGURE 6.8 Series loop hot water baseboard heating system.

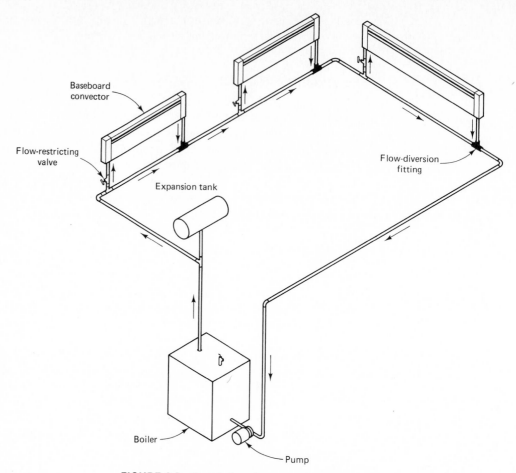

Baseboard
convector

Flow-restricting
valve

Expansion tank

Flow-diversion
fitting

Boiler

Pump

FIGURE 6.9   Supply loop hot water heating system.

Supply Loop System

The supply loop design is illustrated in Figure 6.9. In this type of system, a single supply pipe runs around the perimeter of the house, and all the radiators tap off this main line. Each radiator is equipped with a flow-diversion fitting that siphons off some of the water from the supply loop and directs it through the radiator and back into the supply loop. The amount of hot water admitted to each radiator can be adjusted by means of a variable-flow restricting valve located on the intake side of each radiator. This enables individual room heating to be varied by adjustment of this valve. If heat is not required in a specific room or section of the home, the appropriate radiators can be shut down entirely. This type of system, like its series loop counterpart, is a single-zone system, operated from one centrally located thermostat. Zoned heating systems, which offer somewhat more flexibility in hydronic heating than do single-zone systems, are variations of the simple supply loop system.

Zoned Series Loop System

Of all the baseboard hot water systems currently in use, the system that is the most popular and flexible is the multizoned series loop type of system illustrated in Figure 6.10. Note that in a two-zoned heating system, each zone contains its own thermostat and separate circulating pump to control water flow. This type of system allows separate sections of the house to be heated independently of one another and can significantly reduce fuel con-

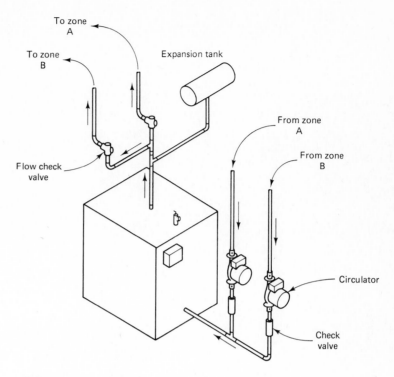

FIGURE 6.10 Two-zone hot water baseboard heating system.

sumption compared to a single-zone system, since heat is delivered only where it is needed. Although Figure 6.10 illustrates the use of separate circulators to control the flow of water to each heating zone, a modification of this system employs a single circulator and electrically operated zone valves (Figure 6.11) to control water flow. Zone valve systems and systems that use separate circulators are similar in nature.

Hot Water Radiant Floor Heating

Radiant floor heating is an adaptation of hot water baseboard design. Rather than using finned baseboard units for heat dispersal, a continuous coil of copper tubing is embedded throughout a concrete floor. The heating loop begins and ends at the boiler, as it does in a series loop system. As the hot water circulates throughout the loop of embedded copper tubing, the concrete floor is heated, and this heat radiates to the interior of the building (Figure

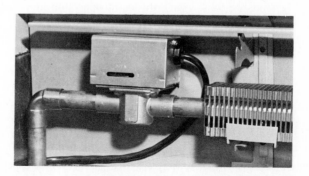

FIGURE 6.11 Electrically operated zone valve. (Courtesy of Flair Mfg. Corp.)

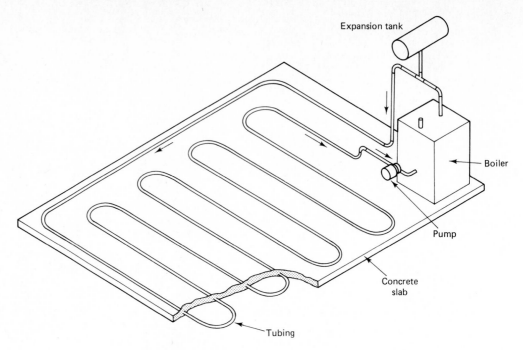

FIGURE 6.12   Hot water radiant floor heating.

6.12). The concrete slab should be well insulated. Should the insulation under the concrete be inadequate, the earth surrounding the concrete floor will act as a massive heat sink, continually drawing off heat from the radiant floor, greatly reducing the operating efficiency of the system.

One characteristic of radiant floor heating is its ability to heat the interior of a building evenly with maximum comfort to the building inhabitants. All system components other than the boiler are hidden in the floors. Due to the thermal mass of the concrete floors, this type of heating system is slow to respond to temperature changes and cannot quickly cycle from a cold start. Once interior temperatures are achieved, however, floor temperatures usually average between 80 and 90°F, and the thermal mass of the heated concrete will radiate heat to the building interior long after the boiler has shut off.

Sequence of Boiler Operation and System Nomenclature

The hot water boiler is the heart of any hydronic heating system. Although there are a wide variety of water heating boilers available, they all function in essentially the same manner. A simplified pictorial of a typical baseboard hot water boiler and piping is illustrated in Figure 6.13. This illustration focuses on the basic arrangement of the boiler controls and plumbing connections. The sequence of operation in a single-zone system is as follows. When the room thermostat calls for heat, the relay that ignites the burner mechanism closes, initiating ignition; assuming that a flame has been established in the burner, a flame or heat-sensing mechanism allows fuel ignition to continue; heat from ignition is transferred to the water in the boiler; when the water temperature in the boiler reaches the low-limit set point on the aquastat, the electrical control to the circulator closes, and hot water begins to circulate from the boiler through the baseboard system.

The burner mechanism will continue to operate until either the water in the boiler reaches the high-limit temperature setting on the aquastat, or the room thermostat in the house has been satisfied. The high-limit aquastat setting prevents the boiler from overheating, and is usually set 20°F above the low-limit set point. The low-limit setting serves as an on–off switch for the circulator, preventing water circulation through the baseboard if the boiler water temperature is too low.

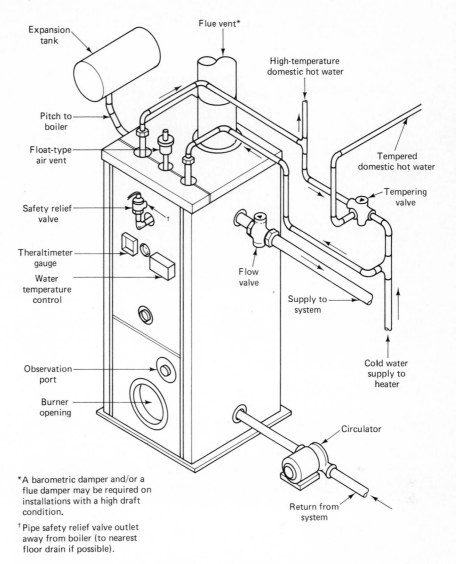

Expansion tank

Flue vent*

High-temperature domestic hot water

Pitch to boiler

Float-type air vent

Safety relief valve

Theraltimeter gauge

Water temperature control

Observation port

Burner opening

Tempered domestic hot water

Tempering valve

Flow valve

Supply to system

Cold water supply to heater

Circulator

Return from system

*A barometric damper and/or a flue damper may be required on installations with a high draft condition.

†Pipe safety relief valve outlet away from boiler (to nearest floor drain if possible).

FIGURE 6.13 Typical piping for hot water boiler. (Courtesy of Burnham Corp.)

A typical baseboard hot water heating system is a combination of two separate electrical circuits: the circulator circuit, dependent for operation on boiler water temperature; and the burner mechanism with associated safety features. The room thermostat serves to bridge the circuits electrically, serving to ignite the boiler and control water circulation through the baseboard. For example, if the thermostat calls for heat and the boiler water temperature is above the low-limit aquastat setting, the circulator will come on, moving hot water through the baseboard units. Additionally, should the boiler water temperature be below the high-limit setting on the aquastat, the burner will begin ignition of fuel. In normal operation, the boiler will maintain internal water temperature between the low- and high-limit aquastat settings, even though the thermostat does not call for heat. This feature keeps the boiler on standby, with a ready supply of heated water. Normal aquastat settings are between 130 and 160°F for the low limit, with high-limit settings 20°F above the low-limit set point. A typical aquastat used in hydronic heating applications is illustrated in Figure 6.14.

The sequence of operation for a multizoned hot water heating system is basically the same as for that of a single-zone system. Modification is made for the boiler to operate in response to the additional thermostats that are placed in the circuitry. The reader is referred to Figure 6.10 for a pictorial of a multizoned heating system. The boiler automatically maintains internal water temperature in the manner described for single-zoned systems. On a call

 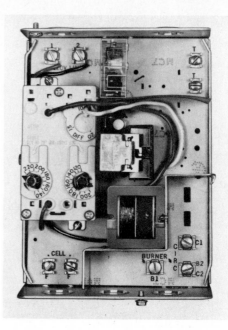

FIGURE 6.14 Aquastat used in hot water heating applications. (Courtesy of Honeywell Corp.)

for heat from any of the room thermostats, the appropriate circulator or zone valve will energize to provide hot water circulation assuming that the boiler water is above the low-limit aquastat setting. Each circulator or zone valve is controlled by a specific thermostat. The circulators are all independent of one another.

Zone valves are often used in place of separate circulators in such heating systems since they are less expensive to install, and use only one circulator to move all the water in the heating system. A zone valve is an electrically operated water valve placed in each heating loop. When the thermostat in a particular zone calls for heat, the zone valve opens, clearing the way for water circulation through the heating loop. Most zone valves incorporate a time-delay mechanism that energizes the circulator only after the valve is fully open. In very large heating systems, zone valves cannot be used because one circulator is too weak to move the amount of hot water required for the total heating system, assuming that all zones are operating. In these larger systems, separate circulators are employed.

Electric Hot Water Boilers

In addition to gas- and oil-fired boilers, electric boilers are available for hydronic heating. An electric boiler is illustrated in Figure 6.15.

Electric boilers contain a series of heating elements similar to those found in hot water heaters. The ability of these resistance elements to heat water instantly allows these boilers to be relatively small in size (an average electric boiler is less than 2 ft long and 1 ft deep). The electrical capacity required to run these units is substantial (often 100-ampere service is required for the boiler), and adequate wiring is a necessity. Since no fossil fuels are oxidized in these units, maintenance costs associated with their operation are minimal. The capacity of the heating elements within the boiler offer instantaneous hot water within the heating system, and their cold-start characteristics are quite good. Rising electrical utility rates in many parts of the United States can make these units costly to operate.

Solid-Fuel and Multifuel Boilers

The popularity of solid-fuel and multifuel heating units has risen dramatically within the past several years (Figure 6.16). Solid-fuel boilers are usually larger than their fossil-fuel counterparts, due to the large fireboxes required for fuel loading and extended burning

FIGURE 6.15 Electric boiler. (Courtesy of Burnham Corp.)

times. Also, the larger fireboxes allow for additional surface area within the combustion chamber required for efficient heat-transfer since the operating temperatures of solid-fuel fireboxes are sometimes lower than those of conventionally fueled boilers.

The operating sequence for solid-fuel units is similar to that of fossil-fuel systems. Internal water temperature is maintained either by immersion draft regulators or draft blowers that increase or decrease the amount of combustion air entering the fuel chamber. A typical immersion aquastat is illustrated in Figure 6.17. Many solid-fuel boilers are installed as add-on units to existing fossil-fuel heating systems. This type of installation offers the flexibility of using solid fuel for the majority of the heating load, allowing the fossil-fuel system to act

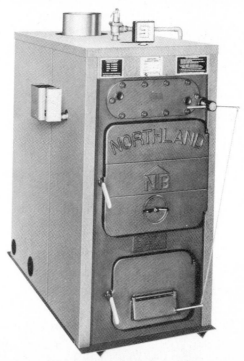

FIGURE 6.16 Solid-fuel hydronic boiler. (Courtesy of Northland Boiler Corp.)

FIGURE 6.17 Immersion-type aquastat for maintaining combustion air. (Courtesy of Ammark Corp.)

as an automatic backup, operating only when needed. This type of add-on installation is illustrated in Figure 6.18.

Note from Figure 6.18 that as the water returns from the baseboard heating units, it is first routed through the solid-fuel boiler. The heated water moves from this unit through the fossil-fuel boiler and then through the baseboard heating system. Thus the solid-fuel boiler preheats the water and a series piping arrangement is established. The fossil-fuel boiler acts independently of the solid-fuel unit, coming on only when the entering water is below the high-limit set point on the aquastat. Although Figure 6.18 shows a typical series hookup, many other types of plumbing arrangements are possible. Manufacturers' specifications and requirements should be consulted prior to undertaking this type of installation.

Since the firebox in all solid-fuel boilers is water cooled, the interior walls of the combustion chamber rarely rise above 212°F. This temperature is below the condensation point

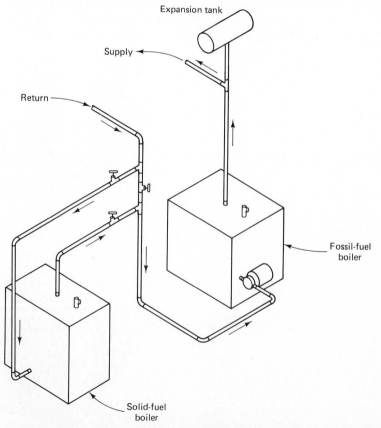

FIGURE 6.18 Add-on solid-fuel boiler. The solid-fuel boiler preheats the return water in a series plumbing arrangement.

for creosote; thus all solid-fuel boilers are subjected to varying amounts of creosote deposition and should be cleaned at frequent intervals when wood is used as the fuel. Although coal does not form creosote, soot that has built up should periodically be removed from both the heat exchanger surfaces of the boiler and all chimney and flue pipes. Following time-tested installation and safety practices, these systems can be highly cost-effective. To help determine the savings from such an installation, the reader is referred to the comparative fuel data calculations in Table 1.2.

Steam Hot Water Boiler

Central steam heating systems have been in use for many years. A steam boiler is similar in appearance to a hot water boiler but functions in a different manner (Figure 6.19). A typical steam heating system contains a closed loop of water that is heated and turned into steam. The steam rises through a single pipe into the room radiators. In the radiators, the steam releases heat into the room, condenses back into water, and returns to the boiler. A single pipe is generally used to supply both the steam to the radiators and carry the condensate back to the boiler. A continual slope on all horizontal pipe runs is necessary to ensure that positive water flow occurs on the return flow from all the radiators.

Steam heating systems do not require a circulating pump, since the steam, being lighter than air, automatically rises in the piping system to the radiators. It returns, gravity fed, via the same route. A glass tube is built into the side of the boiler jacket to indicate boiler water level. Periodically, the water level may drop due to steam leaks and venting from the radiators. A cold water feed pipe is then opened manually to bring the level of the water in the boiler back up to a suitable point, midway on the glass-tube water-level indicator. A low-water cutoff is built into the electrical controls that govern the operation. This control is designed to shut down the burner mechanism should the boiler water drop to a dangerously low level.

Steam heating boilers can be equipped to burn fuel oil, gas, wood, or solid fuel. Tankless coils can be inserted into these units for domestic hot water production. Steam

FIGURE 6.19 Steam boiler. (Courtesy of Burnham Corporation.)

heating systems are generally more expensive to operate and maintain than their baseboard hot water counterparts due to the higher water temperatures involved in the production of steam.

Electric Resistance Heating

Electricity is an adaptable energy source, and as such is used to provide the power for a variety of heating applications: hot air heating, hot water heating, and perhaps the most popular of all electric heating applications, resistance heating. There are two common varieties of electric resistance heating: unit resistance baseboard heaters and radiant ceiling heating.

Electric Resistance Baseboard Heating

Electric resistance baseboard heaters are sold in a variety of lengths to fit different-sized rooms. Installation of these systems is relatively simple since no plumbing connections or ductwork is needed. The unit heater is installed along the base of the wall and wired into the electrical breaker panel together with the thermostat for the unit.

A separate thermostat for each room is usually employed with electric baseboard units, making it a multizoned heating system. A typical electric baseboard unit is illustrated in Figure 6.20. When the unit is turned on, electricity passes through the central heating element. As the element heats up, the fins that are built around the heating element radiate heat into the room. The baseboard unit is constructed so that air enters through the bottom, is heated as it passes over the finned elements, and flows out the top of the unit. Natural convection keeps the air currents moving over the heating element. When the room temperature reaches the thermostat setting, the baseboard heater is turned off.

Electric baseboard heating is virtually maintenance-free and highly reliable. The unit nature of the heating systems allows for zoning flexibility since each room in the home is, in effect, a separate heating zone. Given proper sizing and distribution of the heaters, the heat tends to be even and well distributed.

The rising utility costs of electricity have made electric baseboard heating systems very expensive to operate in many areas of the country. During the 1950s and 1960s, when the cost of electricity was very low, most utilities offered financial incentives, in the form of lowered utility rates during the heating season, to those homeowners who installed electric heating. Due to these incentives and the generally low installation costs of electric heating, it grew rapidly during this time period. Recent escalating utility costs, however, have cut the number of newly constructed homes offering electric heating. In fact, many all-electric homes have converted to gas, oil, or central solid-fuel heating systems in an effort to lower escalating utility costs. In many cases, it is often less expensive to install a complete new fossil-fuel heating system than to continue to operate the electric baseboard heating system.

Radiant Electric Ceiling Heat

The second popular method of electric resistance heating involves the use of resistance heating cable that is embedded in the ceiling. In some systems, rather than embedding the

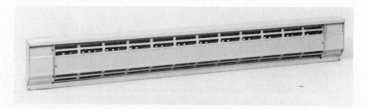

FIGURE 6.20 Electric baseboard heating unit. (Courtesy of Climate Control, a unit of Snyder General Corp.)

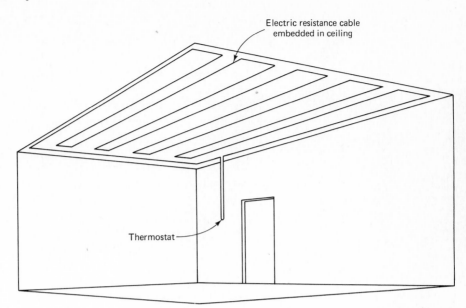

Electric resistance cable
embedded in ceiling

Thermostat

FIGURE 6.21 Electric radiant heat in ceiling.

heating cable in the ceiling panels, it is laid on top of the ceiling material. A typical resistance ceiling layout is illustrated in Figure 6.21. In radiant ceiling systems, the room thermostat controls the operation of the heating cables, and the zoning characteristics of ceiling heat are the same as for resistance baseboard applications. The heat from the ceiling radiates downward into each room, providing even heating with a great deal of comfort. A minimum of 6 in. of insulation above each ceiling is required to ensure that the heat from the cables radiates downward into each room rather than into the room above. The installation costs for this system are somewhat higher than for unit electric baseboard heaters but less than for conventional forced hot air or hydronic systems.

HEAT PUMPS

The use of heat pumps for central heating and cooling has grown rapidly during the past 10 years. A heat pump is a mechanical device that is similar in operating principle to the common household refrigerator. Using a working fluid such as Freon, a heat pump has the ability to transfer heat from a cool environment, outdoor air for example, into a warmer environment such as the home. This is opposite to the natural direction of heat movement; thus the system must be powered by electricity.

A compressor is used to force the working fluid through what is known as a *refrigeration cycle.* The operating characteristics of the refrigeration cycle enable the heat pump to be used to provide both space heating and air conditioning utilizing the same piece of equipment. Let's examine both the heating and cooling cycle of a heat pump to understand the operational characteristics of this increasingly popular technology.

Heating Cycle

The heating cycle of a common heat pump is illustrated schematically in Figure 6.22. The cycle begins when the compressor activates, bringing the Freon gas to a high pressure. This increase in pressure is accompanied by an increase in temperature to approximately 140°F. The high-temperature gas is then circulated either through pipes directly exposed to room air, or through a forced-air heating unit which blows room air over the heated pipes. As the heat is transferred from the gas to the room air, the gas cools and begins to condense into a liquid. Although the Freon has lost a good deal of heat to the room air, it is still very hot.

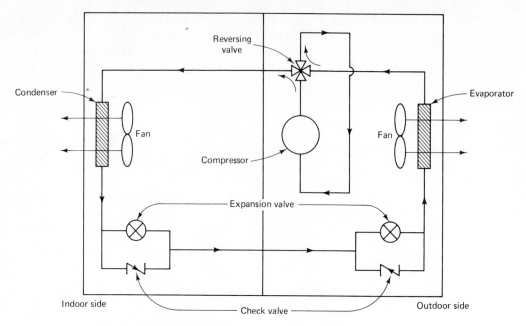

FIGURE 6.22   Heating cycle of a heat pump.

At this point the gas travels through an expansion valve, resulting in a significant drop both in temperature and pressure. The low-pressure Freon liquid travels through a set of evaporator coils located outdoors. Since the temperature of the Freon is now below 0°F, it absorbs heat from the outside air. It is this ability to absorb atmospheric heat that allows the heat pump to operate at very high efficiencies. Outside air, even at low temperatures, contains a considerable amount of usable heat. A heat pump is engineered to take advantage of this available energy. As the Freon absorbs heat from the outside air, it begins to vaporize. The heated gas is now fed back into the compressor, and begins another heating cycle. The amount of heat transferred from the Freon gas to the interior of the home is several times the amount of energy required to run the compressor and associated equipment of the heat pump; thus the operating efficiency of the system, called the *coefficient of performance* (COP), is greater than 100%. Since the efficiency of the heat pump is a function of outside air temperature, the COP will vary depending on the time of year that it is in use. Manufacturers' literature should be consulted to obtain various COPs at different outside air temperatures. The heating cycle of a heat pump is opposite to that of the cooling cycle, which will be examined next.

Cooling Cycle

In the cooling cycle of the heat pump, the flow pattern of the working gas is reversed from that of the heating cycle. The working schematic of the cooling cycle of the heat pump is illustrated in Figure 6.23. We pick up the cooling cycle as the Freon gas travels through the expansion valve, resulting in low-temperature and low-pressure liquid Freon, and then through coils exposed to room air. Warm air in the room is circulated over the cooler Freon-filled coils; heat from the room is absorbed by the coils and the air is cooled. The Freon begins to vaporize as its temperature increases, and it travels back to the compressor. The compressor increases both the temperature and pressure of the Freon gas after which it travels to a heat sink, usually a set of coils located outside the house. Here, heat gained from both the room air and the action of the compressor is radiated to the atmosphere. As the temperature of the gas drops, it begins to condense into the liquid state once again and travels through the expansion valve where the cycle repeats.

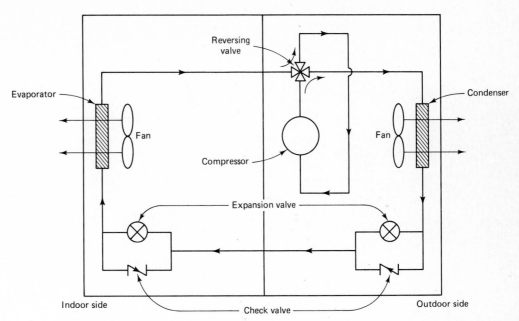

FIGURE 6.23 Cooling cycle of a heat pump.

The operating cost of heat pumps can depend on electric utility rates. These costs should be investigated thoroughly to determine operating costs of heat pumps for specific localities.

It should be kept in mind that the COP of a heat pump drops off significantly in severely cold climates, and manufacturers' specifications must be consulted to help determine which sites are applicable for such installations. The maintenance costs for these systems tend to be the same as for other refrigeration-type devices and maintenance must be performed by a skilled service person.

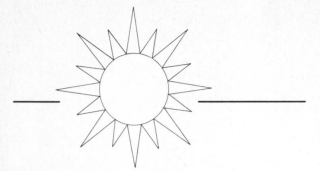

7

Solar
Space Heating
System Design

DESIGN LOGIC

Solar heating engineers are sometimes asked to design a system for a homeowner which will provide 100% of the energy required for space heating and domestic hot water. Many homeowners would naturally like to eliminate the necessity of purchasing conventional energy to run their space heating and hot water systems, since they feel that although they may own their own home, they must "rent" their energy. It is often thought that a solar heating system, designed to provide 100% of those energy needs, would end this continuously increasing energy rent. The inadvisability of 100% solar system sizing will be examined in detail throughout this chapter. It is the responsibility of the solar engineer or system designer to explain to the novice why, in almost all cases, a solar heating system should not be designed to provide 100% of the energy requirements. It is normally much more cost-effective to design a system that interfaces with the existing auxiliary heat source in the home to provide for a meaningful percentage of both space heating and domestic hot water energy requirements.

In attempting to design a system to provide for 100% energy contributions, plans must be made for very long sunless periods, in addition to extremely cold weather conditions that are likely to exist for extended periods of time. These two conditions necessitate excessively large collector arrays and storage systems. Because weather conditions cannot be controlled, system design must take into account worst-case predictions, resulting in a solar system with an almost infinite number of solar collectors together with adequately sized storage. Figures 7.1 and 7.2 illustrate the problems encountered when trying to do too much with a solar energy system, resulting in excess size of the collector array and storage capacity.

Note in Figure 7.1 that the energy output of the solar system falls to its lowest point when the space heating requirement is at its peak. Conversely, large amounts of surplus heat are generated during the warmer months when space heating is needed the least. This surplus heat generation cannot be considered as a contribution to the heating system, or as a savings to the homeowner, unless it can be utilized in some way. System A in Figure 7.1 graphically demonstrates the design of a solar space heating system incorporating excessive numbers of collectors generating large amounts of surplus energy.

The key to successful solar design, therefore, is to minimize the surplus heat generated. Although it may be impressive for a solar system owner to display to friends and neighbors a storage tank at an elevated temperature, it is much more cost-effective for the storage tank to contain just enough energy to meet the owner's needs. Sizing a system that will generate less surplus energy during the warm months means, of course, that an auxiliary, or backup,

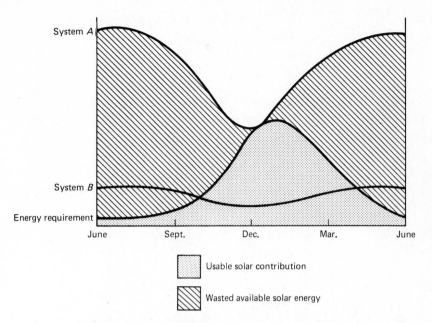

FIGURE 7.1 Sizing for maximum use of solar system.

energy source will be required for portions of the colder months. System B in Figure 7.1 adequately meets the prerequisite for small surplus heat generation and cost-effective heating.

It should be remembered that a solar heating system cannot be oversized as inexpensively as can conventional heating systems. Since all solar system components are quite expensive, oversizing the collector array or storage capability of the system leads to high installed system costs. System A in Figure 7.1 is analogous to a farmer who purchases a very large truck for hauling an annual harvest to market only one or two months each year. The large truck has no other tasks to perform for the remainder of the year. System B in Figure 7.1, on the other hand, is analogous to the farmer who purchases a small pickup truck to perform many smaller tasks on a year-round basis, but must lease a large truck for a short period of time to move the crop to market. The equipment must be fully utilized throughout the year to be cost-effective.

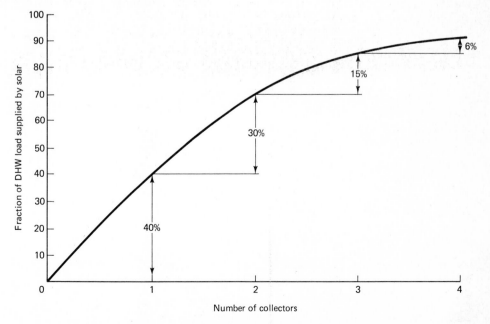

FIGURE 7.2 Collector area versus contribution.

The principle of solar system design for maximum utilization of energy absorbed will be maintained throughout this chapter. Sizing techniques minimizing surplus energy will be discussed. Space heat distribution methods and system controls will be presented with an emphasis on system logic designed for maximum solar output while minimizing backup fuel usage. Alternative uses for surplus solar energy available during the warmer months will be discussed as well as the use of solar storage and control systems designed to facilitate the most efficient use of solid-fuel appliances during the heating season.

SIZING TECHNIQUES

Up through the mid-1970s, a solar-heated house was thought to require thousands of gallons of water for storing the solar energy captured by the collector array. Many houses were built over or around huge storage tanks or rock bins. Others had these storage systems buried under their yards, only to be disappointed by the resulting poor performance of the solar heating system. The number of solar collectors in these systems was often based on the size of the house rather than the heating requirement of the house. Control systems were, and in many cases continue to be, both inadequately designed and troublesome.

Each heating project needs to be investigated carefully prior to undertaking any system design work. Since solar technology is so heavily dependent on site-specific characteristics, each system will vary greatly. Poor potential situations for solar should be immediately recognized as such. For example, all too often the person who wants a solar heating system has an uninsulated and drafty home and is desperate for relief from high fuel bills. The conscientious solar engineer will educate this person regarding the need to insulate the home before installing a solar heating system. Upgrading insulation and heating equipment usually makes more economic sense than installing a larger than necessary solar heating system in these situations. Methods of accomplishing these tasks relating to basic energy conservation and upgrading techniques are discussed more fully in Chapter 10. Only after a homeowner has done everything within reason to decrease building heat loss and increase the efficiency of a conventional energy system can solar be considered as a cost-effective option.

Fuel Consumption History

The first step in designing a solar system for a specific application is to determine the heating load for the home or building. This load is normally composed of two energy consumption figures: the building heat loss and the domestic hot water load. The domestic hot water load is normally the smaller of the two energy requirements and is also the easiest to calculate. The national average consumption of 20 gallons of hot water per person per day is used to determine this load. The building heat loss can be determined in two different ways. One method is to calculate the heat loss of every wall, window, door, floor, and ceiling of the home. The other, and much easier method, is to compare past fuel consumption records with degree-day records for that location. The fuel consumption method, if done properly, is the most accurate because it is based on actual energy requirements and not on theoretical assumptions of heat loss. The life-style of the family and many other factors which cannot be included in a theoretically calculated heat loss analysis, but which do have a significant impact on the energy load, will be reflected in the fuel consumption figures.

An example of the fuel consumption method is as follows.

Example 7.1

A family of four living near Philadelphia, Pennsylvania, burns No. 2 fuel oil for both space heat and domestic hot water. They have a 15-year-old boiler with a separate oil-fired domestic hot water heater. The water heater is 5 years old and incorporates a high-speed flame retention burner. The hot water temperature is set at 130°F. The oil equipment is serviced yearly. The service record identifies the efficiency of both oil burners at 70%. Approximately 1000 gallons of No. 2 fuel oil was used during the past

12 months. The degree-day total for that area (see Appendix 3) was 5400. What is the heat loss of the home?

Solution

Before the space-heating load of the home can be determined, the amount of oil used for production of domestic hot water must be calculated and subtracted from the total amount of oil consumed. This can be done by using the following simple calculations:

(1) No. people $\times$ 20 gal $\times$ 8.3 lb/gal $\times$ set temp. $-$ incoming temp. $=$ no. Btu/day

(2) $\dfrac{\text{No. Btu/day}}{\text{efficiency of unit}} \div$ no. Btu/gallon fuel oil $\times$ no. days used $=$ DHW oil consumption

Therefore,

$$4 \text{ people} \times 20 \text{ gal} \times 8.3 \text{ lb/gal} \times (130°\text{F} - 55°\text{F}) = 49{,}980 \text{ Btu/day}$$

and

$$\frac{49{,}980 \text{ Btu}}{0.60 \text{ efficiency}} \div 140{,}000 \text{ Btu/gallon No. 2 fuel oil} \times 365 \text{ days/yr} = 217 \text{ gal oil/yr}$$

Since 217 gallons of fuel oil was used to heat the domestic hot water, 783 gallons of oil was used to heat the house. The heat loss of the home can now be calculated as follows:

$$\frac{\text{No. gallons oil consumed} \times \text{Btu/gal of fuel} \times \text{efficiency of unit}}{\text{no. degree-days for area}}$$

$$= \text{heat loss in Btu/degree-day (DD)}$$

Substituting the oil consumption figure from our previous calculations, we get

$$\frac{783 \text{ gal No. 2 fuel oil} \times 140{,}000 \text{ Btu/gal} \times 0.55 \text{ efficiency*}}{5400 \text{ degree-days}}$$

$$= 11{,}165 \text{ Btu/DD} \div 24 \text{ hours} = 465 \text{ Btu/°F-hr}$$

In order to use the fuel consumption method for calculating building heat loss, the Btu content of various fuels and the overall efficiency of their respective heating units must be known. This information is available in Chapter 1. It is important to note, however, that the efficiency of a tankless coil hot water heater changes drastically for the worse during the summer months. Because the entire boiler must be kept hot 24 hours a day solely to provide for domestic hot water production capability, the efficiency of a tankless coil system which averages 55% during the winter months drops to an average of 15% during the summer months. This fact must be considered when a heat load is being calculated from fuel consumption figures.

If the family in Example 7.1 had a tankless coil for domestic hot water production instead of a separate oil-fired hot water heater, the calculation would be made in two separate operating modes: one for the production of hot water during the winter months and one for hot water production during the nonheating season, as follows:

49,980 Btu required per day $\div$ 0.55 efficiency $\times$ 275-day heating season $\div$ 140,000 Btu/gal
= 179 gallons of oil used for domestic hot water during the 9-month heating season

49,980 Btu required $\div$ 0.15 efficiency $\times$ 90 days $\div$ 140,000 Btu/gal
= 214 gallons of oil used for domestic hot water production during the nonheating season
of 3 months

---

*The overall efficiency of the heating unit is lower than the burner efficiency due to flue and boiler jacket losses. Similar reductions in efficiency are apparent in hot air heating systems as well.

From the figures above, it is easy to understand why a solar domestic hot water system is an excellent investment for a family with a tankless coil hot water heater. The solar contribution to the system is highest when the auxiliary system efficiency is lowest.

Heat Loss Calculations

In the case of new construction, or when a change in ownership occurs in a home or building, the fuel consumption method of determining building heat loss may not be possible. In new construction, the architect or heating contractor will normally have performed a building heat loss analysis in order to size a conventional heating system. If attempts to obtain this information fail and no fuel consumption records are available, a calculated building heat loss must be performed. The Btu rating of any existing heating equipment should not be relied on to judge the true heat loss of the house, since heating contractors often oversize the boiler or furnace to allow for construction errors or poor heating system performance. Since solar engineers do not have the option of oversizing a heating system, the true heat loss of the building must be carefully calculated.

The heat loss of the building is normally expressed in Btu per hour and is based on some temperature differential between outside ambient air and indoor comfort levels. For example, in Philadelphia, this design temperature differential is 70°F, while in Albany it is 85°F. The design temperature differential is usually based on an indoor temperature of 70°F minus the normal coldest ambient air temperature (0°F in Philadelphia, and −15°F in Albany). The Btu per hour figure can be converted as follows:

$$50{,}000 \text{ Btu/hr with a } 70°F \ \Delta T^* = \frac{50{,}000 \text{ Btu/hr}}{70°F} = 714 \text{ Btu/°F-hr}$$

To predict monthly or yearly fuel requirements, the heat loss can be converted to Btu per degree-day.

Example 7.2

A new home is being built in Boston, Massachusetts, which has a calculated heat loss of 60,000 Btu per hour with a 85°F design temperature differential. The furnace to be installed in the home will burn natural gas, priced at \$1.00 per therm. The average yearly number of degree-days near the construction site is 5634. The expected overall efficiency of the new space heating system is 65%. Assuming an average winter with steady gas prices, how much will it cost to heat the house the first year?

Solution

We first find the heat loss in Btu per degree-day, using the following:

$$\frac{\text{heat loss of building per hour}}{\text{design temp. diff.}} \times 24 \text{ hours} = \text{heat loss, Btu/DD}$$

Therefore,

$$\frac{60{,}000 \text{ Btu/hr}}{85°F} \times 24 \text{ hours} = 16{,}941 \text{ Btu/DD}$$

To convert from heat loss in Btu/DD to amount of energy required per year to heat the home, we use the following formula:

$$\frac{\text{heat loss Btu/DD} \times \text{no. degree-days per year}}{\text{efficiency of heating system} \times \text{Btu/therm of gas}} = \text{therms required per year}$$

---

*The Greek capital letter delta (Δ) is used to denote difference.

$$\frac{16{,}941 \text{ Btu/DD} \times 5634 \text{ DD/yr}}{0.65 \text{ efficiency} \times 100{,}000 \text{ Btu/therm}} = 1468 \text{ therms required per year}$$

$$1468 \text{ therms} \times \$1 \text{ per therm} = \$1468 \text{ for first-year heating costs}$$

In Example 7.2 we assumed that the heat loss calculations were available for use. In many cases, however, they are not and the building heat loss must be calculated before the figures above can be obtained. As discussed in Chapter 1, R values, the resistance to the flow of heat of various objects, are the mathematical reciprocals of K values. When calculating a building heat loss, the cumulative conductive properties of many different building materials must be derived. The term used to describe how well a wall, window, or floor conducts heat is the overall coefficient of heat transmission, or the U value. A smaller U value denotes a higher R value and therefore less heat loss. U and R values are expressed in Btu/hr-ft²-ΔT(°F). If the U value or R value is known for a wall, its heat loss can be quickly calculated.

Example 7.3

A 20 ft $\times$ 9 ft wall which has no windows has an R value of 20. What is the heat loss of the wall for a design temperature differential of 70°F?

Solution

$$20 \text{ ft} \times 9 \text{ ft} \times \frac{1}{20} \times 70 = 630 \text{ Btu/hr heat loss}$$

Example 7.4

The wall in Example 7.3 now has two windows added to it. The windows measure 3 ft $\times$ 4 ft and have a U value of 0.65. What is the heat loss of the wall and window installation combined, based on a design temperature differential of 70°F?

Solution

$$\text{Walls: } [(20 \text{ ft} \times 9 \text{ ft}) - (2 \times 3 \text{ ft} \times 4 \text{ ft})] \times \frac{1}{20} \times 70 = 546 \text{ Btu/hr}$$

$$\text{Windows: } 2 \times 3 \text{ ft} \times 4 \text{ ft} \times 0.65 \times 70 = 1092 \text{ Btu/hr}$$

$$564 + 1092 = 1638 \text{ Btu/hr heat loss}$$

Air Infiltration

Another significant source of heat loss from a building is infiltration, the direct loss of heat due to the exchange of air into and out of the building. Warm air that leaks through the cracks around windows and doors must be replaced by new unheated air from outside the building. The heat capacity (C) of air is 0.018 Btu/ft³-°F. If a house is maintained at a temperature of 68°F while the outside air is 38°F, every cubic foot of air that is exchanged will result in a heat loss of 0.54 Btu. The exchange of air, Q, is measured in cubic feet per foot of crack per hour.

Half of the total crack length around all the windows and doors in the home should be used, because for each location where cold air enters the home, there must be a separate location where the warm air escapes, and the air exchange should not be counted twice. Table 7.1 lists Q values for selected wind speeds and types of windows. For doors, use the same chart but double the Q value. The perimeter of a window or door is normally used as the crack length, but if the frame is not tightly sealed to the wall or jamb, it must also be considered.

Table 7.1   AIR INFILTRATION THROUGH WINDOWS: *Q* VALUES (Ft³/hr-ft)[a]

	Wind Speed	
	15 mph	*20 mph*
Weather-stripped	25	35
Non-weather-stripped	40	60

[a]*For wood double-hung windows.*
SOURCE: ASHRAE, Handbook of Fundamentals, 1981.

Let's see how the values in Table 7.1 are used to determine air infiltration heat losses in an average setting.

Example 7.5

A room has two windows on the west wall and two windows on the east wall. All four windows have average fitting and weather stripping and measure 3 ft × 4 ft. The room is maintained at a temperature of 68°F and the outside air is 0°F. If there is a prevailing wind of 15 mph from the west, what is the infiltration heat loss for the room?

Solution

To determine infiltration heat loss, we use the formula

$$\text{length of crack} \times Q \text{ value} \times \Delta T \times C = \text{infiltration heat loss}$$

Substituting the values into our equation from the discussion above, we get

$$2 \times 3 \text{ ft} \times 4 \text{ ft} \times 24 \times 68°F \times 0.018 = 705 \text{ Btu/hr heat loss}$$

By using the *R* values of common building materials and air films listed in Table 1.1, the total heat loss from all surfaces of an average building can be calculated. After adding infiltration losses, the building heat loss is complete.

Sizing the Collector Array

Once the space heat and domestic hot water energy loads are known, the number of collectors can be determined. A careful analysis should be made at the installation site to check for possible shading, orientation, angle of inclination, and length of all pipe runs. These items were discussed in Chapter 5 and should be reviewed prior to undertaking any extensive solar installation or design problem. Time should not be wasted in designing a system that cannot be installed due to undesirable conditions presented by any of the factors noted above. These and many other variables are discussed in Chapter 8, which covers the *f*-chart method of solar system sizing. The *f*-chart is the most widely accepted method of sizing a collector array.

Selection of Size and Type of Storage

A common mistake made by solar system designers is to size the storage system prior to sizing the collector array. The prevailing philosophy has been to try to provide a house with enough heat storage to "carry over" for 1 or 2 days of cloudy and extremely cold weather, without paying attention to what effect this might have on the overall performance of the solar system. Some large storage systems were even installed with the hope of efficiently storing solar energy collected during the summer to be utilized the following heating season.

By November, the storage in these systems was usually depleted due to storage heat loss and building heating demand. The large storage volume would then prevent the collector array from reheating the storage to a usable temperature until the following spring. Clearly in these situations, the collector array, system storage, and building heating load were not engineered to work in harmony.

It should be noted that not all large storage systems experience such severe design and operating problems, but it is important to remember that storage is an integral part of system design. To be engineered properly, system performance should dictate storage size. Small storage capacities have been proven to provide better overall system performance than large ones. Although small storage systems will quickly become depleted during cold weather, they can also be heated to a useful working temperature within a relatively short period of time on the following sunny day. During severely cold and sunny days, the solar energy absorbed by the collectors will be transferred almost immediately to the living space, with little, if any, energy going into storage. During milder weather conditions, however, the storage heat may last for several days. The most significant accomplishment of this type of system design lies in the fact that a contribution, however small or large, can consistently be made to the energy requirements of space heat and hot water production by the solar energy system. In order for the solar system to pay for itself, it must make a real contribution to meeting these energy requirements.

The proper amount of storage for solar space heating systems ranges from 1.25 to 1.8 gallons of water per square foot of gross collector area. The type of space heat distribution system should influence the amount of storage designed into the solar system. Low-temperature distribution systems such as forced air or radiant floor heating will generally require larger solar storage systems, while high-temperature baseboard systems, for example, normally require smaller amounts of storage necessary for achieving higher working temperatures. One exception to these storage size parameters is where the solar-heated storage is used to power a liquid-source heat pump. Liquid-source heat pumps can extract heat from storage until the storage water temperature is chilled down to approximately 45°F. In this type of application, the size of the solar storage should be doubled, to a range of between 2.5 and 4 gallons of water per square foot of gross collector area.

The argument continues within the solar industry between those who prefer air-cooled collectors to liquid-cooled collectors. Rock storage is normally used with air-cooled collectors, while water is normally used as the storage medium in liquid-cooled collector applications. Rock storage systems require over three times the volume of water storage systems to store an equal amount of heat. Water is the most commonly used storage medium in the solar industry.

The type of storage tank to be used in a specific system application depends on many considerations. Retrofit applications normally require a group of small tanks. This is due to restrictions imposed by the width of existing doorways within the home. Interestingly, a group of standard 120-gallon solar hot water heater tanks, the most commonly utilized storage tank in larger systems, will usually be less expensive than one tank of equivalent volume and pressure rating. Atmospheric pressure tanks are used for some applications to reduce cost, and can also be site-built to allow entry through standard-sized doorways.

EFFECTS OF CONVENTIONAL HEATING SYSTEMS ON SOLAR SPACE HEATING SYSTEM DESIGN

When dealing with new construction a solar engineer or system designer is free to make recommendations to the homeowner as to what type of auxiliary heating system will be the most advantageous based on the potential solar system design. Prime considerations include the ability of the system to provide even, responsive heating; initial system installed cost; long-term operating costs of the system; and how well the auxiliary system will interface with the solar heating system with respect to system control, cost, and efficiency of operation. Even though the vast majority of solar heating systems are retrofit installations, the auxiliary heating system is still a primary consideration, since the distribution system of the auxiliary heater is a major factor in determining the overall performance of the solar heating system. Many homes cannot make use of solar heating systems without first making major changes in the existing heating distribution systems. In fact, one of the most common reasons that a house is unsuited for an interfacing solar heating system is that it has a high-temperature natural convection heat distribution system, such as electric baseboard or steam radiation.

Forced convection and radiant floor heating distribution systems facilitate the high heating performance required for a cost-effective solar application. A forced convection distribution system that possesses the ability to heat a house directly from solar storage with temperatures ranging between 75 and 85°F enables the system not only to operate efficiently, but also to provide a meaningful contribution to the overall heating load as well.

Not all electric baseboard and steam radiation systems are unsuited for solar energy retrofit applications, however. Small fan convection units can easily be added to many of these types of systems to facilitate heat delivery from solar storage. The heat delivery capacity of any natural convection system is a product of the amount of surface area available for heat transfer, the temperature differential between the transfer surface and the room, and the amount of time that the heat transfer takes place. If a solar system is to operate at low temperatures in order to maximize the solar contribution to the heating load of the building, the surface area for heat transfer as well as the amount of time for that heat transfer to occur must be at a maximum level. One method of accomplishing this is the use of two-stage heating. This concept increases the time for low-temperature system operation.

Since natural and forced convection systems rely on the movement of air to transfer heat efficiently, anything that aids in this movement of air will increase heat delivery, hence system efficiency. To this end, air filters should be replaced at frequent intervals. Baseboard and radiation units should be kept free of dust and dirt. Furniture and draperies should be placed well away from air vents and baseboard units where possible so as not to hinder convection. Radiator covers and baseboard louvered vents can significantly decrease heat transfer and should be avoided.

DESIGN FOR THE APPLICATION

Basic System Logic

Figure 7.3 clarifies the operation of the collector and demand heating loops of a typical solar space heating system. This system logic has some interesting features: collector loop to main storage heat transfer is accomplished by a very efficient single-wall heat exchanger arrangement; secondary heat transfer to domestic hot water occurs in a double-wall separation be-

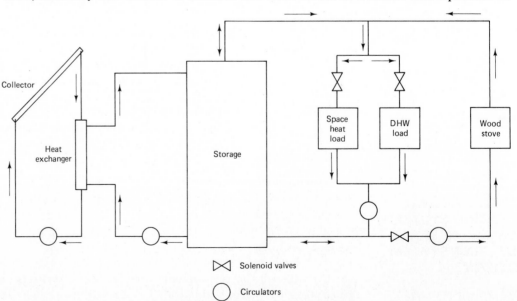

FIGURE 7.3 Logic of solar space heating.

tween the collector-loop antifreeze and the potable water in the home. In addition, the domestic hot water can be heated 24 hours a day and on cloudy days when the main storage contains available heat. If storage tanks with internal heat exchangers are used, the operation of one or both of the demand loops during a sunny period will cause storage water to be circulated across all three of the collector-loop heat exchangers. This establishes a counterflow single-pass heat exchange arrangement which increases the efficiency of heat transfer. Let's examine now in more detail how a solar system interfaces with standard types of heating systems.

Solar with Forced-Air Heat

Forced-air heating systems utilize the principle of forced convection for heat transfer. Because these systems do not rely on natural convection, low-temperature heating sources can be used for interior space heating. This ability to utilize lower-than-normal temperatures for space heating makes solar well suited for interfacing with forced warm air heating systems.

Figure 7.4 depicts a typical solar space heat and domestic hot water system interfaced with a standard fossil-fuel backup heating system. Three 120-gallon tanks are used for main solar storage, with one 80-gallon tank used for domestic hot water. All four tanks are stone lined with internal raised copper fin heat exchanger coils. This quantity of storage, together with the amount of heat exchanger surface area contained within the tanks, is well sized for between 220 and 290 ft^2 of gross collector surface area.

The collector-loop differential temperature controller, located on panel A, operates the circulators (1) when the collectors are warmer than the bottom of the coolest storage tank. The collector loop operates at a pressure that normally ranges between 10 and 30 psi. As the storage water within the three storage tanks (11) is heated, it will automatically be circulated to the forced-air heat exchanger (20) if there is a demand for space heat within the building. The aquastat controller (C) monitoring the storage temperature will prevent first-stage heat if the water is not warm enough to make a significant heating contribution. In addition, the main solar storage water will also be circulated through the raised fin copper heat exchanger in the bottom of the domestic hot water tank (18) whenever the main solar storage is warmer than the bottom of the domestic hot water tank. Circulator 13 is used for both the space heat and domestic hot water demand loops. The water in the main storage tanks and demand loops is normally between 10 and 15 psi.

A variation of the standard solar system interface with forced warm air heating is illustrated in Figure 7.5. This system uses potable water as the storage medium. Main storage for the space heating is designed so that the storage tanks (1) are plumbed in series with the auxiliary domestic hot water heater. This type of system is simpler in operation than the system shown in Figure 7.4, and is also less expensive to install. However, certain precautions must be taken in the design of this system if it is to function properly. Since fresh potable water is constantly being introduced into the space heating system, all active components must be manufactured from stainless steel, copper, or brass. A brass check valve (13) and potable water expansion tank (17) must be incorporated into the system to prevent expansion of the heated water back into the cold water supply lines. The expansion tank (17) will prevent the buildup of excessive pressure that can occur within the system during periods of bright sunshine with no domestic hot water usage. In this design, the entire heating demand loop and storage system operates at house pressure, which normally ranges between 30 and 70 psi.

Solar with Baseboard Hot Water Heat

Hot water baseboard heating systems are one of the most common natural convection heating applications. A hot water baseboard heating system with solar interface is illustrated in Figure 7.6. Until such solar system designs were developed, baseboard and other high-

EQUIPMENT LIST

PLUMBING

PART NO.	DESCRIPTION
1	Circulator #UP26-64
2	Set Flanges #51.97 56
3	Air Purger #440
4	Flush and Drain Valve
5	Pressure Gauge 0-60 P.S.I.
6	Pressure Relief Valve 50 P.S.I.
7	Air Vent #122
8	Expansion Tank #S-30
9	Check Valve 3/4"
10	Fill and Drain Valve
11	120 Gallon Maximum Insulation Storage Tank with Heat Exchanger
12	Pressure and Temperature Relief Valve
13	Circulator #UP20-42
14	Air Purger #440
15	Expansion Tank #SX-30V
16	Manual Ball Valves 3/4"
17	Solenoid Valve #3SX821009HW
18	80 Gallon Maximum Insulation Storage Tank with Heat Exchanger
19	Tempering Valve 1/2"
20	Forced Air Heat Exchanger
21	Existing Forced Air Furnace

ELECTRICAL

PART NO.	DESCRIPTION
A	Solar Control Monitor and Relay Terminal
B	Thermostat #T675A1540
C	Remote Bulb Controller #T675A1417
D	To Collector Sensor
E	STO. Sensor
F	Aux. 2 Sensor
G	Aux. 3 Sensor
H	Aux. 4 Sensor
J	Existing Furnace Control
K	150°F Upper Limit-Wire in Series with Aux. 1 Sensor
L	190°F Upper Limit-Wire in Series with Storage Sensor

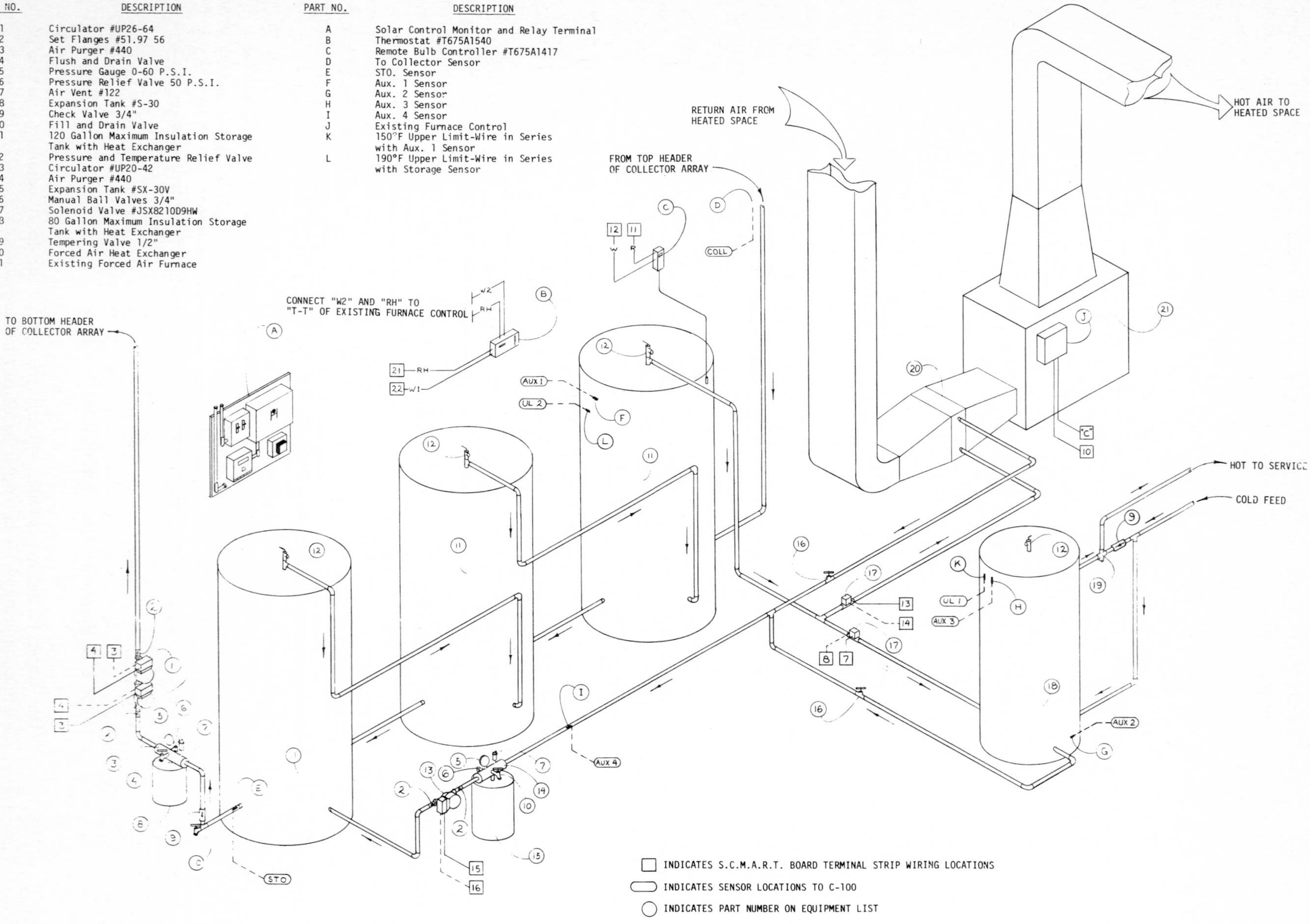

☐ INDICATES S.C.M.A.R.T. BOARD TERMINAL STRIP WIRING LOCATIONS

◯ INDICATES SENSOR LOCATIONS TO C-100

◯ INDICATES PART NUMBER ON EQUIPMENT LIST

FIGURE 7.4 Solar forced-air space heating system. (Courtesy of Heliotherm.)

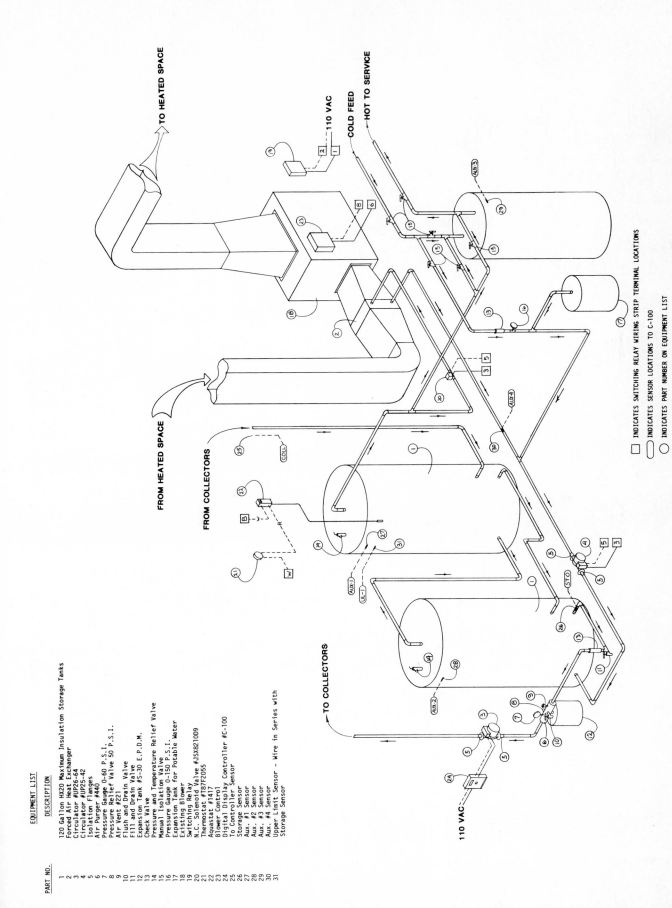

FIGURE 7.5 Potable water storage in forced warm air application. (Courtesy of Heliotherm.)

☐ INDICATES SWITCHING RELAY WIRING STRIP TERMINAL LOCATIONS

⬭ INDICATES SENSOR LOCATIONS TO C-100

◯ INDICATES PART NUMBER ON EQUIPMENT LIST

TO HEATED SPACE

FROM HEATED SPACE

FROM COLLECTORS

TO COLLECTORS

110 VAC

110 VAC

COLD FEED

HOT TO SERVICE

159

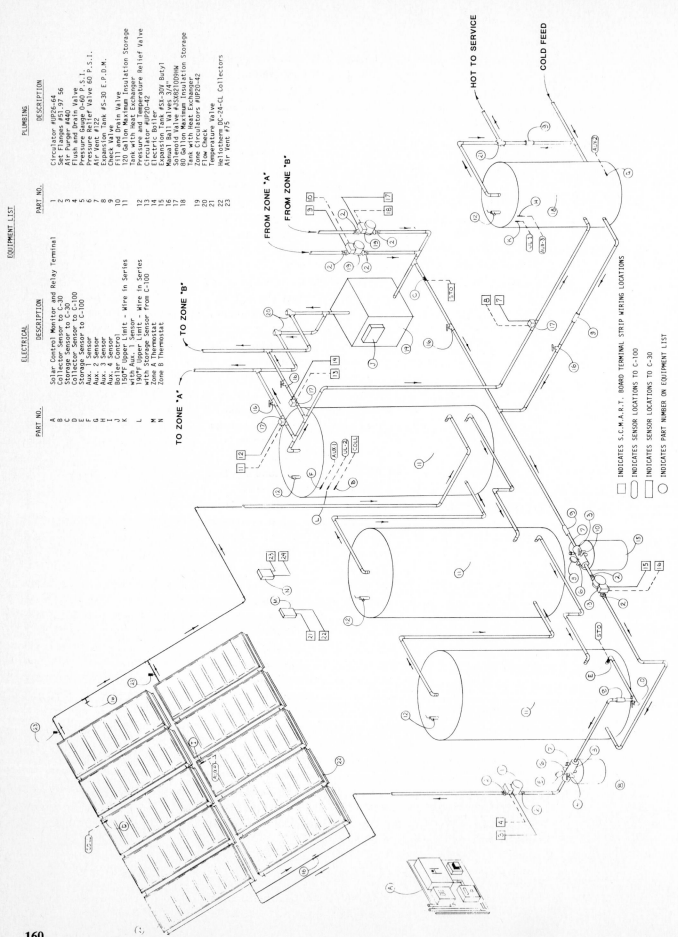

FIGURE 7.6 Two-stage solar heat with auxiliary boiler. (Courtesy of Heliotherm.)

EQUIPMENT LIST

ELECTRICAL

PART NO.	DESCRIPTION
A	Solar Control Monitor and Relay Terminal
B	Collector Sensor to C-30
C	Storage Sensor to C-30
D	Collector Sensor to C-100
E	Storage Sensor to C-100
F	Aux. 1 Sensor
G	Aux. 2 Sensor
H	Aux. 3 Sensor
I	Aux. 4 Sensor
J	Boiler Control
K	150°F Upper Limit - Wire in Series with Aux. 1 Sensor
L	190°F Upper Limit - Wire in Series with Storage Sensor from C-100
M	Zone A Thermostat
N	Zone B Thermostat

PLUMBING

PART NO.	DESCRIPTION
1	Circulator #UP26-64
2	Set Flanges #51.97 56
3	Air Purger #440
4	Flush and Drain Valve
5	Pressure Gauge 0-60 P.S.I.
6	Pressure Relief Valve 60 P.S.I.
7	Air Vent #122
8	Expansion Tank #S-30 E.P.D.M.
9	Check Valve
10	Fill and Drain Valve
11	120 Gallon Maximum Insulation Storage Tank with Heat Exchanger
12	Pressure and Temperature Relief Valve
13	Circulator #UP20-42
14	Electric Boiler
15	Expansion Tank #SX-30V Butyl
16	Manual Ball Valves 3/4"
17	Solenoid Valve #JSX821009HW
18	80 Gallon Maximum Insulation Storage Tank with Heat Exchanger
19	Zone Circulators #UP20-42
20	Zone Circulators
21	Flow Check
22	Temperature Valve
23	Heliotherm DC-24-CL Collectors
24	Air Vent #75

HOT TO SERVICE

COLD FEED

FROM ZONE "A"

FROM ZONE "B"

TO ZONE "A"

TO ZONE "B"

□ INDICATES S.C.M.A.R.T. BOARD TERMINAL STRIP WIRING LOCATIONS

◯ INDICATES SENSOR LOCATIONS TO C-100

◯ INDICATES SENSOR LOCATIONS TO C-30

◯ INDICATES PART NUMBER ON EQUIPMENT LIST

160

temperature natural convection heat distribution systems were not practical for a solar interface arrangement. The key to successful utilization of hot water baseboard heating as a backup to solar hot water heating relies on three design factors that must be incorporated into heating system design: the use of a maximum amount of finned high-efficiency baseboard units, a well-insulated dwelling, and two-stage logic for the heating system. Solar, the first-stage heating mode, is activated by separate thermostats M and N (two separate heating zones), which are set at least 3°F above the second-stage (backup) thermostats. This type of arrangement will allow the solar system to be used first, on initial demand for heat. The auxiliary backup system is energized only if an inadequate supply of solar heat causes the house temperature to fall 3°F below the first-stage thermostat setting. When the second-stage backup heating is activated, the solar system is disengaged from the baseboard loop. In this manner, all available solar heat is utilized to ensure a maximum contribution from the system. Lower baseboard temperatures will provide a significant portion of the required space heat if used with adequate surface area and sufficient operating time. The use of lower boiler temperatures for effective space heating is dramatically demonstrated when homeowners find that they can usually lower the high-limit setting on a fossil-fuel hydronic boiler by as much as 60°F with no apparent loss of interior comfort or temperature levels.

An interesting advantage of the two-stage heating concept in Figure 7.6 is that the solar contribution does not have to supply the entire heating demand of the home. Rather, it can be used to provide whatever contribution the solar storage temperature and finned baseboard units will allow prior to energizing the backup second-stage heating system. When the backup heating system is activated, it will remain in operation only until the second-stage thermostat setting has been satisfied. Once this has taken place, first-stage solar heating will resume. Thus we find that a successful solar heating system will shift between first- and second-stage heating throughout the day, depending on solar storage temperatures, to ensure maximum solar contribution to the heating load of the house.

In order to prevent hot water in the auxiliary boiler from circulating through, and heating, the solar storage tanks, a differential temperature controller (A) constantly compares the water temperature at sensor B with the water temperature at sensor C. Sensor B measures the water temperature in the hottest solar storage tank; sensor C measures the temperature of the water returning from the baseboard radiation units. First-stage solar heat can be activated only when the top of the hottest solar storage tank is warmer than the water returning from the baseboard radiation. When second-stage heating is activated in response to the closing of the thermostat relay, circulation of water through the boiler begins immediately. Circulation through the solar storage tanks will continue only until the hot water from the backup boiler completes its loop through the baseboard system, heating up sensor C. Once sensor C becomes warmer than the top of the hottest solar storage tank, circulation through the solar system ceases. During the heating cycle, when the house temperature rises above the second-stage backup heater setting, but is still below the first-stage solar setting, a momentary pause occurs in all space heating circulation while the differential temperature controller waits for the baseboard to cool down to a temperature below that of the solar storage tanks. In this way, all the heat supplied by the auxiliary system is transferred to the house before the solar system once again comes on line to continue heating the house.

First-stage solar-heated water is prevented from short cycling through the backup boiler by the use of flow control valves (20) located just above the backup boiler. Flow valve 17 functions both to stop first-stage circulation through the baseboard, and to prevent second-stage circulation from circulating through the solar storage tanks. The collector-loop operation and the domestic hot water demand loop are the same as those shown in Figures 7.4 and 7.5 for forced-air heating systems.

Solar with Individual Fan Convectors

When the space heat distribution system in a home is inadequate for direct solar interface due to high-fluid-temperature requirements or because the house is heated electrically, small fan convectors can be utilized successfully with a solar heating retrofit. These units are forced water-to-air heat exchangers, similar in operating principle to an automobile heater. Each fan convector contains its own water circulating pump, three-speed air distribution fan, and energizing thermostat. A typical fan convection unit is illustrated in Figure 7.7 and the solar system retrofit incorporating these convectors is illustrated in Figure 7.8. Fan convectors can be used to heat a whole house, or can be used when only a portion of a house is to be solar heated. In addition, these units offer a great deal of system flexibility in that they can easily be added to the solar system at a later date as the system expands to take on larger portions of the heating load.

The main solar storage tanks are filled with potable water. Thus, in addition to supplying the water for space heating, the solar storage tanks also act as a domestic hot water preheating system. The entire system, with the exception of the collector loop, operates at street or well-pump pressure, ranging between 30 and 70 psi.

A thermostat switch (21) prevents the fan convectors from operating when the storage water is too cool to make a significant heating contribution to the dwelling. This low-limit switch also prevents cool drafts from the unit which would occur under low-temperature operating conditions. The proper temperature setting for this low-limit thermostatic switch (21) is between 85 and 110°F.

Solar with Radiant Heat

Radiant heating systems operate at the lowest distribution temperatures of all forced-liquid heating systems; thus solar is ideally suited for interfacing with radiant applications. The temperature of the water circulating through the radiant floor distribution system is rarely required to be above 85°F.* We have continually examined how solar is best suited for heating systems which provide an almost continuous flow of heat to the living space at low operating temperatures. In these cases, the interior temperature of the dwelling is maintained by a slow, continuous heat transfer from the entire surface area of the floor to the room interiors. The system schematic for a radiant heat application is the same as for a hot water baseboard system, illustrated in Figure 7.6.

FIGURE 7.7 Forced-air convection unit used with solar heating application. (Courtesy of Turbonics, Inc.)

*The assumption is made that the radiant slab is poured on an insulated foundation.

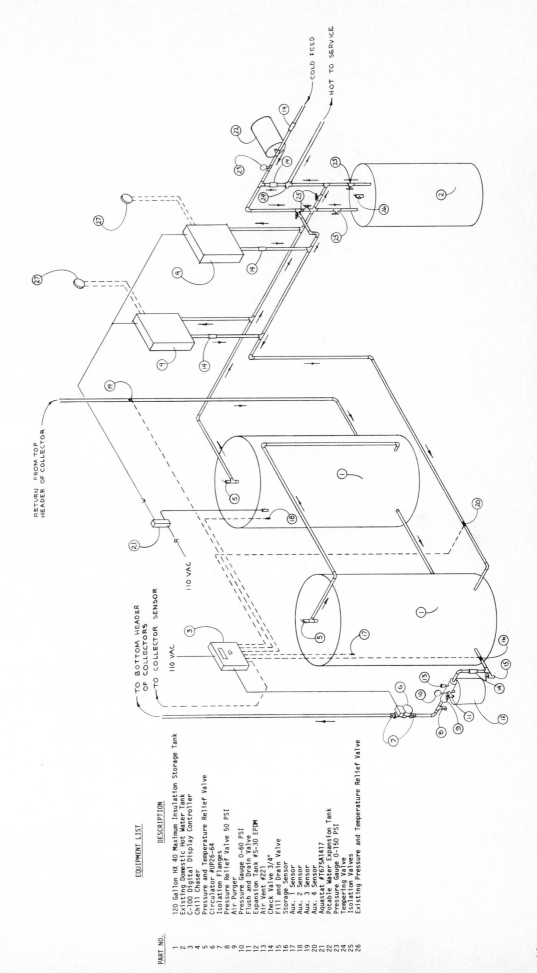

FIGURE 7.8 Partial space heat and domestic hot water system with forced-air convectors.

EQUIPMENT LIST

PART NO.	DESCRIPTION
1	120 Gallon HX 40 Maximum Insulation Storage Tank
2	Existing Domestic Hot Water Tank
3	C-100 Digital Display Controller
4	Chill Chaser
5	Pressure and Temperature Relief Valve
6	Circulator #UP26-64
7	Isolation Flanges
8	Pressure Relief Valve 50 PSI
9	Air Purger
10	Pressure Gauge 0-60 PSI
11	Flush and Drain Valve
12	Expansion Tank #S-30 EPDM
13	Air Vent #221
14	Check Valve 3/4"
15	Fill and Drain Valve
16	Storage Sensor
17	Aux. 1 Sensor
18	Aux. 2 Sensor
19	Aux. 3 Sensor
20	Aux. 4 Sensor
21	Aquastat #T675A1417
22	Potable Water Expansion Tank
23	Pressure Gauge 0-150 PSI
24	Tempering Valve
25	Isolation Valves
26	Existing Pressure and Temperature Relief Valve

163

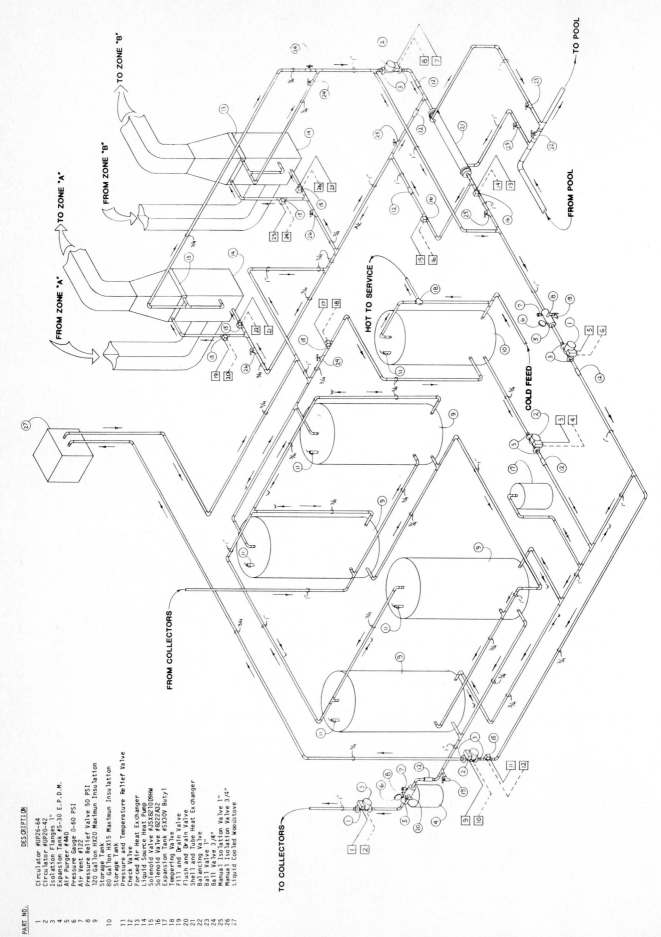

DESCRIPTION

PART NO.	
1	Circulator #UP26-64
2	Circulator #UP20-42
3	Isolation Flanges 1"
4	Expansion Tank #S-30 E.P.D.M.
5	Air Purger #440
6	Pressure Gauge 0-60 PSI
7	Air Vent #122
8	Pressure Relief Valve 50 PSI
9	120 Gallon HX20 Maximum Insulation
10	80 Gallon HX15 Maximum Insulation
11	Storage Tank
12	Pressure and Temperature Relief Valve
13	Check Valve
14	Forced Air Heat Exchanger
15	Liquid Source Heat Pump
16	Solenoid Valve #JSX821009HW
17	Solenoid Valve #822A32
18	Expansion Tank #SX30V Butyl
19	Tempering Valve
20	Fill and Drain Valve
21	Flush and Drain Valve
22	Shell and Tube Heat Exchanger
23	Balancing Valve
24	Ball Valve 1"
25	Ball Valve 3/4"
26	Manual Isolation Valve 1"
27	Manual Isolation Valve 3/4"
	Liquid Cooled Woodstove

FIGURE 7.9 Solar with liquid-source heat pump.

164

Solar-Assisted Heat Pump Systems

An air-to-air heat pump system requires the same solar system design as that for any central forced-air heating system. Figure 7.3, which illustrates basic solar system design logic, is applicable to air-to-air heat pump applications as well. The distribution duct systems used for heat pumps are usually well suited for the distribution of solar heat since they are sized for lower-temperature air heating encountered in heat pump installations.

Liquid-to-air heat pumps are also well suited for solar retrofit applications as they possess the ability to extract heat efficiently from the solar storage water or from well water down to as low as 45°F. Figure 7.9 illustrates the design of a solar and liquid-source heat pump system. Two separate heat pumps (14) are used for zone control within the house. The system is designed to provide solar heat to the house through the forced-air heat exchangers (13) before the heat pump compressor cycle is required. In this type of system design, solar energy can eliminate the electricity that would otherwise be used to operate the heat pump until the solar storage water temperature drops below 85°F. The heat pump compressors will then draw on the water storage until the tank temperature drops below 45°F. As tank temperatures fall below this point, electrically operated strip heaters are used for emergency heating. The water-jacketed woodstove (27) is used during cold weather to help keep the water storage as warm as possible. In heat pump applications, one large storage tank is preferable to several smaller tanks in order to keep heat loss from storage to a minimum. Smaller tanks are sometimes used, however, due to lower price or size restrictions.

DESIGN AND FUNCTION OF CONTROL MECHANISMS

Solar space heating system controls should be designed so that they are safe, easy to understand, and assure the maximum solar contribution to the heating system. Standardization of solar space heating control systems is virtually nonexistent within the industry. Only a few solar manufacturers have attempted to design and prepackage a standardized control center for all functions of the space heating system. Figures 7.10 and 7.11 are examples of such space heating control packages. Note that circuit breakers are used to prevent electrical overloads and to act as a disconnect for the system should service be required. A wiring schematic of the entire system should be attached to the control package for the convenience of service technicians. An example of such a wiring schematic is illustrated in Figure 7.12. A digital display of system sensor temperatures is a very helpful feature to incorporate into system design for ease in both monitoring the system performance and in troubleshooting. Another helpful option in the control package is the use of indicator lights to inform the owner and service technician which modes of the heating system are in operation.

A well-designed control package will operate all functions of the solar heating system efficiently without interfering with the operation of the backup space heat and domestic hot water system. Any interconnection with the backup system, other than plumbing connections and bypass valving arrangements, should be avoided or minimized. When a solar system is designed in this manner, a problem with the system will not prevent the auxiliary backup system from taking over the full heating load of the building until repairs to the solar system have been completed. Solar engineers and technicians should prevent the possibility of "no heat" emergencies from developing in a solar-heated house. Most heating contractors are well prepared for such emergencies on a 24-hour basis, but they would need to contact the solar-engineer responsible for the system design if they found that the backup heating system was not controlled in a conventional manner.

A well-designed solar control package will also be easy to install. A terminal strip for all the system electrical components should be numbered or labeled to ensure proper wiring. All relays, differential controllers, transformers, and other peripheral components within the control system should be easily replaceable. Space should be reserved on the control board for future expansion of the system. Often, additional heating zones, such as a pool or a hot tub, are added at a later date.

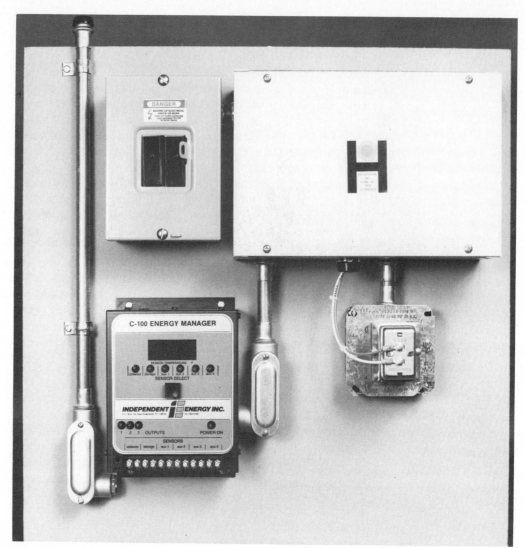

FIGURE 7.10 Solar control, monitor, and relay board, or S.C.M.A.R.T. Terminal. (Courtesy of Heliotherm.)

Multistage Thermostats

The two-stage heating logic discussed in this chapter relies on the use of a special thermostat called a two-stage thermostat. This thermostat contains two separate single-pole single-throw mercury switches that are set to close approximately 3°F from one another. The wiring diagram and pictorial of a typical multistage thermostat are illustrated in Figures 7.13 and 7.14. Two-stage thermostats assure that the auxiliary heating system will be activated only if available solar heat in storage is inadequate to meet the space-heating demands of the dwelling. The 3°F separation between first- and second-stage heat is normally field-adjustable.

Differential Temperature Controllers

Differential temperature controllers play a key role in the proper operation of any solar heating system. Circulators, solenoid valves, and fans must be energized at exactly the right time to prevent the system from wasting valuable heat. Solar controllers have been greatly

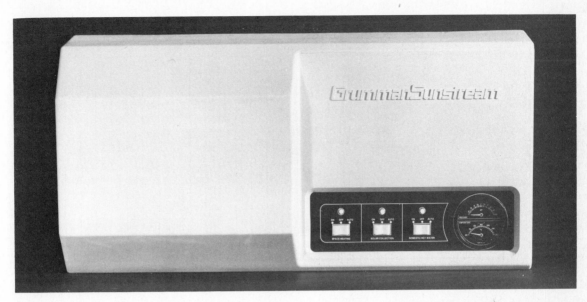

FIGURE 7.11 Control module for solar space heating and domestic hot water. (Courtesy of Grumman Energy Systems Co.)

improved in both function and reliability within recent years. Digital temperature displays, indicator lights, field calibration and programming, proportional outputs, and multiple controlled outputs have all been developed for the solar industry within these solid-state controllers. A solar engineer can now choose from a wide variety of controllers to handle virtually any heating application.

Proportional outputs should be used whenever possible to increase the overall system performance. Proportional controllers can automatically adjust the speed of the circulator to the availability of solar heat incident on the collector array. However, proportional controllers cannot be used to energize relays or solenoid valves since these components require full output for operation. Proportional outputs tied to these components would cause the energizing coils to overheat and be destroyed. For this reason, most of the control functions of a solar space-heating system are controlled by a standard output in the controller circuitry. Only the collector-loop circulators are normally controlled by proportional outputs. Circulators should be checked for compatibility before a proportional output controller terminal is used.

Low differentials can be used to transfer heat from main storage to the building interior, domestic hot water system, or hot tub, since relatively little heat will be lost between storage and demand. Turn-on and turn-off differentials for demand loop operation are normally 8°F and 3°F, respectively. The same low differentials can be used to transfer heat from a woodstove to the solar storage tanks.

Overheat Protection and Heat Rejection Controls

When utilizing a mixture of propylene glycol and water as the collector-loop fluid, it is imperative that fluid circulation within the loop be maintained during periods of system overheating if the chemical integrity of the fluid is to be maintained. The traditional approach to overheat control has been for an upper-limit sensor, wired through the differential controller, to open the electrical circuit to the collector-loop circulator when the upper-limit temperature has been reached within the solar storage tanks. Although this approach will prevent the storage tanks from overheating, it also has a damaging effect on the propylene glycol/water mixture within the collector loop. A 50/50 mixture of glycol and water boils at approximately 240°F. The heat rejection approach discussed above would cause the glycol/water mixture within the solar collectors to boil off due to stagnation of the collector

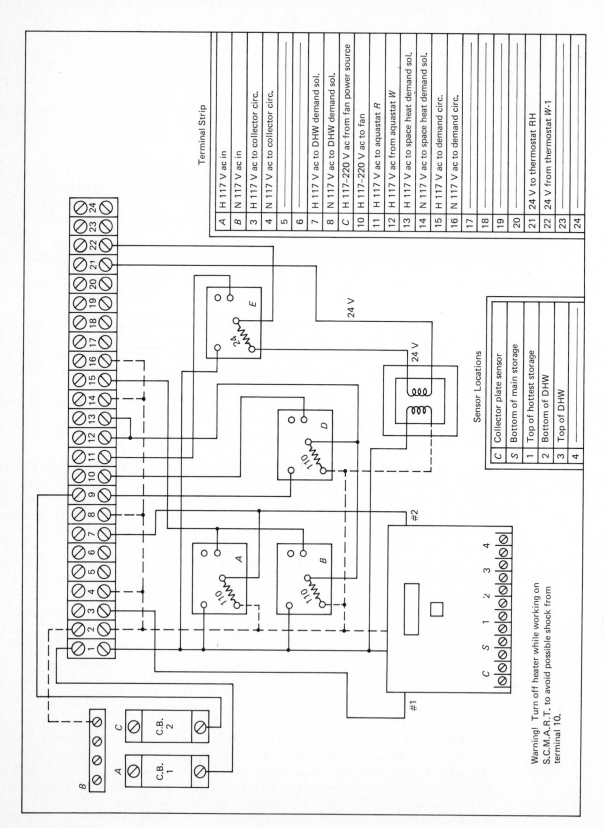

FIGURE 7.12 Wiring schematic for S.C.M.A.R.T. Board. (Courtesy of Heliotherm.)

Terminal Strip

A	H 117 V ac in
B	N 117 V ac in
3	H 117 V ac to collector circ.
4	N 117 V ac to collector circ.
5	
6	
7	H 117 V ac to DHW demand sol.
8	N 117 V ac to DHW demand sol.
C	H 117–220 V ac from fan power source
10	H 117–220 V ac to fan
11	H 117 V ac to aquastat *R*
12	H 117 V ac from aquastat *W*
13	H 117 V ac to space heat demand sol.
14	N 117 V ac to space heat demand sol.
15	H 117 V ac to demand circ.
16	N 117 V ac to demand circ.
17	
18	
19	
20	
21	24 V to thermostat RH
22	24 V from thermostat *W*-1
23	
24	

Sensor Locations

C	Collector plate sensor
S	Bottom of main storage
1	Top of hottest storage
2	Bottom of DHW
3	Top of DHW
4	

Warning! Turn off heater while working on S.C.M.A.R.T. to avoid possible shock from terminal 10.

168

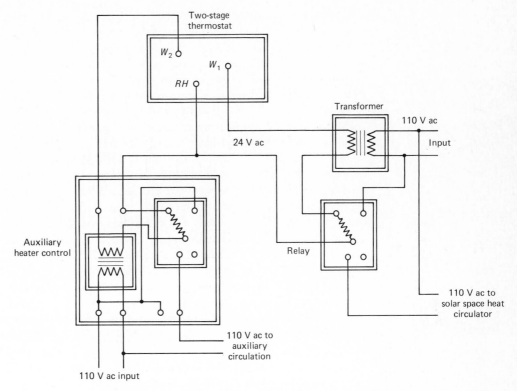

FIGURE 7.13 Wiring diagram of a typical two-stage thermostat.

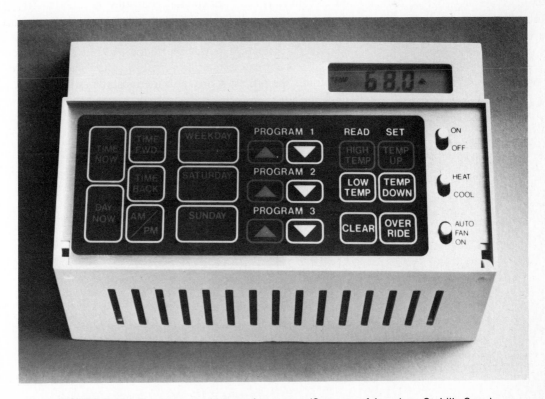

FIGURE 7.14 Solid-state multistage thermostat. (Courtesy of American Stabilis Corp.)

array once the circulation within the loop has stopped. (It should be remembered that most overheating situations occur during very sunny days, when the demand on the system for both space heat and hot water is at a minimum.) In addition to loss of collector-loop fluid, exposure of the glycol/water mixture to elevated temperatures for extended periods of time turns the loop fluid acidic. This can cause deterioration within the collector absorber plates and piping system if the solution is not drained, flushed, and replaced promptly.

An improved, proven solution to this overheating problem is to wire, in series with the collector-loop storage sensor, an upper-limit sensor which opens with a temperature rise at approximately 190°F. A wiring diagram for this is illustrated in Figure 7.15. The upper-limit sensor is located at the top of the hottest solar storage tank. If the hottest tank reaches or exceeds 190°F, the upper-limit sensor will break the connection between the collector-loop storage sensor and the differential controller. When this occurs the differential controller interprets this open switch as a very cold solar storage tank temperature. Any warm reading at all from the collector sensor will cause the controller to energize the loop circulator at full speed (assuming a proportional output connection from the fluid loop circulator to the controller), regardless of the available solar radiation. The system, in essence, will run in reverse to reject heat. Heated water from the solar storage tank will heat the collector fluid within the storage tank heat exchanger. As this heated fluid is circulated through the solar collector array, heat will be radiated from the surface of the collector absorber plate to the cooler, ambient air. This action will continue until the solar storage tanks have cooled to a temperature below 190°F. As long as the collector loop fluid is kept circulating, the system will not boil, since the collection efficiency of the collectors will be very low.

The domestic hot water tank illustrated in Figure 7.7 is prevented from overheating by an upper-limit sensor that is set to open at 150°F. This sensor is identified as K in Figure 7.7. This sensor is similar in operation to the upper-limit switch described previously, except for activating at a lower temperature setting. This sensor is wired in series with the sensor Aux-1 at the top of the hottest solar storage tank. When the domestic hot water temperature reaches or exceeds 150°F, the circuit is broken between the sensor Aux-1 and the domestic hot water controller. The differential controller interprets this open switch as a very cold temperature in main solar storage. Thus any warm temperature at all at sensor Aux-2 will prevent circulator 13 and flow valve 17 from becoming energized.

Another important overheat protection and rejection approach is necessary when a water-jacketed wood stove is interfaced with a solar heating system. The schematic in Figure 7.16 illustrates the proper wiring of a thermostatic switch used to dump excess heat from a solid-fuel appliance.

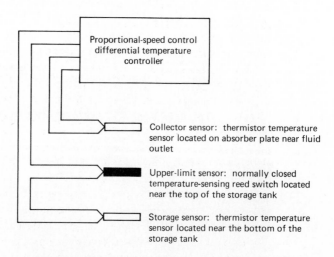

FIGURE 7.15 Wiring schematic for upper-limit sensor.

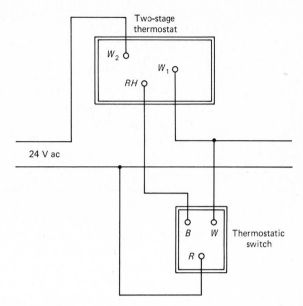

FIGURE 7.16 Wiring schematic for thermostatic overheat protection.

When the main storage which is heated by the wood or coal appliance reaches or exceeds approximately 190°F, the single-pole, double-throw high-limit control will eliminate the possibility of second-stage heat and force the first-stage heating system to activate. Hot water from main storage and the solid-fuel appliance will be distributed throughout the house. First-stage heat distribution to the house will continue until the main solar storage tank cools below 190°F. If the homeowner continues to burn solid fuel, the house will simply become warmer.

ASSOCIATED FUNCTIONS OF SOLAR SPACE HEATING SYSTEMS

The need to utilize a solar heating system as fully as possible to make it cost-effective has been a central theme of this chapter. The control and storage portions of the solar system should both be designed to put surplus and wasted solar heat to use, thus making the solar system as cost-effective as possible. Reducing the energy requirements of domestic hot water heaters, swimming pools, and hot tubs, and interfacing with solid-fuel boilers are associated functions of solar space heating systems which can increase their cost-effectiveness.

Domestic Hot Water

Domestic hot water heating capability is almost always included in solar space heating system design. Figures 7.4 through 7.9 illustrate this design function. The control logic of these solar systems is designed so that the main solar storage tanks are rarely more than a few degrees warmer than the domestic hot water tank. In this way the domestic hot water loop extracts heat from solar storage when available. In fact, a large hot water demand within the home will prevent a solar system from making a contribution to the space heating requirement. This can occur because the domestic hot water is a lower-temperature demand than the space-heating demand and will siphon off heat from the main solar storage tanks when available. The system should be designed in this way since the collectors will be operating at low temperatures whenever possible. Low-temperature operation, as we have discussed, is the most efficient operating mode for the collector array. A good solar designer attempts to develop a low-temperature collector operating mode in the initial system design. For example, if the domestic hot water and main solar storage tanks illustrated in Figure 7.4 are both 140°F, the collector array should be designed to operate between 145 and 150°F. On the other hand, if the domestic hot water tank temperature is 100°F and the main solar storage tanks are 140°F, the system designer should incorporate features that will ensure a transfer of heat from main storage to domestic hot water, thus lowering the operating temperature of the collector array. An upper-limit sensor, such as sensor K in Figure 7.4, is used to stop heat transfer to the domestic hot water tank at temperatures above 150°F.

Swimming Pools, Spas, and Hot Tubs

Swimming pools, spas, and hot tubs are, in essence, very large heat sinks. As such, they represent an excellent end use of the surplus solar energy available during the nonheating season. These bodies of water require, however, very large energy inputs to effect increases in temperature. Full research into the exact energy requirements of each specific system should be undertaken before designing the solar space-heating system. For example, a typical swimming pool measuring 15 ft × 30 ft with an average depth of 5 ft requires the Btu equivalent of 1 gallon of No. 2 fuel oil, burned at 100% efficiency, for each 1°F rise in temperature.

Example 7.6

$$5 \text{ ft} \times 15 \text{ ft} \times 30 \text{ ft} \times 7.48 \text{ gal/ft}^3 \times 8.33 \text{ lb/gal} \times 1 \text{ Btu/°F}$$
$$= 140{,}194 \text{ Btu/°F temperature rise}$$

In this instance, a 240-ft² collector array can contribute only enough energy to raise the pool temperature less than 2°F on the average sunny summer day in the northeastern United States. Sizing requirements for swimming pools are discussed in more detail in Chapter 9. However, this example serves to illustrate the enormous Btu requirements inherent in some applications.

Spas and hot tubs are much easier than swimming pools to solar-heat because they are generally smaller in size and the heat loss from them can be controlled to a large extent. A shell-and-tube heat exchanger is normally used to transfer heat from the main solar storage tanks or collector loop to the spa or hot tub. Figure 7.9 illustrates the use of such a shell-and-tube heat exchanger for this purpose. In addition to the solar heat availability, in this system the spa or hot tub can also be heated with waste heat from the home air-conditioning system. Since domestic hot water is usually more essential than pool, spa, or hot tub heating requirements, a thermostatic switch is often used to prevent solar heating of these auxiliary appliances until the domestic hot water has reached the desired temperature.

Solid-Fuel Boilers and Furnaces

A typical solid-fuel stove must be properly located in a home if it is to heat efficiently. Forced convection of heated air with these units is sometimes used to aid in heat distribution throughout the home. In all too many cases, however, the room in which the stove is located is very warm while the rest of the house is relatively cool. The use of solid-fuel boilers and furnaces interfaced with the solar storage and auxiliary space-heating distribution system can greatly increase the efficiency with which wood and coal are burned. Common interfaces of solid-fuel boilers and furnaces with standard heating systems were examined in detail in Chapter 6. When these units are installed in conjunction with solar space-heating systems, excess energy from the solid-fuel units can be efficiently stored in the solar storage tanks to be distributed throughout the house as first-stage heat during periods of diminished solar system performance. Figures 7.3 and 7.9 illustrate how heat from the solid-fuel boiler is transferred to the solar storage tanks, and from there to supply end-use heating demands.

In these systems, the solid-fuel appliance is used to heat the house and domestic hot water, as well as the pool, spa, or hot tub. The mating of solid fuel with solar applications is an ideal one, since the solid-fuel appliances can make up for low system output during the winter months while utilizing the solar system controls and storage facilities.

With an understanding of system design in hand, we can now move on to an examination of procedures for determining exact system sizing requirements, and the resulting cost analysis calculations.

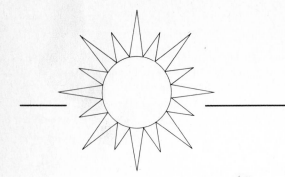

8

Thermal Performance and Economic Analysis of Solar Energy Systems

A relatively small number of solar systems have well documented performance records. Of those that do, a number were used in conjunction with detailed computer simulations to develop a simplified and reliable method of predicting the thermal performance and economic feasibility of solar installations. This effort resulted in a widely accepted method known as the *f*-chart. The *f*-chart was developed at the University of Wisconsin–Madison by Beckman, Duffie, and Klein in the mid-1970s. The *f*-chart is a set of curves that represent the fraction of the space heat and/or domestic hot water load which will be supplied by a proposed solar heating system. The *f*-chart curves are a function of two dimensionless parameters, *X* and *Y,* and are calculated by the following equations:*

$$X = \frac{\text{collector heat loss}}{\text{heating load}} = \frac{AF'_R U_L (212 - \tilde{T}_A)\,\Delta t}{L_{\text{SH}}\,\text{DD} + L_{\text{DHW}}} \tag{8.1}$$

and

$$Y = \frac{\text{absorbed insolation}}{\text{heating load}} = \frac{AF'_R\,(\overline{\tau\alpha})\bar{H}_t\Delta t}{(L_{\text{SH}}\,\text{DD} + L_{\text{DHW}})24} \tag{8.2}$$

where

A = square feet of collector area
$F'_R U_L$ = absolute value of the slope of the collector efficiency curve
$\tilde{T}_A$ = monthly average ambient air temperature
Δt = number of hours each month
L_{SH} = Btu/DD building design heat loss
DD = total degree-days for the month

*Equations (8.1) through (8.5) are excerpted from W. A. Beckman, S. A. Klein, and J. A. Duffie, *Solar Heating by the f-Chart Method*, Wiley-Interscience, New York, 1977.

L_{DHW} = Btu/month domestic hot water load

$F_R{}'(\overline{\tau\alpha})$ = intercept value of the collector efficiency curve

$\overline{H}_t$ = Btu/ft²-day average insolation incident on the collectors

Once the values for the X and Y parameters are know, they can be used to calculate the solar fraction, f. For liquid-cooled collectors, this equation is

$$f = 1.029Y - 0.065X - 0.245Y^2 + 0.0018X^2 + 0.0215Y^3 \qquad (8.3)$$

For air-cooled collectors, this equation is

$$f = 1.040Y - 0.065X - 0.159Y^2 + 0.00187X^2 - 0.0095Y^3 \qquad (8.4)$$

These two *f*-charts are illustrated in Figures 8.1 and 8.2. The solar fraction f is the portion of the *monthly* heating load that will be supplied by the proposed solar system. This value can be found by calculating the values of X and Y for the load and solar system in question and then locating the intersection of X and Y. The actual Btu contribution from the solar system for a particular month is found by multiplying the solar fraction f by the monthly heating load. The portion of the *yearly* load that will be supplied by the solar system is known as the solar fraction F. This figure can be found by dividing the sum of the monthly Btu contributions by the total yearly load.

Equations (8.1) and (8.2) are true for solar systems that provide both space heating and domestic hot water. If a solar system is being designed which will supply only domestic hot water, it is logical to assume that the efficiency of the collectors, and therefore the contribution of the system, will improve due to the lower operating temperatures of the collector array. Since the dimensionless parameter X is a factor of the heat loss from the solar collector, X will obviously be affected by the type of solar system application in question. For example, colder incoming water temperatures from the street or deep well will decrease the value of X. When an *f*-chart is being calculated for a domestic hot water system, X can be corrected by multiplying it by a correction factor, X_C/X. This factor can be calculated by

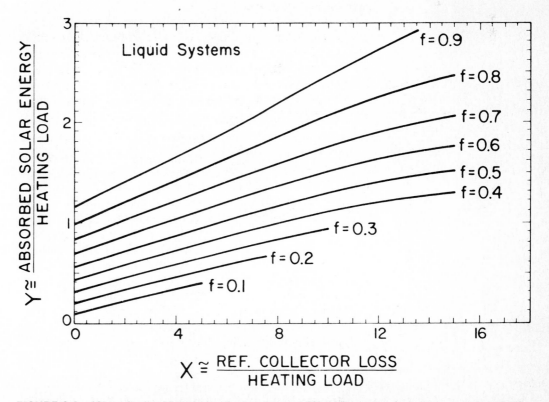

FIGURE 8.1 *f*-Chart for liquid-cooled collector systems. (W. A. Beckman. S. A. Klein, and J. A. Duffie, *Solar Heating Design: By the f-Chart Method,* Wiley-Interscience, New York, 1977.)

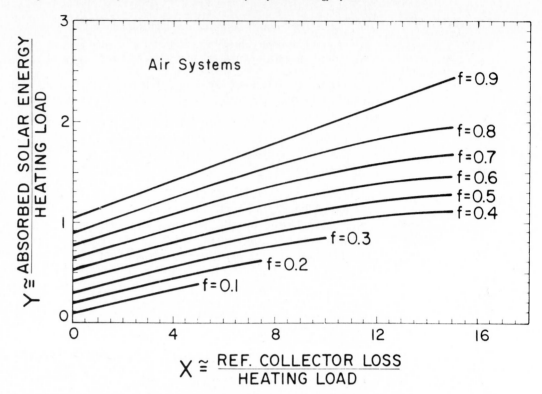

FIGURE 8.2   *f*-Chart for air-cooled collector systems. (W. A. Beckman, S. A. Klein, and J. A. Duffie, *Solar Heating Design: By the f-Chart Method,* Wiley-Interscience, New York, 1977.)

the following equation:

where

$$\frac{X_C}{X} = \frac{11.6 + 1.18T_W + 3.86T_m - 2.32\tilde{T}_A}{212 - \tilde{T}_A}$$
(8.5)

T_W = hot water set temperature, °F
T_m = incoming water temperature, °F

The solar fraction F can be found only after the 12 monthly solar fractions f have been calculated. The collector instantaneous efficiency curve will provide the values for the intercept $F_{R'}(\tau\alpha)$ and for the slope $F_{R'}U_L$. All the other variables that determine X and Y are related to the installation site and specific application of the proposed solar system. Factors such as the efficiency of the collector loop heat exchanger, the size of solar storage, the temperature requirement, and the method of contribution to the heating load all affect the performance of the system. Although this information is readily available, it is a somewhat tedious process to calculate X and Y for each month. A computer program was developed at the University of Wisconsin–Madison which greatly simplifies these calculations. The *f*-chart computer program contains weather and insolation data for hundreds of geographical locations in the United States. Only the location of the installation site and information about the specific application needs to be entered by the computer operator to obtain an *f*-chart readout.

f-Chart Variables

The six major categories of variables that affect the thermal performance of a solar system and which can be changed by design are as follows.

Collector Parameters

C1. *Collector area*: The total collector aperture area used in your system. The same area (gross or net) as used in determining C2 and C3 should be supplied here.

C2. *FR-UL product*: If collector performance data are plotted as collector efficiency versus $(T_i - T_a)/I$, a straight line will usually fit the data points well. FR-UL, the product of the heat removal factor FR and the overall loss coefficient UL, is the negative of the slope of this line. A collector test procedure for determining FR-UL and FR-TAU-ALPHA is described in ASHRAE Standard 93-77. FR-UL is assumed to be based on aperture area.

If the data are plotted with either the outlet temperature or the average temperature in place of the inlet temperature, it is necessary to calculate FR-UL and FR-TAU-ALPHA.

The data used to construct the plot of efficiency versus $(T_i - T_a)/I$ must be taken using the same collector flow rate as is expected in the installation. For liquid heating solar collectors, data taken at different flow rates may be modified.

C3. *FR-tau-alpha* (normal incidence): The collector heat removal factor FR times the transmittance–absorptance product TAU-ALPHA (also called optical efficiency) is the y-intercept of the straight-line efficiency plot described above.

C4. *Concentration ratio*: The ratio of collector aperture area to absorber area. Used for CPCs and imaging concentrators only.

C5. *CPC acceptance half-angle*: CPC collectors accept all radiation incident within the acceptance half-angle of the optical axis.

C6. *Number of covers*: Number of transparent covers over the collector absorber. Parameters C6 through C8 are used to estimate incidence angle effects unless a nonzero-incidence-angle modifier constant (parameter C9) is supplied.

C7. *Index of refraction*: For glass covers, use 1.526.

C8. *Extinction coefficient × length*: Product of the extinction coefficient of the cover material and the thickness of each cover.

C9. *Incidence angle modifier constant*: The incidence angle modifier $K(t)$ is the transmittance–absorptance product at incidence angle t divided by the transmittance–absorptance product at normal incidence. Parameter C9, the incidence-angle modifier constant, is the number b which provides the best fit in the equation

$$K(t) = 1 - b\,\frac{1}{\cos t - 1}$$

The constant b can be determined experimentally as described in ASHRAE Standard 93-77. If a value of b of zero is specified, parameters C6 to C8 are used to consider incidence angle effects.

C10. *Collector flow rate × specific heat/area*: Collector capacitance rate per unit area. In air-based systems, this parameter is used to correct for variations in storage stratification. In systems with collector-storage heat exchangers, it is used to account for heat exchanger effects.

C11. *Tracking axis* (1 = EW, 2 = NS, 3 = 2 − axis): Three tracking modes are available for imaging collectors: east-west axis, north-south axis, and dual-axis tracking.

C12. *Collector slope*: The angle between the plane of the collector aperture and the ground (horizontal) (may have 12 monthly values).

C13. *Collector azimuth*: The deviation of the normal to the collector surface from the local meridian; the zero point directly facing the equator, west positive, and east negative. At present (Version 4.0), only zero azimuth collectors may be used for liquid-based systems.

C14. *Ground reflectance*: Reflectance of horizontal surface in front of collector. This is used to estimate the total solar radiation incident on flat-plate collectors (may have 12 monthly values).

C15. *Incidence angle modifiers* (10, 20, 30, 40, 50, 60, 70, 80 degrees): Incidence modifiers for the plane which passes through the collector axis and is normal to the collector aperture.

Collector-to-Storage Transfer Parameters

T1. *EPS-CMIN of collector-store HX/collector area*: Effectiveness of collector-store heat exchanger $\times$ the minimum of the two capacitance rates (mass flow rate $\times$ specific heat) through the heat exchanger $\div$ collector area. In air-based combined space and water heating systems, this parameter refers to the heat exchanger from the collector outlet duct to the water preheat tank.

T2. *UA of collector inlet pipe or duct*: Overall loss coefficient U of inlet pipe or duct $\times$ surface area A.

T3. *UA of collector outlet pipe or duct*: Overall loss coefficient U of outlet pipe or duct $\times$ surface area A.

T4. *Collector duct leak rate* (percent): Percent of collector outlet duct flow rate which is leakage. The collector is assumed to be under negative pressure (i.e., all leaks are into the ducts rather than out).

T5. *Duct leak location* (1 = inlet, 2 = outlet, 3 = both): The total leakage may be assumed to occur in the inlet duct, the outlet duct, or may be evenly distributed between the two ducts.

Storage Unit Parameters

S1. *Tank capacity/collector area*: Tank energy storage capacity per unit collector area (storage mass $\times$ specific heat $\div$ collector area). For liquid-based combined space and water heating systems, this is the capacity of the space heating tank plus the capacity of the domestic water preheat tank if one is used. When two tanks are to be accounted for, parameters S2 and S3 must be set so that the correct tank UA will be calculated. For air-based combined space and water heating, this is the capacity of the water preheat tank.

S2. *Storage unit height/diameter ratio*: Ratio of height of tank referred to by parameter S1 to tank diameter.

S3. *Heat loss coefficient*: Overall loss coefficient of tank referred to by parameter S1.

S4. *Environment temperature* (-1000 for TENV = TAMB): Temperature to which tank losses occur. If -1000 is specified, the environment temperature is set equal to the average monthly ambient temperature.

S5. *Hot water auxiliary tank UA*: Overall loss coefficient $U \times$ surface area A of hot water auxiliary tank. The auxiliary tank temperature is assumed to be the hot water set temperature (parameter L4).

S6. *Hot water auxiliary tank environment temperature*: Temperature to which auxiliary tank losses occur.

S7. *Rock bed capacity/collector area*: Energy storage capacity (mass times specific heat) of rock bed per unit collector area.

S8. *Phase-change volume/collector area* ($\times$ 1000): Volume of phase-change unit $\div$ collector area $\times$ 1000. The volume of phase-change material is the phase-change unit volume $\times$ 1 $-$ the void fraction (parameter S10).

S9. *Phase-change material density*: Density of the phase-change material used in the storage device (*not* the average density of the storage device).

S10. *Void fraction*: Fraction of the phase-change unit volume which does not contain phase-change material.

S11. *Solid-phase specific heat*: The specific heat of the phase-change material when it is in solid phase.

S12. *Liquid-phase specific heat*: The specific heat of the phase-change material when it is in liquid phase.

S13. *Heat of melting*: Energy absorbed by melting of unit mass of phase-change material.

S14. *Melting temperature*: Temperature at which phase-change material melts.

Delivery Device Parameters

D1. *EPS-CMIN of load heat exchanger*: Effectiveness of load heat exchanger $\times$ the minimum of the two capacitance rates (mass flow rate $\times$ specific heat) through the heat exchanger.

D2. *Minimum temperature for HX operation*: Minimum storage temperature for operation of load heat exchanger pump and fan.

D3. *Delivery heat pump number*: Heat pump performance data are indexed by heat pump numbers. This parameter is the number corresponding to the performance data for the heat pump, which delivers heat from storage to the load.

D4. *Minimum heat pump absorber temperature*: Minimum storage temperature (heat pump absorber temperature) required for heat pump operation.

D5. *Heat pump bypass temperature*: Maximum storage temperature for heat pump operation. When storage is above this temperature, heat is delivered through the load heat exchanger and the heat pump is bypassed.

Load Parameters

L1. *Building UA*: Overall building loss coefficient $U \times$ exterior surface area A. The total building load for a month is the UA times the monthly degree-days [based on 18.3°C (65°F) base temperature]. Degree-days may be listed using LIST DATA or LIST DDAYS and entered or changed using ENTER or CHANGE.

L2. *Room temperature*: The temperature of the unheated air entering the load heat exchanger.

L3. *Hot water use*: Total average daily mass demand (total mass demand for month ÷ number of days in month).

L4. *Hot water set temperature*: Desired temperature of heated water. This temperature is used to calculate total water heating loads and also affects the auxiliary tank heat losses (see parameter S5 description).

L5. *Water main's temperature*: Temperature of cold water supplied to water heating system.

L6. *Total process or space heating load*: Total average daily energy demand.

L7. *Hours per day*: Number of hours per day that the load operates. It is assumed that the time distribution of the load is not significant. Only the monthly capacitance rate (hourly capacitance rate $\times$ hours of operation over the month) is considered in the analysis.

L8. *Load return temperature*: Load heat exchanger cold side inlet temperature; the temperature of the fluid returned from the load to the heat exchanger.

Auxiliary Parameters

A1. *Auxiliary fuel type* (1 = gas, 2 = elec, 3 = oil): The type of fuel used for space heating or process heating auxiliary. Use 1 for gas auxiliary, 2 for electric, and 3 for oil. If some other fuel is used, select one of these three and enter the appropriate cost using the command FLDATA. This parameter is used for the purchased energy summary and fuel cost calculations only.

A2. *Auxiliary device efficiency*: Average annual auxiliary device efficiency. The total fuel consumption is the total energy demand ÷ the consumption efficiency.

A3. *Hot water auxiliary fuel* (1 = gas, 2 = elec, 3 = oil): Type of fuel used for domestic water heating auxiliary (see A1).

A4. *Auxiliary water heater efficiency*: Average annual auxiliary water heater efficiency. Fuel consumption is energy demand ÷ efficiency.

```
COLLECTOR PARAMETERS
   C1. COLLECTOR AREA ...................................   72.00 FT2
   C2. FR-UL PRODUCT ...................................    0.72 BTU/HR-FT2-DEG F
   C3. FR-TAU-ALPHA (NORMAL INCIDENCE) .................    0.75
   C6. NUMBER OF COVERS ................................    1.00
   C7. INDEX OF REFRACTION .............................    1.53
   C8. EXTINCTION COEFFICIENT X LENGTH (KL)............    0.04
   C9. INCIDENCE ANGLE MODIFIER CONSTANT ...............    0.13
   C10. COLLECTOR FLOW RATE * SPECIFIC HEAT/AREA ......   12.60 BTU/HR-FT2-DEG F
   C12. COLLECTOR SLOPE ................................   45.00 DEGREES
   C13. COLLECTOR AZIMUTH ..............................    0.00 DEGREES
   C14. GROUND REFLECTANCE .............................    0.20

COLLECTOR-STORE TRANSFER PARAMETERS
   T1. EPS-CMIN OF COLLECTOR-STORE HX/COLLECTOR AREA ..   10.00 BTU/HR-FT2-DEG F
   T2. UA OF COLLECTOR INLET PIPE OR DUCT ..............    0.00 BTU/HR-DEG F
   T3. UA OF COLLECTOR OUTLET PIPE OR DUCT .............    0.00 BTU/HR-DEG F

STORAGE UNIT PARAMETERS
   S1. TANK CAPACITY/COLLECTOR AREA ....................   13.90 BTU/DEG F-FT2
   S2. STORAGE UNIT HEIGHT/DIAMETER RATIO ..............    2.00
   S3. HEAT LOSS COEFFICIENT ...........................    0.09 BTU/HR-FT2-DEG F
   S4. ENVIRONMENT TEMPERATURE (-1000 FOR TENV=TAMB) ..   68.00 DEG F
   S5. HOT WATER AUXILIARY TANK UA .....................    7.58 BTU/HR-DEG F
   S6. HOT WATER AUX TANK ENVIRONMENT TEMPERATURE .....   68.00 DEG F

LOAD PARAMETERS
   L3. HOT WATER USE ...................................   80.00 GALLONS/DAY
   L4. HOT WATER SET TEMPERATURE .......................  130.00 DEG F
   L5. WATER MAINS TEMPERATURE .........................   55.00 DEG F

AUXILIARY PARAMETERS
   A3. HOT WATER AUXILIARY FUEL (1=GAS,2=ELEC,3=OIL) ..    2.
   A4. AUXILIARY WATER HEATER EFFICIENCY ...............    1.00


                          FUEL COST

              ELEC (BLOCK RATE STRUCTURE)
                   ANNUAL INFLATION RATE   15.0 %
                   BLOCK               1
                   COST ($/MMBTU)     23.4
                   MAX. USE (MMBTU)10000.0


****************   FCHART  ANALYSIS   (VERSION 4.0)   ****************

PHILADELPHIA    PA      LATITUDE 39.5


                      THERMAL PERFORMANCE

              HT      TA    HWLOAD    QU    QLOSS FDHW
           (MMBTU)(DEG-F)(MMBTU)  (MMBTU)(MMBTU)
      JAN   2.06   32.0    1.90    0.80   0.05 0.37
      FEB   2.23   33.8    1.72    0.83   0.06 0.46
      MAR   2.88   42.8    1.90    1.10   0.08 0.52
      APR   3.06   53.6    1.84    1.18   0.10 0.60
      MAY   3.29   62.6    1.90    1.31   0.11 0.62
      JUN   3.32   71.6    1.84    1.36   0.12 0.68
      JUL   3.40   77.0    1.90    1.47   0.12 0.70
      AUG   3.32   75.2    1.90    1.45   0.13 0.70
      SEP   3.02   68.0    1.84    1.32   0.11 0.66
      OCT   2.85   57.2    1.90    1.22   0.10 0.60
      NOV   2.12   46.4    1.84    0.90   0.07 0.46
      DEC   1.82   35.6    1.90    0.73   0.05 0.36
      YR   33.37   54.7   22.39   13.67   1.10 0.56


                  PURCHASED ENERGY SUMMARY

                   GAS    ELECTRIC   OIL    TOTAL
      USE (MMBTU)  0.00     9.83    0.00     9.83
      COST ($)     0.00   230.37    0.00   230.37
```

FIGURE 8.3 f-Chart for domestic hot water systems.

A5. *Auxiliary heat pump number*: Heat pump performance data are indexed according to heat pump number. Set this parameter to the number of the heat pump performance data set which corresponds to the ambient source (auxiliary) heat pump.

f-Chart for Domestic Hot Water Systems

The *f*-chart computer program contains more than 80 possible input parameters. Not all of these inputs are used for sizing a domestic hot water system, however. Many of the inputs are used for sizing space heating systems and factoring variables that are different from one solar collector to another. The required input parameters, and the resulting thermal analysis of a typical solar domestic hot water system for a family of four living in Philadelphia, Pennsylvania, are shown in Figure 8.3. In this sample computer run, 72 ft^2 of collector area will provide 56% of the total yearly domestic hot water requirement. The peak contribution of the solar system is achieved during the months of July and August, while the lowest occurs in December, yielding only 36% of the domestic hot water contribution to the family. HT, the solar energy incident on the collector array; HW LOAD, the hot water load; QU, the actual solar contribution; and Q LOSS, the solar heat that is lost from piping and storage, are listed in millions of Btu. TA is the average monthly ambient air temperature.

f-Chart for Space Heating Systems

The *f*-chart computer program shown in Figure 8.4 is localized for a family of four living in Boston, Massachusetts. The thermal analysis illustrates the operation of a 240-ft^2 solar collector array designed to provide both space heat and hot water in a home with a heat loss of 10,320 Btu/DD. In this example, the 240 ft^2 of collector area will provide 56% of the total yearly hot water requirement and 43% of the total yearly combined space heating and domestic hot water requirement. FNP represents the fraction of the total load that is nonpurchased energy. SH LOAD is the space heating load. DELETE is the total change in internal energy of solar storage. Note that this relatively large collector array is able to supply 100% of the total heating and hot water load during the months of July and August. Ground reflectance, parameter C-14, was input into the computer program on a monthly basis so that the higher ground reflectance caused by snow during December, January, and February would be taken into consideration in the performance of the solar system.

f-Chart for Process Heat Systems

The required input parameters and resulting thermal analysis of a 720 ft^2 collector array for process heat are shown in Figure 8.5. This system is designed to preheat 1000 gallons of water daily for a factory located in New York City. In this thermal analysis, the 720 ft^2 of collector surface area will provide 49% of the total yearly process hot water requirement. The total monthly process load, parameter L6, changes from month to month because the incoming water temperature in the factory is greatly affected by ambient air temperature. In this application, 1000 gallons of street water from the New York City water system is heated to 110°F daily. The energy requirement needed to accomplish this can be calculated as follows:

January:	$(110°F - 38°F) \times 8.33 \times 1000 = 600$ MBtu
February:	$(110°F - 38°F) \times 8.33 \times 1000 = 600$ MBtu
March:	$(110°F - 44°F) \times 8.33 \times 1000 = 550$ MBtu
April:	$(110°F - 54°F) \times 8.33 \times 1000 = 466$ MBtu
May:	$(110°F - 65°F) \times 8.33 \times 1000 = 375$ MBtu

June: $(110°F - 75°F) \times 8.33 \times 1000 = 292$ MBtu

July: $(110°F - 79°F) \times 8.33 \times 1000 = 258$ MBtu

August: $(110°F - 79°F) \times 8.33 \times 1000 = 258$ MBtu

September: $(110°F - 75°F) \times 8.33 \times 1000 = 292$ MBtu

October: $(110°F - 65°F) \times 8.33 \times 1000 = 375$ MBtu

November: $(110°F - 54°F) \times 8.33 \times 1000 = 466$ MBtu

December: $(110°F - 43°F) \times 8.33 \times 1000 = 558$ MBtu

```
COLLECTOR PARAMETERS
  C1. COLLECTOR AREA ................................. 240.00 FT2
  C2. FR-UL PRODUCT .................................    0.72 BTU/HR-FT2-DEG F
  C3. FR-TAU-ALPHA (NORMAL INCIDENCE) ...............    0.75
  C6. NUMBER OF COVERS ..............................    1.00
  C7. INDEX OF REFRACTION ...........................    1.53
  C8. EXTINCTION COEFFICIENT X LENGTH (KL)...........    0.04
  C9. INCIDENCE ANGLE MODIFIER CONSTANT .............    0.13
  C10. COLLECTOR FLOW RATE * SPECIFIC HEAT/AREA .....   12.60 BTU/HR-FT2-DEG F
  C12. COLLECTOR SLOPE ..............................   57.00 DEGREES
  C13. COLLECTOR AZIMUTH ............................    0.00 DEGREES
  C14. GROUND REFLECTANCE ...........................
        0.70    0.70    0.20    0.20    0.20    0.20
        0.20    0.20    0.20    0.20    0.20    0.70


COLLECTOR-STORE TRANSFER PARAMETERS
  T1. EPS-CMIN OF COLLECTOR-STORE HX/COLLECTOR AREA ..  10.00 BTU/HR-FT2-DEG F
  T2. UA OF COLLECTOR INLET PIPE OR DUCT .............    0.00 BTU/HR-DEG F
  T3. UA OF COLLECTOR OUTLET PIPE OR DUCT ............    0.00 BTU/HR-DEG F

STORAGE UNIT PARAMETERS
  S1. TANK CAPACITY/COLLECTOR AREA ..................   12.50 BTU/DEG F-FT2
  S2. STORAGE UNIT HEIGHT/DIAMETER RATIO ............    2.00
  S3. HEAT LOSS COEFFICIENT .........................    0.09 BTU/HR-FT2-DEG F
  S4. ENVIRONMENT TEMPERATURE (-1000 FOR TENV=TAMB) ..  68.00 DEG F
  S5. HOT WATER AUXILIARY TANK UA ...................    7.58 BTU/HR-DEG F
  S6. HOT WATER AUX TANK ENVIRONMENT TEMPERATURE .....  68.00 DEG F

DELIVERY DEVICE PARAMETERS
  D1. EPS-CMIN OF LOAD HEAT EXCHANGER ............... 1200.00 BTU/HR-DEG F
  D2. MINIMUM TEMPERATURE FOR HX OPERATION ..........   80.00 DEG F

LOAD PARAMETERS
  L1. BUILDING UA ................................... 430.00 BTU/HR-DEG F
  L2. ROOM TEMPERATURE ..............................   68.00 DEG F
  L3. HOT WATER USE .................................   80.00 GALLONS/DAY
  L4. HOT WATER SET TEMPERATURE ..................... 130.00 DEG F
  L5. WATER MAINS TEMPERATURE .......................   50.00 DEG F

AUXILIARY PARAMETERS
  A1. AUXILIARY FUEL TYPE (1=GAS,2=ELEC,3=OIL) ......    3.
  A2. AUXILIARY DEVICE EFFICIENCY ...................    0.55
  A3. HOT WATER AUXILIARY FUEL (1=GAS,2=ELEC,3=OIL) ..   2.
  A4. AUXILIARY WATER HEATER EFFICIENCY .............    1.00


                  FUEL COST


       ELEC (BLOCK RATE STRUCTURE)
             ANNUAL INFLATION RATE  15.0 %
             BLOCK             1
             COST ($/MMBTU)    23.4
             MAX. USE (MMBTU)10000.0

       OIL  (BLOCK RATE STRUCTURE)
             ANNUAL INFLATION RATE  10.0 %
             BLOCK             1
             COST ($/MMBTU)    7.9
             MAX. USE (MMBTU)10000.0
```

FIGURE 8.4 f-Chart for space heat systems.

```
**************** FCHART ANALYSIS (VERSION 4.0) ****************

BOSTON          MA      LATITUDE 42.2


                        THERMAL PERFORMANCE

         HT     TA    SHLOAD  HWLOAD    QU    QLOSS   DELTE  FNP  FDHW
       (MMBTU)(DEG-F)(MMBTU) (MMBTU)  (MMBTU)(MMBTU) (MMBTU)
  JAN   6.86   30.2   11.46   2.00    2.98   0.07    0.00  0.22 0.31
  FEB   7.50   32.0    9.99   1.81    3.25   0.06    0.00  0.27 0.31
  MAR   8.73   37.4    8.60   2.00    3.73   0.07    0.00  0.35 0.31
  APR   8.94   48.2    5.07   1.94    3.97   0.07    0.00  0.56 0.31
  MAY   9.88   59.0    2.25   2.00    3.60   0.18    0.12  0.77 0.52
  JUN  10.09   68.0    0.28   1.94    2.55   0.36    0.12  0.94 0.93
  JUL  10.30   73.4    0.00   2.00    2.41   0.44   -0.03  1.00 1.00
  AUG   9.78   71.6    0.07   2.00    2.49   0.41    0.00  1.00 1.00
  SEP   9.75   64.4    0.78   1.94    2.81   0.36   -0.07  0.93 0.90
  OCT   9.14   55.4    3.10   2.00    3.72   0.20   -0.13  0.72 0.54
  NOV   6.12   44.6    6.13   1.94    2.92   0.07    0.00  0.35 0.31
  DEC   6.17   33.8   10.24   2.00    2.76   0.07    0.00  0.22 0.31
  YR  103.25   51.5   57.98  23.60   37.18   2.34    0.00  0.43 0.56


                   PURCHASED ENERGY SUMMARY

                    GAS    ELECTRIC   OIL    TOTAL
      USE (MMBTU)  0.08  -  10.30   66.25 -  76.56
      COST ($)     0.00    241.63   520.74  762.37
```

FIGURE 8.4 (*continued*)


```
COLLECTOR PARAMETERS
  C1. COLLECTOR AREA .................................... 720.00 FT2
  C2. FR-UL PRODUCT ....................................    0.72 BTU/HR-FT2-DEG F
  C3. FR-TAU-ALPHA (NORMAL INCIDENCE) ..................    0.75
  C6. NUMBER OF COVERS .................................    1.00
  C7. INDEX OF REFRACTION .............................    1.53
  C8. EXTINCTION COEFFICIENT X LENGTH (KL)..............    0.04
  C9. INCIDENCE ANGLE MODIFIER CONSTANT ...............    0.19
  C10. COLLECTOR FLOW RATE * SPECIFIC HEAT/AREA .......   12.60 BTU/HR-FT2-DEG F
  C12. COLLECTOR SLOPE ................................   45.00 DEGREES
  C13. COLLECTOR AZIMUTH .............................     0.00 DEGREES
  C14. GROUND REFLECTANCE ...........................     0.20

COLLECTOR-STORE TRANSFER PARAMETERS
  T1. EPS-CMIN OF COLLECTOR-STORE HX/COLLECTOR AREA ..   10.00 BTU/HR-FT2-DEG F
  T2. UA OF COLLECTOR INLET PIPE OR DUCT .............   15.00 BTU/HR-DEG F
  T3. UA OF COLLECTOR OUTLET PIPE OR DUCT ...........    15.00 BTU/HR-DEG F

STORAGE UNIT PARAMETERS
  S1. TANK CAPACITY/COLLECTOR AREA ..................    10.00 BTU/DEG F-FT2
  S2. STORAGE UNIT HEIGHT/DIAMETER RATIO ...........     2.00
  S3. HEAT LOSS COEFFICIENT ........................     0.09 BTU/HR-FT2-DEG F
  S4. ENVIRONMENT TEMPERATURE (-1000 FOR TENV=TAMB) ..  68.00 DEG F

DELIVERY DEVICE PARAMETERS
  D1. EPS-CMIN OF LOAD HEAT EXCHANGER ............... 2274.72 BTU/HR-DEG F
  D2. MINIMUM TEMPERATURE FOR HX OPERATION .......... 111.00 DEG F



     LOAD PARAMETERS
       L6. TOTAL PROCESS OR SPACE HEATING LOAD .............
            600.00  600.00  550.00  466.00  375.00  292.00
            258.00  258.00  292.00  375.00  466.00  558.00          MBTU/DAY
       L7. HOURS PER DAY ..................................     8.00
       L8. LOAD RETURN TEMPERATURE .......................   105.00 DEG F

     AUXILIARY PARAMETERS
       A1. AUXILIARY FUEL TYPE (1=GAS,2=ELEC,3=OIL) .......    3.
       A2. AUXILIARY DEVICE EFFICIENCY ...................     0.55
```

FIGURE 8.5 *f*-Chart for process heat systems.

```
                         FUEL  COST

        ELEC  (BLOCK RATE STRUCTURE)
               ANNUAL  INFLATION  RATE   15.0 %
               BLOCK                1
               COST  ($/MMBTU)      18.6
               MAX. USE  (MMBTU)10000.0


        OIL   (BLOCK RATE STRUCTURE)
               ANNUAL  INFLATION  RATE   10.0 %
               BLOCK                1
               COST  ($/MMBTU)       8.9
               MAX. USE  (MMBTU)10000.0




   ***************   FCHART  ANALYSIS   (VERSION 4.0)   ***************

   NEW YORK(CEN PK)NY     LATITUDE 40.5


                      THERMAL  PERFORMANCE

             HT      TA     LOAD      QU    QLOSS  FNP
           (MMBTU)(DEG-F) (MMBTU)  (MMBTU)(MMBTU)
      JAN   18.65   32.0   18.60    4.56   0.42 0.22
      FEB   20.26   33.8   16.80    4.94   0.39 0.27
      MAR   26.99   41.0   17.05    6.87   0.46 0.37
      APR   29.20   51.8   13.98    7.71   0.48 0.52
      MAY   32.67   62.6   11.63    8.59   0.53 0.69
      JUN   31.63   71.6    8.76    7.96   0.55 0.85
      JUL   32.87   77.0    8.00    8.17   0.60 0.94
      AUG   31.39   75.2    8.00    8.03   0.60 0.93
      SEP   28.75   68.0    8.76    7.89   0.55 0.83
      OCT   26.71   59.0   11.63    7.93   0.52 0.64
      NOV   18.12   48.2   13.98    5.14   0.42 0.34
      DEC   15.60   35.6   17.30    3.91   0.41 0.20
      YR   312.86   54.7  154.47   81.69   5.93 0.49


                 PURCHASED ENERGY SUMMARY

                  GAS     ELECTRIC    OIL      TOTAL
   USE (MMBTU)    0.00       0.00   143.10    143.10
   COST ($)       0.00       0.00  1273.63   1273.63
```

FIGURE 8.5 (*continued*)

Note from the monthly thermal contribution figures that the solar system's maximum contribution during the summer months still allows the collectors to operate at full efficiency with no surplus energy going to waste.

**ECONOMIC
AND PAYBACK
ANALYSIS
OF SOLAR ENERGY
SYSTEMS**

Installation Costs for Solar Versus Conventional Heating

The use of a solar energy system does not eliminate the need for the conventional heating system. An auxiliary, backup system, sized to provide the maximum possible heating load, must remain operative. Thus in new building construction, the cost of a solar space heat and domestic hot water system must be added to the cost of the conventional heating units.

The cost of a typical solar heating system is between 5 and 10 times greater than the cost of the related auxiliary system. Since no money is saved by virtue of the solar installation, the energy savings accrued over the operating life of the solar system must justify its purchase. Tax credits and other financial incentives also help to make solar economically attractive. In some areas of the country, the combination of federal and state income tax credits reduce the net purchase price of the solar system by more than 50%. Other ways to decrease the cost of a solar system are to either reduce its size and/or self-install it. For example, a proposed space heat and domestic hot water system can be modified to provide reduced energy contribution, and will also result in greater energy savings per dollar invested since the solar collectors will operate at lower, hence more efficient temperatures. Thirty to forty percent of the total installed cost of any solar system is due to labor and materials. Labor costs can be significantly reduced if the system purchaser possesses the skills necessary to accomplish the installation. Many solar system dealers are willing to provide on-site supervision and inspection of owner-installed systems at little or no additional charge.

Many people postpone the purchase of a solar system in the belief that the price of equipment will decrease with increased component production output. Television sets and hand-held calculators are often cited as examples of dramatic price reductions resulting from mass-production techniques and full market penetration. However, the solar collector manufacturing industry is not as labor intensive as was the infant electronics industry. Solar collectors are essentially material intensive. The prices of most of the components are directly affected by the price of conventional energy sources. For example, copper and aluminum, two metals used extensively in most flat-plate collectors, require a great deal of energy to extract and refine, making them relatively expensive. Rather than the price of solar collectors decreasing with automated production, the price has inflated in a similar manner to the price of energy. Thus the initial cost of a solar heating system is considerably higher than its conventional counterpart. A decision not to invest in solar, however, is simply not cost-effective because the home or business owner will pay for the energy system many times over in the form of higher energy bills and lost tax and financial incentives. In short, it pays to buy solar.

Comparative Costs of Delivered Energy

The degree to which a solar system is cost-effective is largely determined by its energy production. Also, the durability of the system components and the degree of reliability they possess are critical to the long-term energy contribution of the system. The more energy the solar energy system produces, the greater will be its return on investment. A long operating system life results in a low cost per unit of delivered energy. For example, if a solar domestic hot water system has a net cost of $1800 after the federal and state tax credits, and provides the owner with 11,600,000 Btu per year, the expected operating life of the system can be used to determine the cost per delivered Btu.

Example 8.1

Ten-year system life:

$$\frac{\$1800}{10 \times 11.6 \text{ MMBtu}} = \$15.52/\text{MMBtu}$$

Fifteen-year system life:

$$\frac{\$1800}{15 \times 11.6 \text{ MMBtu}} = \$10.34/\text{MMBtu}$$

Twenty-year system life:

$$\frac{\$1800}{20 \times 11.6 \text{ MMBtu}} = \$7.76/\text{MMBtu}$$

It is evident that the longevity of the system is just as important as the collector performance or installed system cost when considering delivered cost of energy.

The cost of delivered solar energy can be compared to the cost of delivered energy of conventional fuels. Table 8.1 lists common fuels and their respective delivered costs per MMBtu. The efficiency of the conventional system must be included in each calculation to assure accurate comparisons. The listed comparisons are all relatively simple. Initial cost of the conventional heating system, inflation of conventional fuel prices, and the opportunity costs associated with a solar investment are not considered in these calculations. These factors will be discussed as part of the *f*-chart life-cycle analysis that follows.

Table 8.1   COST OF DELIVERED ENERGY OF SELECTED FUEL SOURCES

Fuel	Cost/Unit	Btu/Unit	Efficiency	Cost/MMBtu
Electricity	$0.12/kwh	3,413	0.95	$37.01
Fuel oil	1.20/gallon	140,000	0.60	14.29
Natural gas	0.80/therm	100,000	0.60	13.33
Wood	125.00/cord	24,000,000	0.50	10.42
Coal (anthracite)	120.00/ton	25,000,000	0.60	8.00
Solar	1800.00/system	11,600,000/yr[a]	1.00[b]	7.76[c]

[a]*Based on an average of 665 Btu/ft^2/day.*

[b]*The inefficiencies in the energy systems have been considered in the figure for yearly energy contribution.*

[c]*Based on 20-year operational life.*

f-Chart Life-Cycle Economic Analysis

The decision to purchase a solar energy system should be based on a comparison of the costs associated with a conventional heating system and those of a solar system with an auxiliary backup. Initial system costs as well as the costs over the expected life of the solar system must be considered for both options. These costs should be expressed in terms of their value at the time of the solar investment. In other words, an economic analysis should be based on life-cycle costs and should also consider the time value of money.

An *f*-chart computer program will perform, in addition to a solar thermal analysis, an economic analysis for both the conventional heating system and the solar system with auxiliary backup, and then compare the two. The initial and future costs associated with system ownership and operation considered in this program are:

1. Down payment
2. Cost of capital (mortgage principal and interest)
3. Insurance and maintenance costs of the system
4. Property taxes
5. Auxiliary fuel costs (decreases with solar system)
6. State and federal tax credits
7. State and federal tax deductions
8. Depreciation (commercial/industrial systems only)
9. System resale value

These costs are computed for each year of the analysis, and are based on the results of the thermal performance analysis and the economic parameter values input into the computer program. The yearly sum of these costs is listed as the net annual cash flow for the year. This figure is discounted at the specified market discount rate (parameter E15) in order to determine its present worth. The sum of the yearly discounted net cash flows is the present worth of the system life-cycle costs. This figure is the most significant calculation by the economic computer program. A comparative life-cycle economic analysis represents the difference in energy costs and total costs for the solar and conventional systems. These differences in costs are expressed in undiscounted, present worth, and annualized terms.

The economic input parameters are described below.

Economic Parameters

E1. *Economic output detail* (1, 2, or 3): Three levels of detail are available for the economic analysis output. Use 1 for short output, 2 for more complete economic indicators, and 3 for annual cash flow summaries.

E2. *Reference or comparison system* (1 or 2): The "economics package" can be used to determine the life-cycle cost of purchasing and operating a system, or for determining costs and savings relative to some reference system. Set this parameter to 1 to get life-cycle cost information ("reference system") only. When E2 = 2, the current system is compared against the most recently run reference system.

E3. *Calculate rate of return* (yes = 1, no = 2): If E3 = 1, the rate of return on the current system investment relative to the reference system can be calculated. Set parameter E3 to 2 to avoid this calculation.

E4. *Income-producing building* (yes = 1, no = 2): Use 1 for commercial economics, 2 for residential. Commercial buildings have different federal tax credits, income tax deductions, and depreciation deductions.

E5. *Depreciation* (straight line = 1, declining balance = 2, sum of the years' digits = 3, none = 4): Three types of depreciation for commercial buildings are offered: straight line, declining balance, and sum of the years' digits. If E = 4, depreciation is not considered.

E6. *Consider federal tax credits* (yes = 1, no = 2): Set this parameter to 1 to get federal tax credits; use 2 to ignore federal tax credits.

E7. *Length of analysis*: Number of years of life cycle. The system is assumed to be sold and any remaining mortgage paid off in the final year of the analysis.

E8. *Tax-creditable system base cost*: In general, some of the system costs will be eligible for tax credits and some will not. In addition, some of the costs result from purchasing components which are being OPTIMIZED or LOOPed while the rest of the initial investment is independent of these components. This parameter should be set to the total system cost which is eligible for tax credits and which is independent of any components OPTIMIZEd or LOOPed.

E9. *Non-tax-creditable system base cost*: System cost which is not eligible for tax credits and which is independent of any components on LOOP or OPTIMIZE commands (see E8).

E10. *Annual increase in purchased energy demand*: Annual percent increase in demand for purchased energy. Usually, this increase is a result of degradation in solar system performance. Warning: The purchased energy requirement after N years is $(1 + E7/100)N \times$ the first year's requirement. If E7 is 15%, the purchased energy requirement will double in 5 years.

E11. *Term of mortgage*: Length of mortgage (in years). If the system is financed by cash outlay, set E11 to 1 and set E12 to 100.

E12. *Down payment* (percent of original investment): Percent of total system cost which will be paid initially.

E13. *Mortgage annual interest rate*: Interest rate on mortgage.

E14. *Resale value* (percent of original investment): Cash received for resale in final year of analysis as percentage of original system cost. For commercial buildings, the cash received is assumed to be the maximum of the resale value and the value to which the investment has been depreciated.

E15. *Annual nominal (market) discount rate*: The nominal discount rate is the annual rate of return of the best alternative investment. This is the real rate of return plus the general inflation rate. Since extra cash can be used to pay back the mortgage, the minimum rate of return (discount rate) is the mortgage interest rate.

E16. *Extra insurance and maintenance in year 1* (percent of original investment): All miscellaneous costs which cannot be entered elsewhere should be included here together with insurance and maintenance costs. Costs are as a percent of original investment (total system cost).

E17. *Annual percent increase in above expenses*: Annual percent increase in insurance, maintenance, and miscellaneous costs covered by parameter E16.

E18. *Effective federal-state income tax rate*: Since state income tax paid is deductible on federal returns, the effective federal-state income tax rate is

$$[F + S - (F \times S)] \times 100$$

where F and S are the federal and state tax rates, respectively (between 0 and 1; not percent). These tax rates are assumed constant throughout the period of analysis. For states that allow income tax deductions for federal income tax paid, the effective tax rate is

$$[F + S - 2(F \times S)] \times \frac{100}{1 - (F \times S)}$$

E19. *True property tax rate per dollar of original invest*: Since property taxes are paid on assessed value rather than actual cost, the true property tax rate is

$$\text{tax rate on assessed value} \times \frac{\text{assessed value}}{\text{system cost}}$$

E20. *Annual percent increase in property tax rate*: Annual percent increase in the true property tax rate (supplied as parameter E19).

E21. *State credit in tier one*: A two-tier state tax credit system is assumed. One rate is applied to costs up to some maximum investment and a second rate is applied to additional costs up to a second maximum. The state tax credit received is parameter E21 times the minimum of the system cost and E22, plus E23 times the remaining system cost above E22 but not in excess of E24.

E22. *State credit tier one break*: Maximum investment credited at the rate given by parameter E21.

E23. *State credit in tier two*: Tax credit rate on investment in excess of the cost specified for parameter E22.

E24. *State credit tier two break*: Cost above which no additional state tax credit is received (see E21).

E25. *Useful life for depreciation purposes*: The length of time over which the commercial system is to be depreciated to zero value. If this is greater than the length of the analysis (parameter E7), this value may affect the estimated resale value (see E14).

E26. *Percent of straight-line depreciation rate* (declining-balance depreciation): The federal government allows several rates for writing off investments when the declining-balance method is used. These rates are expressed as percentages of the straight-line depreciation rate (e.g., set E26 to 200 for double-declining-balance depreciation).

The economic parameters for the three-collector domestic hot water system are listed as follows:

E1.	Economic output detail (1, 2, or 3)	3.00
E2.	Reference or comparison system (1 or 2)	2.00
E3.	Calculate rate of return (yes = 1, no = 2)	1.00
E4.	Income-producing building (yes = 1, no = 2)	2.00
E5.	Depreciation: straight line = 1, declining balance = 2, sum of the years' digits = 3, none = 4	4.00
E6.	Consider federal tax credits (yes = 1, no = 2)	1.00
E7.	Length of analysis	20.00 years
E8.	Tax-creditable system base cost	$4500.00
E9.	Non-tax-creditable system base cost	0.00
E10.	Annual increase in purchased energy demand	0.00%/yr
E11.	Term of mortgage	1.00 year
E12.	Down payment (% of original investment)	100.00%
E13.	Mortgage annual interest rate	14.00%
E14.	Resale value (% of original investment)	100.00%
E15.	Annual nominal (market) discount rate	8.00%
E16.	Extra insurance maintenance in year 1 (% of original investment)	1.00%
E17.	Annual % increase in above expenses	10.00%
E18.	Effective federal-state income tax rate	35.00%

E19. True property tax rate per dollar of original investment   0.00%

E20. Annual % increase in property tax rate   6.00%/yr

E21. State credit in tier one   0.00%

E22. State credit tier one break $10,000.00

E23. State credit in tier two   0.00%

E24. State credit tier two break $10,000.00

E25. Useful life for depreciation purposes   20.00 years

E26. Percent of straight-line depreciation rate
(declining-balance depreciation)   150.00%

The computer program shown in Figure 8.6 is calculated with parameter E2 as "1" for the reference or conventional system. The collector area, C1, is input as "0" and the system cost, E8, is input as "0." Once the life-cycle costs of a conventional heating system are calculated, the solar system or current system parameters are input into the computer program. Parameter C1 is then changed to 72 ft^2, E2 is changed to comparison, and E8 is changed to a system cost of $4500. The resulting life-cycle cost analysis and comparative life-cycle economic analysis for this three-collector domestic hot water system are given in Figure 8.6.

In addition to a year-by-year life-cycle analysis of the current system (a solar system with auxiliary back-up), a year-by-year cost comparison between the current system and the reference system (conventional system only) is calculated. Costs are broken down into system and energy categories for both systems. The column labeled NET SAVE is the difference between the sum of the first two columns (net cost for the conventional system) and the sum of the second two columns (net cost for the solar system with auxiliary backup). Net savings are also expressed in present dollars in the next-to-last column and as a running sum of present dollars in the last column.

The *break-even point* or *payback* is the year in which the cumulative undiscounted net savings equal zero. The net savings reach zero before the energy savings in this particular example, because the tax credits are included only in the net savings.

Return on Investment

One method used to determine the attractiveness of an investment option is to calculate its return on investment (ROI). The ROI percentage is a means of comparing an investment in a solar system to other investment opportunities.

The ROI is a well-understood and widely used method for calculating relative benefits. For example, the interest rate on a savings account or Treasury certificate is an ROI investment percentage. Simple ROI calculations are not intended to replace a thorough *f*-chart economic analysis, but merely to act as assists in this area. It is often useful to quickly calculate the ROI to assess the economic viability of solar energy systems.

In general terms, ROI is determined by dividing the energy costs saved the first year by virtue of the operation of the solar system, by the net cost of the solar energy system. Net cost in this instance is defined as the installed purchase price of the system minus all the federal and state tax credits:

$$\text{return on investment} = \frac{\text{first year's energy saving}}{\text{net system purchase price}} \qquad (8.6)$$

Example 8.2

Assume that a solar domestic hot water system costs $3500 to install, saves the purchaser $360 in electric hot water heating costs the first year of operation, and is eligible for combined federal and state tax credits of 55%. What is the ROI of the solar investment?

LIFE CYCLE COST ANALYSIS OF CURRENT SYSTEM

INITIAL COST OF SYSTEM	$	4500.
DOWN PAYMENT	$	4500.
TOTAL TAX CREDITS	$	1800.
ANNUAL MORTGAGE PAYMENT	$	0. FOR 1 YEARS

	****INFORMATION****			******SYSTEM COSTS*******				*ENERGY*	***INDICATORS****		
	END OF YR	IN-TEREST	COMM DEPREC	LOAN	MAINT	PROP	TAX CREDIT	NET FUEL	NET CASH	PW OF CASH	CUMU-LATIVE
YR	PRINC	PAID	DEDUCT	PAYMT	& INS	TAX	RESALE	COST	FLOW	FLOW	PW
0	0	0	0	4500	0	0	1800	0	2700	2700	2700
1	0	0	0	0	45	0	0	230	275	255	2955
2	0	0	0	0	49	0	0	265	314	270	3225
3	0	0	0	0	54	0	0	305	359	285	3510
4	0	0	0	0	60	0	0	350	410	302	3811
5	0	0	0	0	66	0	0	403	469	319	4130
6	0	0	0	0	72	0	0	463	536	338	4468
7	0	0	0	0	80	0	0	533	613	357	4825
8	0	0	0	0	88	0	0	613	700	378	5204
9	0	0	0	0	96	0	0	705	801	401	5605
10	0	0	0	0	106	0	0	810	917	425	6029
11	0	0	0	0	117	0	0	932	1049	450	6479
12	0	0	0	0	128	0	0	1072	1200	477	6955
13	0	0	0	0	141	0	0	1233	1374	505	7461
14	0	0	0	0	155	0	0	1417	1573	535	7996
15	0	0	0	0	171	0	0	1630	1801	568	8564
16	0	0	0	0	188	0	0	1875	2063	602	9166
17	0	0	0	0	207	0	0	2156	2362	639	9804
18	0	0	0	0	227	0	0	2479	2707	677	10482
19	0	0	0	0	250	0	0	2851	3101	719	11200
20	0	0	0	0	275	0	4500	3279	-945	-202	10997
TOTAL		0	0	4500	2575	0	6300	23601	24379	10997	
PW OF TOT		0	0	4500	996	0	2765	8265	10997		

	UNDISCOUNTED	ANNUALIZED	PRESENT WORTH
ENERGY COST	23601	842	8265
TOTAL COST	24379	1120	10997

COMPARATIVE LIFE CYCLE ECONOMICS

	REFERENCE SYSTEM*		***CURRENT SYSTEM****		*******SAVINGS*******		
	SYSTEM	ENERGY	SYSTEM	ENERGY	NET	PW OF	CUM.
YR	COST	COST	COST	COST	SAVE	SAVE	PW
0	0	0	2700	0	-2699	-2699	-2699
1	0	525	45	230	249	231	-2468
2	0	603	50	265	289	248	-2220
3	0	694	54	305	335	266	-1955
4	0	798	60	350	388	285	-1670
5	0	918	66	403	449	305	-1364
6	0	1055	72	463	519	327	-1037
7	0	1214	80	533	601	351	-686
8	0	1396	88	613	695	376	-311
9	0	1605	96	705	804	402	90
10	0	1846	106	810	929	430	521
11	0	2122	117	932	1074	461	981
12	0	2441	128	1072	1241	493	1474
13	0	2807	141	1233	1433	527	2001
14	0	3228	155	1417	1655	564	2564
15	0	3712	171	1630	1911	603	3167
16	0	4269	188	1875	2207	644	3811
17	0	4909	207	2156	2547	688	4499
18	0	5646	227	2479	2939	736	5235
19	0	6493	250	2851	3392	786	6021
20	0	7467	-4224	3279	8413	1805	7826
TOTAL	0	53748	777	23601	29371	7826	
PW TOT	0	18824	2732	8265	7826		

	UNDISCOUNTED	ANNUALIZED	PRESENT WORTH
ENERGY SAVING	30147	1075	10559
TOTAL SAVING	29371	797	7826

CUMULATIVE UNDISC. ENERGY SAVINGS EQUAL INCREASED
INVESTMENT IN 9 YEARS

CUMULATIVE UNDISC. NET SAVINGS EQUAL ZERO
(BREAK EVEN POINT) IN 7 YEARS

RATE OF RETURN ON ADDITIONAL INVESTMENT IS 21.9 %

FIGURE 8.6 Life-cycle cost analysis.

Solution

$$\text{ROI} = \frac{\text{first year savings (\$360)}}{\text{net system cost (\$3500} - \$1925 = \$1575)}$$

$$= 22.8\%$$

Once the ROI has been determined, the economic attractiveness of the solar investment quickly becomes obvious. A rate of return on investmenof 22.8% is difficult to achieve elsewhere.

There are limitations to the effectiveness of ROI calculations. Factors such as escalating or deescalating fuel costs, maintenance costs of the solar system, and unpredictable inflation rates are not taken into account in simple ROI calculations. Increases in alternate or backup fuel rates will result in equivalent increases in ROI. As electric rates increase by 10%, the ROI increases by 10% as well. Increases in tax credits and other financial incentives all increase the ROI of the solar system.

Third-Party Financing

A major barrier to the widespread use of solar energy is the high initial cost of the associated equipment. Even large solar systems used for production of domestic or process hot water which operate at very high efficiencies are often not purchased due to the magnitude of the capital outlay involved. An additional objection voiced by many potential system purchasers is that the length of the payback period to recoup initial investment is too long. Many businesses will not seriously consider a project unless it can be shown to have less than a 5-year payback.

In these instances, the purchase of large solar systems is made easier by third-party financing, which has had a major impact on the solar industry. Solar heating systems that are otherwise unaffordable are now possible for many applications. There are two general categories of third-party financing: leasing and shared savings agreements. Leasing arrangements are common in many business applications and were instrumental in fostering the growth of the computer industry. Shared savings agreements were first developed for the energy management industry but are quickly becoming a significant concept in the marketing of solar energy systems. With either type of third-party financing, the end user of the solar energy avoids the high initial cost of the system. Also, no risks are taken by the end user regarding the true performance and long-term reliability of the system. Because the end user does not own the system, the tax credits and depreciation allowances cannot be taken by him. The rationale for leasing a solar system or entering into a shared savings agreement is primarily that the dollar value of the energy saved by the solar system is greater than the contract payments for it. In fact, such financing offers the end user immediate savings, no capital outlay, and little or no risk. In addition, many contracts offer the end user an option to buy the solar system at its fair market value at the conclusion of the contract period. If the end user decides not to purchase the system, the contract can be either renewed or the solar system removed. If the system is removed, the collectors can be used elsewhere. The formation of a "used collector market" will be an outgrowth of increased thirty-party financing arrangements and will allow people to purchase good-quality solar equipment at reduced prices.

Leasing

The two common types of leases are the true lease and the financing lease. With a *true lease*, the lessor is considered as the owner of the solar system for income tax purposes. The lessee in this instance can claim tax credits only for the lease payments. The lessor is eligible for the investment tax credits and existing federal energy tax credits, accelerated depreciation allowance, and any state tax incentives. True leases are short term, and always much shorter than the operating life of the solar system.

Financing leases have much higher payments than true leases. In a financing lease, the entire cost of the system in addition to interest is paid by the lessee. The lessee can purchase the system at the end of the lease term for a nominal sum. This sum is generally written into the original lease agreement. The lessee in this instance is considered the owner of the solar system for income tax purposes.

Shared Savings Agreements

Shared savings agreements appear to be a highly promising area for third-party financing arrangements in the solar industry. The risks in shared savings agreements are reduced for all the parties concerned. The end user does not own the solar system but rather, agrees to provide a location for the solar system and to purchase all the energy that it produces. The purchase price for this energy is generally between 75 and 90% of the current market value of utility-generated heat or power and is paid to the investors who own the solar system. The end user assumes no risk since poor solar system performance cannot cost more than the backup conventional fuel that the solar system is designed to replace. The arrangement of a typical shared savings agreement is illustrated in Figure 8.7.

The investors in this arrangement are often represented by an administrative group known as an *energy supply company* (ESC). The ESC can be thought of as a microutility. The ESC arranges for the design, installation, and maintenance of the solar system. The major risks for the investors in this arrangement lie in two areas: solar system performance and future conventional energy prices. The solar system must perform as predicted for the life of the agreement or the investors will lose income equal to the value of uncollected solar energy that the system was supposed to produce. Since the agreement is predicted on continually escalating prices of conventional energy, a decline in these prices can adversely affect income produced by the arrangement. Decreasing conventional energy prices will lower the payments paid to the ESC by the lessee of the solar system.

To minimize these risks, experienced professionals must be employed to design, install, and maintain the solar system. In addition, conservative inflation rates should be used to project future conventional energy costs.

Many of the elements illustrated in Figure 8.7 can be combined under larger umbrella organizations. For example, the collector manufacturer can also supply all the solar system components and provide for long-term maintenance of the system operation. In some instances the system installer may also provide the maintenance service required. Because of the importance of uninterrupted system operation, durability of system components, proper design, and necessity of immediate servicing for any malfunction, special warranties are required for a good shared savings agreement. Spare parts, an alarm system for pressure and flow malfunctions, and redundancy of critical segments and components of the system are recommended.

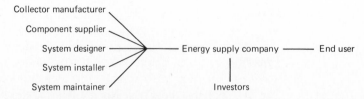

FIGURE 8.7   Organizational chart of a shared savings agreement.

I.E.S. Solar Contracting and Design
8 Post Office Square
Lynnfield, Mass. 01940
(617) 598-9700

SHARED SOLAR SAVINGS SYSTEM

USER'S PROJECTED CASH FLOW

SAMPLE PROGRAM ONLY

25	# of Heliotherm collectors, model CEC24
50000	Installed system cost
5000	Current cost for energy being supplemented
2750	Estimated value of energy produced by IES/KFG solar system
2475	Price paid to IES/KFG for that energy @ 90%
5%	Escalation of fuel costs years 2 + 3
10%	Escalation of fuel costs years 4–10

PROJECTED SAVINGS FROM THE IES/KFG SOLAR SYSTEM

YEAR	1	2	3	4	5	6	7	8	9	10
Energy cost without IES/KFG solar system	5000	5250	5513	6064	6670	7337	8071	8878	9766	10742
Value of energy produced by IES/KFG system	2750	2888	3032	3335	3669	4035	4439	4883	5371	5908
Price paid for that energy	2475	2599	2729	3002	3302	3632	3995	4395	4834	5317
Savings that year	275	289	303	334	367	404	444	488	537	591
Cumulative savings	275	564	867	1200	1567	1971	2415	2903	3440	4031
Projected fair market value selling price	45000	40000	35000	30000	25000	20000				

ASSUME THAT THE USER PURCHASES THE SYSTEM IN YEAR 6.

YEAR	1	2	3	4	5	6	7	8	9	10
Purchase price of the IES/KFG solar system that year						20000				
Value of energy saved from the system						4035	4439	4883	5371	5908
Cumulative value of energy savings						4035	8474	13357	18728	24637

FROM THIS YOU CAN SEE THAT:

A) DURING THE FIVE YEARS PRIOR TO PURCHASING THE SYSTEM, THE USER GETS 1971 IN FREE SAVINGS.

B) THE CUMULATIVE SAVINGS FROM THE SYSTEM IN YEARS 6 THROUGH 10 IS 24637, ON A PURCHASE PRICE IN YEAR 6 OF 20000.

FIGURE 8.8   End user's projected cash flow: shared solar savings system. (From I.E.S. Solar Contracting and Design, Lynnfield, Mass.)

The advantages to the end user in shared savings agreements focus on immediate reduction in energy bills coupled with no investment and minimal risk. The advantages to the system investors focus on tax advantages accruing from federal, state, and depreciation allowances as well as income from the sale of energy produced by the solar system. Since income is derived from the ongoing sale of solar energy, 12-month applications with higher collector efficiency such as domestic or process hot water are preferred. The minimum investment in this type of arrangement is usually in the neighborhood of $25,000.

Figure 8.8 is a computer printout highlighting the end user's projected cash flow in a typical shared savings agreement. Figure 8.9 is a printout illustrating the investors' projected cash flow. Both printouts demonstrate the attractiveness of this form of third-party financing, and why it is bound to become an integral component of solar financing in the immediate future.

I.E.S. Solar Contracting and Design
8 Post Office Square
Lynnfield, Mass. 01940
(617) 598-9700

SHARED SOLAR SAVINGS SYSTEM

OWNER'S CASH FLOW

SAMPLE
PROGRAM
ONLY

##COLLECTORS	25ENTER
SYSTEM COST	50000ENTER
I'S TAX BRCK	0.50
SYSTEM SAVNG	2750ENTER
%ESC OF SVNG	0.10ENTER
FED ITC	5000
SOLAR ITC	7500
STATE ITC	0
TOTAL ITC	12500
DEPRECIATION	
YEARS	5
AMOUNT$$/YR	10000
$$ DOWN	10000ENTER
$$ FINANCED	40000
#YRS OF LOAN	10
INTEREST RAT	0.14ENTER

YEAR	1	2	3	4	5	6	7	8	9
DOWN PAYMENT	10000								
LOAN BALANCE	40000	36000	32000	28000	24000	20000	16000	12000	8000
INT. PAID	5600	5040	4480	3920	3360	2800	2240	1680	1120
PRIN.REDUCTI	4000	4000	4000	4000	4000	4000	4000	4000	4000
TAX CONSEQUENCES (PRE-TAX DOLLARS)									
PROJECTED INCOME AT 90%	2475	2723	2995	3294	3624	3986	4385	4823	5305
LESS DEPRECIATION	7500	11000	10500	10500	10500	0	0	0	0
LESS INTEREST PAID	5600	5040	4480	3920	3360	2800	2240	1680	1120
TAXABLE INCOME/SHELTER	−10625	−13318	−11985	−11126	−10236	1186	2145	3143	4185
TAX CREDITS	12500								
PRE-TAX VALUE OF CREDITS (ASSUMES 50% BRACKET)	25000								
TOTAL TAX SHELTER AVAIL	35625	13318	11985	11126	10236	0	0	0	0

CASH FLOW

CASH REQUIRED									
PROJECTED INCOME AT 90%	2475	2723	2995	3295	3624	3986	4385	4823	5305
LESS DOWN PAYMENT	−10000								
LESS PRINCIPAL REDUCTN	−4000	−4000	−4000	−4000	−4000	−4000	−4000	−4000	−4000
LESS INTEREST PAID	−5600	−5040	−4480	−3920	−3360	−2800	−2240	−1680	−1120
NET CASH AVAILABLE	0	0	0	0	0	0	0	0	0
NET CASH REQUIRED	17125	6318	5485	4626	3736	2814	1855	857	93
TAX SHELTER AVAILABLE FROM THAT CASH	35625	13318	11985	11126	10236	0	0	0	0
SHELTER TO CASH RATIO	2.08	2.11	2.18	2.41	2.74	0.00	0.00	0.00	0.00
CUMULATIVE SHELTER TO CASH RATIO	2.08	2.09	2.11	2.15	2.21	2.05	1.96	1.92	1.92

FIGURE 8.9 Investor's (owner's) projected cash flow in a shared solar savings system. (From I.E.S. Solar Contracting and Design, Lynnfield, Mass.)

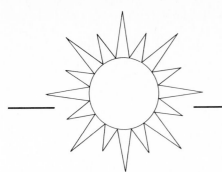

9

Solar Swimming Pool Heating

The heating of swimming pool water represents a highly cost-effective application for solar energy systems. The costs associated with conventional fuel pool heating are often substantial. Due to the large quantities of water contained in most residential and commercial swimming pools, the fossil-fuel heating units in these systems commonly have higher Btu ratings than those of residential heating systems. In fact, the annual cost for pool heating in many sections of the United States is often equal to or higher than the costs for space heating the home.

The plumbing and heating systems in most swimming pools are modifications of standard domestic hot water applications. In most pools, the water is circulated between the pool and a filter utilizing a high-volume fractional or multihorsepower circulating pump. A heating unit is inserted into the piping system after the filter in a series-type arrangement pictured in Figure 9.1.

Swimming pool heating systems can consume prodigious amounts of energy. A simple example will emphasize this point.

Example 9.1

A swimming pool measuring 15 ft × 30 ft with an average depth of 5½ ft loses an average of 8°F in 1 day if the pool heater is turned off when the water reaches a temperature of 78°F. The pool utilizes an oil-fired hot water heater, and we wish to determine how much fuel oil will be needed to replace the heat loss that has occurred.

Solution

(1) No. cubic feet in pool × 7.48 gal/ft³ = no. of gallons of water in the pool

$$15 \text{ ft} \times 30 \text{ ft} \times 5\tfrac{1}{2} \text{ ft} = 2475 \text{ ft}^3$$

$$2475 \text{ ft}^3 \times 7.48 \text{ gal/ft}^3 = 18{,}513 \text{ gal water in the pool}$$

(2) No. gallons in pool × 8.33 lb/gal × no. degrees of temp. change = no. Btu lost

$$18{,}513 \text{ gal} \times 8.33 \times 8°F = 1{,}233{,}700 \text{ Btu of heat lost}$$

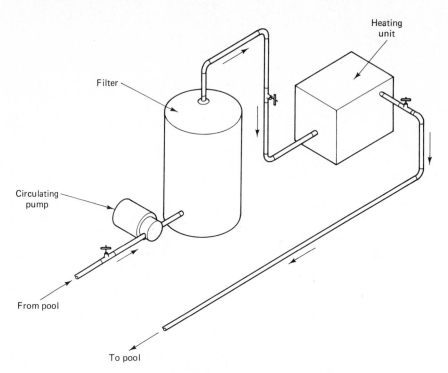

FIGURE 9.1 Plumbing arrangement of residential swimming pool.

(3) Btu lost ÷ 98,000 (fuel oil burned at 70% efficiency) = no. gallons of fuel oil required to make up heat lost = 12.6 gal fuel oil equivalent

Under these conditions, the swimming pool heater will consume 12.6 gallons of fuel oil to make up for the heat loss. Fuel consumption of this magnitude can result in excessively high costs to maintain the heat of the pool over the course of the normal swimming season.

Pool heaters of 400,000 Btu/hr output or higher are commonly used. Thus the rated output of many heaters is larger than the residential space heating system. High-capacity Btu units are generally employed for pool heating rather than smaller units since a rapid temperature increase in the pool water is a desirable feature of the system. The very large thermal mass of most pools requires large quantities of energy for even small increases in water temperature. Many hours or even days of heater operation are often necessary to bring a pool gradually up to the desired temperature.

In conventional pool systems the circulating pump pulls water from the bottom of the pool, pushes it through the filter, and then moves it back to the upper sides of the pool. These pumps are sized to move between 10 and 50 gallons of water per minute in the average swimming pool. Thus the pool in Example 9.1, with a volume of 18,513 gallons of water, requires approximately 5 hours of operation to move the entire volume of water once through the heating unit. Length of operation time of both the circulators and heater is usually controlled by an automatic timing switch.

While the mechanics of swimming pool heating are simple, the solar designer must take into account the magnitude of the pool's energy requirement when sizing a system. For example, smaller solar collector arrays are sometimes employed to heat an entire house than to heat the family swimming pool.

The solar collectors used in pool heating applications can be either flat-plate collectors or special unglazed, uninsulated collectors specifically designed for pool heating applications. Collector design as it applies to pool heating systems will now be discussed.

**SOLAR POOL
COLLECTOR
DESIGN
FUNDAMENTALS**

The type of solar collector to be used in a pool heating application depends on the length of time that the system will be in operation throughout the year and the climatic conditions to which the system will be exposed. For example, an indoor swimming pool located in a cold climate which is to be solar heated throughout the year requires the use of standard flat-plate collectors in conjunction with an antifreeze closed-loop system design. On the other hand, when a pool is located in either a warm climate not subject to freezing temperatures, or in a cold climate in which the pool is in use only during the warmer summer months, a specially designed pool heating collector that is simpler in design and material configuration than the standard flat-plate collector can be used. These pool collectors are also much less expensive than typical flat plates since no glazing, insulation, or collector box is required. In fact, the only component of a swimming pool collector is the absorber plate. This type of collector is illustrated in Figure 9.2. The design of this type of collector is based on the principle that as long as the ambient air temperature is equal to or greater than the pool water temperature, there is no reason to try to prevent heat loss from the absorber panel. In fact, ambient temperatures warmer than the pool water will help to warm the pool water by transferring heat to the exposed absorber panels, eliminating the need for all collector glazing, insulation, and boxing.

Absorber Materials and Design

The most commonly used materials in pool collector absorbers are ethylene propylene diene monomer (EDPM) and polycarbonate plastics (polypropylene and polyethylene). These materials are excellent choices for this application due to their low cost, high durability, and

FIGURE 9.2 Copolymer plastic absorber. (Courtesy of Sun-Glo Solar Ltd.)

corrosion resistance. Corrosion resistance in a solar-heated swimming pool is of primary concern in the selection of a flat-plate collector. For example, pool water is rarely circulated directly through copper absorber plates because the low pH of the pool water will eventually corrode the absorber flow channels.

In this instance a cupro-nickel (90% copper, 10% nickel) alloy heat exchanger is recommended. No such concern exists when using EPDM or polymer absorber plates; the pool water can be circulated directly through these collectors with no adverse affects. To prolong the useful life of these materials, carbon black and chemical ultraviolet stabilizers are formulated into the absorber to prevent decomposition from prolonged exposure to the sun.

While standard flat-plate collectors are unaffected by small changes in ambient temperature, pool collector efficiency will either increase or decrease rapidly with small changes in temperature. Since the pool collector has no glazing or insulation, its performance is governed by its operating temperature relative to the temperature of its surroundings. The efficiency curve for a typical well-constructed copolymer pool heating panel is illustrated in Figure 9.3. This efficiency curve illustrates the limitations of such a simple collector design: It cannot be expected to provide energy if the ambient air temperature is 50°F or more below the collector fluid temperature. Wind also reduces the performance of the unit since no glazing or insulation is incorporated to prevent convective heat loss.

Working within these design limitations, pool collectors are well suited for heating outdoor swimming pools in use during warm weather. In fact, when the pool water is nearly equal to, or cooler than, the ambient air temperature, pool collectors will outperform a conventional flat-plate collector.

The efficiency curve in Figure 9.3 demonstrates the importance of keeping a swimming pool collector operating at as low a temperature as possible. Any temperature buildup within the collector will significantly reduce performance. A maximum wetted surface area of the absorber plate is necessary for high performance due to the relatively low thermal conductivity of the polycarbonates and rubber compounds used in manufacturing the absorber plate.

Flow Characteristics

Fluid flow rate through the absorber is also very important to maximize collector efficiency. It is recommended that the pool water be circulated through swimming pool collectors at least 3½ times faster than water would normally be circulated through a conventional flat-

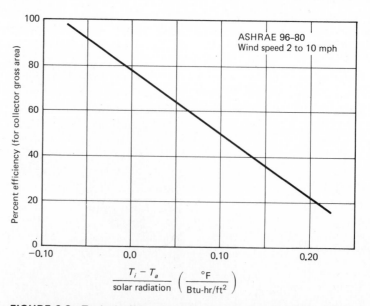

FIGURE 9.3 Typical efficiency curve for a swimming pool collector. Note the steep slope of the curve caused by the lack of glazing and absorber insulation.

plate collector. This increased flow rate is required in order to keep the upper part of the collectors from becoming more than 2 or 3 degrees warmer than the bottom collector inlet. For purposes of comparison, a conventional flat-plate collector has an optimum water flow rate of approximately 0.03 gallon per minute per gross square foot of collector area, while at least 0.10 gallon per minute of water flow rate for swimming pool collectors is recommended.

The relatively poor thermal conductivity of both polycarbonates and EPDM rubber necessitate keeping the absorber plate fluid flow channels relatively thin in order to maximize heat transfer. These thin walls also restrict the pressures under which the collectors can successfully operate. Most pool system operating pressures are much lower than the operating pressures encountered in conventional domestic hot water and space heating systems. The only pressure buildup in a pool collector will be caused by friction or restriction of pool water as it returns from the collector array to the pool. Proper system design and installation procedures will keep this pressure to a minimum.

COLLECTOR ARRAY SIZING TECHNIQUES

In Chapter 8 the widely accepted *f*-chart computer program for sizing solar domestic hot water and space heating systems was discussed. No comparable computer program for sizing solar pool heating systems exists, however. What has emerged through many years of experience are sizing techniques largely based on the surface area of the swimming pool. Accurate pool system sizing akin to computerized programs for domestic hot water and space heating is difficult due to the many variables encountered in swimming pool operation. The most significant of these variables are: surface area of the pool; the use, or lack, of an insulating pool blanket; location of the pool (above or below ground); desired water temperature; exposure of the pool to prevailing winds; exposure of the pool to direct sunlight; how much the pool is used; and specific climatic conditions.

In addition to conduction, convection, and radiation heat losses from a pool, an additional major cause of heat loss is due to evaporation of water from the pool's surface. This evaporative heat loss can often be the largest form of heat loss encountered within the system. The change of phase from liquid to gas requires extremely large amounts of energy. For example, raising the temperature of 1 gallon of water 1°F requires 8.33 Btu, but to evaporate 1 gallon of water at a temperature of 80°F requires 9196 Btu. Of this figure, 8097 Btu is required merely to change the phase of the water from a liquid to a gas. This phenomenon is experienced, for example, when a person steps out of a shower and runs to answer the telephone in another room of the house. In so doing, the person experiences a cold feeling. The biggest reason for experiencing a chill is not the 68°F air temperature of the home, but the water droplets evaporating from the surface of the skin. The faster the person runs, the colder they feel, due to acceleration of this process. The quickest relief from the cold in this instance is to dry off or to cover the body to prevent further evaporation from taking place. The major portion of the heat required for this evaporation comes from the person's body. The remainder of the heat is drawn from the surrounding air. An uncovered swimming pool is in a similar situation, the wind is the major factor that accelerates pool water evaporation. Energy required for this evaporation process comes primarily from the heat within the pool water; the remainder is drawn from the surrounding air. The net result is to decrease the pool water temperature. Clearly, the lack of a pool cover or insulating pool blanket will greatly increase heat loss, due largely to the evaporative process at work. Exposure to the wind and the amount of human activity in the pool, which encourages evaporation by agitating the surface of the pool water, are also important variables affecting heat loss.

Due to these factors, sizing a solar system to heat a swimming pool is a difficult task with relatively little scientific data to use as a guide. Two successful sizing techniques will now be discussed: the surface area method and the fuel consumption method.

Surface Area Method

The surface area of a pool is the most important variable contributing to heat loss. The magnitude of conduction, convection, radiation, and evaporation are all proportional to the surface area of the pool. The surface areas of the most common shapes of swimming pools can be calculated from the information provided in Figure. 9.4. Once the surface area of the pool has been calculated, a number of other variables must be determined relative to sizing the solar collector array effectively. Figure 9.5 considers one suggested method for sizing the system, taking into account the use of pool blankets, prevailing winds, specific environment of the pool, and exposure of the proposed collector array to sunlight.

Figure 9.5 illustrates a rule-of-thumb method for sizing a pool collector array. This type of calculation is valid either for outdoor pools in use during the summer swimming season, or for pools located in warm climates year round. A pool heating system sized in this way can extend the swimming season in northern climates and will result in sizable temperature increases in the pool water in the warmer areas of the country. As with any solar installation, procedures must be followed to ensure solar access for the collectors with a minimum of shading.

A solar collector array area equal to between 50 and 75% of the surface area of the pool is an acceptable sizing ratio for normal pool heating requirements. "Normal" in this instance is defined as raising the temperature of the water in the "comfort zone" of the pool, that is, the top 4 ft of water, approximately 10 to 15°F above ambient temperature during the swimming season.

It should be remembered that in both indoor and outdoor pools, an insulating pool cover is essential to decreased evaporative heat losses from the water. Although indoor pools are not prone to accelerated evaporation losses due to the effects of wind, this is offset by the lack of direct solar gain on the pool's surface.

Fuel Consumption Record Method

Past fuel consumption records for heating a swimming pool can be very helpful in determining the proper size for a solar collector array. To help ensure the accuracy of the analysis, at least two years of fuel records should be examined to discount the effect of weather and seasonal fluctuations on the final sizing calculations. The Btu requirements of an outdoor pool will change dramatically with weather conditions, while that of an indoor pool in a heated natatorium will remain relatively constant.

Once the energy requirement for heating the pool is known, an *f*-chart computer program can be used as the design tool. The pool heating load is treated as a large domestic

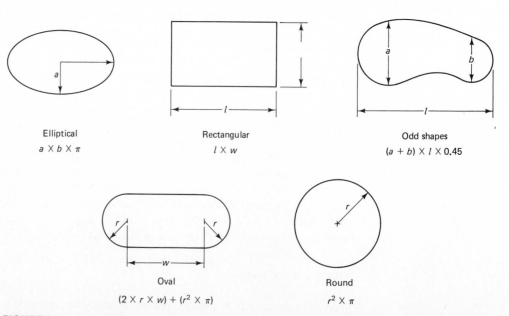

FIGURE 9.4 Methods for determining the surface areas of commonly shaped swimming pools.

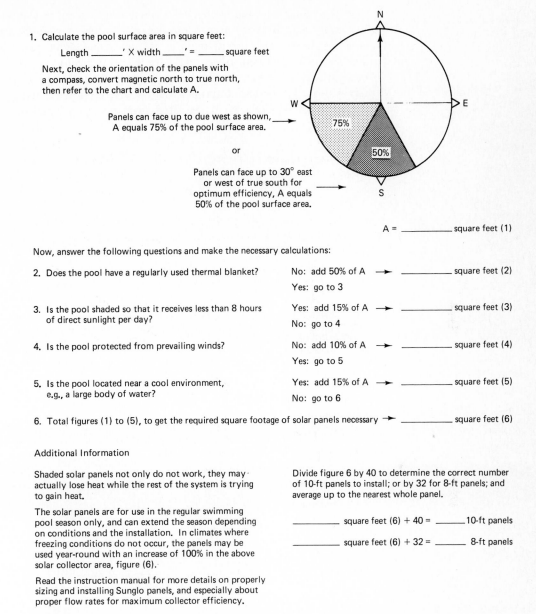

1. Calculate the pool surface area in square feet:

   Length _____' × width _____' = _____ square feet

   Next, check the orientation of the panels with a compass, convert magnetic north to true north, then refer to the chart and calculate A.

   Panels can face up to due west as shown, A equals 75% of the pool surface area. →

   or

   Panels can face up to 30° east or west of true south for optimum efficiency, A equals 50% of the pool surface area. →

   A = _____ square feet (1)

Now, answer the following questions and make the necessary calculations:

2. Does the pool have a regularly used thermal blanket?
   No: add 50% of A → _____ square feet (2)
   Yes: go to 3

3. Is the pool shaded so that it receives less than 8 hours of direct sunlight per day?
   Yes: add 15% of A → _____ square feet (3)
   No: go to 4

4. Is the pool protected from prevailing winds?
   No: add 10% of A → _____ square feet (4)
   Yes: go to 5

5. Is the pool located near a cool environment, e.g., a large body of water?
   Yes: add 15% of A → _____ square feet (5)
   No: go to 6

6. Total figures (1) to (5), to get the required square footage of solar panels necessary → _____ square feet (6)

Additional Information

Shaded solar panels not only do not work, they may actually lose heat while the rest of the system is trying to gain heat.

The solar panels are for use in the regular swimming pool season only, and can extend the season depending on conditions and the installation. In climates where freezing conditions do not occur, the panels may be used year-round with an increase of 100% in the above solar collector area, figure (6).

Read the instruction manual for more details on properly sizing and installing Sunglo panels, and especially about proper flow rates for maximum collector efficiency.

Divide figure 6 by 40 to determine the correct number of 10-ft panels to install; or by 32 for 8-ft panels; and average up to the nearest whole panel.

_____ square feet (6) ÷ 40 = _____ 10-ft panels

_____ square feet (6) ÷ 32 = _____ 8-ft panels

FIGURE 9.5 Method of determining the surface area of the solar collector array for various swimming pools. (Courtesy of Sun-Glo Solar Ltd.)

hot water or process heat load, while the solar collector parameters are entered at the same time. The following example, together with the accompanying computer printout, illustrates this method of system sizing.

Example 9.2

We wish to maintain a swimming pool containing 20,000 gallons of water at 78°F. The heating load for the program is entered as 20,000 gallons per day with a set temperature of 78°F. The street water temperature is entered as 75°F.

Solution

The resulting computer printout (Figure 9.6) provides a monthly quantity of solar energy delivered to the swimming pool water.

The solar contribution in this system will depend on the number of collectors used in the array. The collector area parameter can be changed until the delivered solar energy is equal to the monthly energy requirement determined from past energy consumption records. The use of a good pool blanket (Figure 9.7) is critical to ensuring the validity of this method. Without it the excess solar energy contributed to the pool on a better-than-average solar day will quickly be lost. Caution should be exercised when sizing the collector array because the efficiency of unglazed, uninsulated swimming pool collectors are more severely affected

```
TEN FLATE PLATE COLLECTORS TO HEAT SWIMMING POOL



LIST ALL


COLLECTOR PARAMETERS
  C1. COLLECTOR AREA ................................  240.00 FT2
  C2. FR-UL PRODUCT ................................    0.74 BTU/HR-FT2-DEG F
  C3. FR-TAU-ALPHA (NORMAL INCIDENCE) .............    0.70
  C6. NUMBER OF COVERS ............................    1.00
  C7. INDEX OF REFRACTION ..........................    1.53
  C8. EXTINCTION COEFFICIENT X LENGTH (KL)..........    0.04
  C9. INCIDENCE ANGLE MODIFIER CONSTANT ...........    0.09
  C10. COLLECTOR FLOW RATE * SPECIFIC HEAT/AREA ......  12.60 BTU/HR-FT2-DEG F
  C12. COLLECTOR SLOPE .............................   30.00 DEGREES
  C13. COLLECTOR AZIMUTH ...........................    0.00 DEGREES
  C14. GROUND REFLECTANCE ..........................    0.20

COLLECTOR-STORE TRANSFER PARAMETERS
  T1. EPS-CMIN OF COLLECTOR-STORE HX/COLLECTOR AREA ..  11.50 BTU/HR-FT2-DEG F

STORAGE UNIT PARAMETERS
  S1. TANK CAPACITY/COLLECTOR AREA ................  700.00 BTU/DEG F-FT2
  S2. STORAGE UNIT HEIGHT/DIAMETER RATIO ..........    2.00
  S3. HEAT LOSS COEFFICIENT .......................    0.00 BTU/HR-FT2-DEG F
  S4. ENVIRONMENT TEMPERATURE (-1000 FOR TENV=TAMB) ..  68.00 DEG F
  S5. HOT WATER AUXILIARY TANK UA .................    0.00 BTU/HR-DEG F
  S6. HOT WATER AUX TANK ENVIRONMENT TEMPERATURE .....  68.00 DEG F

LOAD PARAMETERS
  L3. HOT WATER USE ..............................20000.00 GALLONS/DAY
  L4. HOT WATER SET TEMPERATURE ...................   78.00 DEG F
  L5. WATER MAINS TEMPERATURE .....................   75.00 DEG F

AUXILIARY PARAMETERS
  A3. HOT WATER AUXILIARY FUEL (1=GAS,2=ELEC,3=OIL) ..   1.
  A4. AUXILIARY WATER HEATER EFFICIENCY ...........    0.60
?


LIST FLDATA


        FUEL COST

GAS  (BLOCK RATE STRUCTURE)
        ANNUAL INFLATION RATE  10.0 %
        BLOCK           1
        COST ($/MMBTU)       8.0
        MAX. USE (MMBTU)10000.0

ELEC (BLOCK RATE STRUCTURE)
        ANNUAL INFLATION RATE  10.0 %
        BLOCK           1
        COST ($/MMBTU)       6.3
        MAX. USE (MMBTU) 9477.7

TYPE FLDATA TO ENTER OR CHANGE FUEL COSTS
 ?

RUN 220 LONG
```

FIGURE 9.6 *f*-Chart computer printout for swimming pool system.

```
***************   FCHART  ANALYSIS   (VERSION 4.0)  ****************

WASHINGTON       DC      LATITUDE 38.6


                       THERMAL PERFORMANCE

             HT      TA     HWLOAD      QU     QLOSS FDHW
          (MMBTU)(DEG-F)(MMBTU)   (MMBTU)(MMBTU)
    JAN    6.33    32.0    15.52     2.46     0.00 0.15
    FEB    7.19    33.8    14.01     2.99     0.00 0.22
    MAR    9.68    41.0    15.52     4.48     0.00 0.28
    APR   10.86    53.6    15.01     5.75     0.00 0.38
    MAY   12.30    62.6    15.52     7.06     0.00 0.45
    JUN   12.77    71.6    15.01     7.88     0.00 0.52
    JUL   12.79    75.2    15.52     8.14     0.00 0.52
    AUG   12.05    73.4    15.52     7.62     0.00 0.49
    SEP   10.72    66.2    15.01     6.39     0.00 0.43
    OCT    9.53    55.4    15.52     5.13     0.00 0.33
    NOV    6.78    44.6    15.01     3.13     0.00 0.21
    DEC    5.49    33.8    15.52     2.11     0.00 0.14
    YR   116.49    53.6   182.68    63.13     0.00 0.35


               PURCHASED ENERGY SUMMARY

                  GAS      ELECTRIC    OIL     TOTAL
    USE (MMBTU)  199.32      0.00      0.00    199.32
    COST ($)    1594.57      0.00      0.00   1594.57
```

FIGURE 9.6 (*Cont.*)

by wind speed than are conventional flat-plate collectors for which the *f*-chart was developed.

The solar designer must be aware of the many factors that contribute to the heat loss of a pool. Evaporative heat loss, for example, is surprisingly large in this application. Heat losses associated with conduction, convection, and radiation which are insignificant in other types of solar installations are greatly exaggerated in pool systems due to the large surface area of the pool.

Experience has shown these two sizing methods to be accurate in predicting solar system performance.

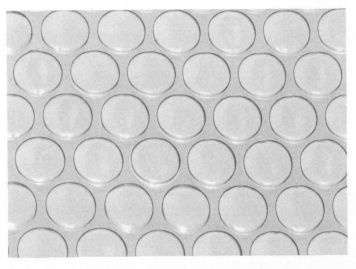

FIGURE 9.7 Insulated pool blanket. (Courtesy of Century Products, Inc.)

COLLECTOR ORIENTATION AND INCLINATION

The optimum angles of solar collector orientation and inclination are determined by the specific applications to which the system is put. Some systems will require solar heat for only a few months during the summer and early fall, whereas others require heat throughout the entire year. The energy load may be relatively constant throughout the year, as with an indoor pool, or it may change significantly at certain times, as is often the case with outdoor pools located in changeable climates. These factors must all be taken into consideration when determining orientation and inclination angles for the collector array.

Indoor swimming pools require heat throughout the entire year, and the heating load in this instance will be relatively constant. This type of application is similar to a domestic hot water demand. In this instance the collector array should face within 20° of true south at an inclination angle within 10° of site latitude. If in addition to the swimming pool, the natatorium is to be solar-heated as well, the collector array should be inclined at site latitude plus 15°, optimizing the array for space heating. In many applications it is better to concentrate on the pool heating load itself rather than on heating the natatorium, keeping the solar collectors at the lower angle of inclination. The pool is the most efficient and cost-effective application for the solar collectors.

Outdoor pools are generally used only during the summer months, and the unglazed, uninsulated type of collector is normally recommended for these applications. Again, these collectors should face as close to true south as possible. The proper inclination of the array is equal to the angle necessary for the sun's rays to strike the collectors at a right angle during periods of maximum solar radiation. For example, in the northeastern United States, where the swimming season is generally restricted to the summer months, the inclination angle would be 20 to 30° from horizontal.

SWIMMING POOL SOLAR SYSTEM DESIGN

We have examined how the energy requirements of the pool are used to size the collector array. The size of the array thus determines the size of all other system components. Piping, valves, pumps, and heat exchangers must all be properly matched to the energy output of the solar collectors. Also, the system may take on auxiliary functions such as heating hot tubs and domestic hot water tanks, which will vary the size and capacity of all system components. The distance between the pool circulating pump and the collectors, as well as the number and size of the collectors, will have an effect on the size of all piping used within the system. The use of an insulating pool blanket is critical to minimize the size of the collector array while maximizing system performance.

As with all solar systems, final design considerations are determined by the uses to which the system is put. Location of the pool, size, and type of the collectors, pool energy requirements, and various site characteristics all contribute to final system design. Indoor pools in use throughout the entire year will ordinarily use conventional flat-plate collectors while outdoor pools will normally use the simpler and less expensive swimming pool collectors. These two basic collector configurations dictate two different types of system designs, known as closed loop and open loop, which will now be discussed in greater detail.

Closed Loop

Closed-loop pool heating systems, as with their domestic hot water counterparts, are employed when the collectors must operate in ambient air temperatures far below internal collector fluid temperatures. Conventional liquid flat-plate collectors are used in these systems, mostly for heating indoor pools located in northern climates subject to frequent freezing conditions. A 50/50 mixture of inhibited propylene glycol and water is used in the collector loop. A bronze or cupro-nickel shell-and-tube heat exchanger is recommended for use with chlorinated low-pH pool water for heat transfer. Cupro-nickel is an alloy consisting of 90% copper and 10% nickel which is highly resistant to corrosion (Figure 9.8).

These types of systems are very simple and require no storage tanks; the thermal mass of the pool water is large enough to absorb all the collector output, resulting in small increases in pool water temperature. Figure 9.9 illustrates a simple closed-loop system in

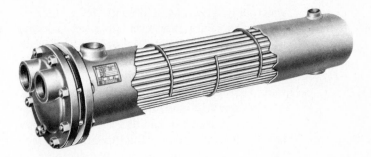

FIGURE 9.8 Shell-and-tube heat exchanger. (Courtesy of ITT Fluid Handling Division.)

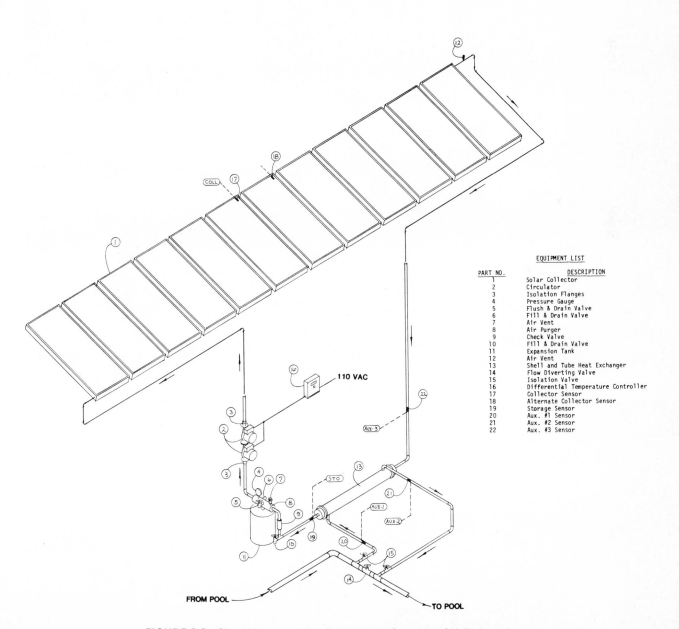

EQUIPMENT LIST	
PART NO.	DESCRIPTION
1	Solar Collector
2	Circulator
3	Isolation Flanges
4	Pressure Gauge
5	Flush & Drain Valve
6	Fill & Drain Valve
7	Air Vent
8	Air Purger
9	Check Valve
10	Fill & Drain Valve
11	Expansion Tank
12	Air Vent
13	Shell and Tube Heat Exchanger
14	Flow Diverting Valve
15	Isolation Valve
16	Differential Temperature Controller
17	Collector Sensor
18	Alternate Collector Sensor
19	Storage Sensor
20	Aux. #1 Sensor
21	Aux. #2 Sensor
22	Aux. #3 Sensor

110 VAC

FROM POOL

TO POOL

FIGURE 9.9 Closed-loop pool heating system. (Courtesy of Heliotherm.)

which only one differential controller is utilized. The controller serves to activate the collector-loop circulator whenever the solar collectors become warmer than the swimming pool water, moving antifreeze through the tubes of the shell-and-tube heat exchanger. At the same time that the collector loop is energized, pool water is circulated through the shell of the heat exchanger, where a transfer of heat takes place from the collector loop to the pool water. This exchange is highly efficient and coincides with the availability of solar energy.

Figure 9.10 illustrates a modification of the closed-loop pool system in which domestic hot water is heated in addition to the pool water. Solenoid valves are employed to divert the flow of the collector loop fluid to either the shell-and-tube heat exchanger for the pool, or to the heat exchanger within the solar domestic hot water storage tank. Thermostatic switches (aquastats) are used to automatically open and close the two solenoid valves responsible for routing the antifreeze fluid. Priority can be established for either of the two heating requirements. If the domestic hot water is given priority it will be solar-heated until it reaches a set point of, for example, 120°F, at which time the antifreeze fluid will be diverted to heat the pool via the shell-and-tube heat exchanger. If the swimming pool is given priority, it will be solar heated until it reaches a set point of, say, 78°F. If the pool water rises above this temperature, the antifreeze fluid will be diverted to heat the domestic hot water. To prevent overheating the domestic hot water, provision is made for the collector-loop fluid to be diverted back to the pool heat exchanger when temperatures in excess of the high-limit set point are reached. Figure 9.11 illustrates the wiring schematic of the priority toggle switch and aquastat used to achieve this high-limit operational logic.

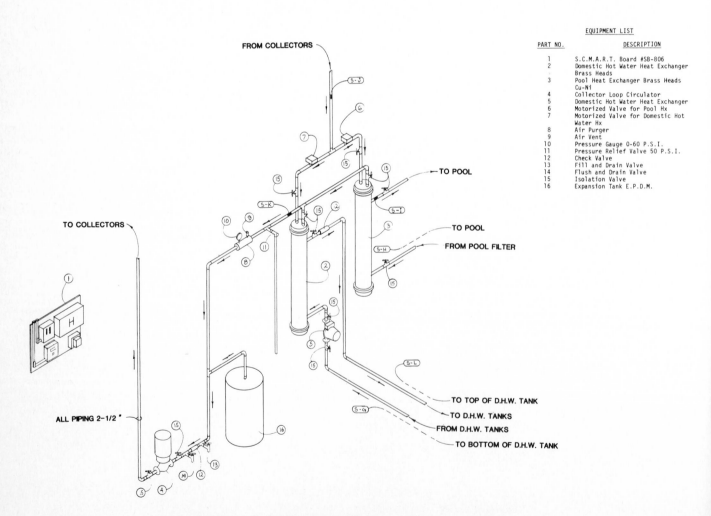

FIGURE 9.10 Closed-loop pool and domestic hot water heating system.

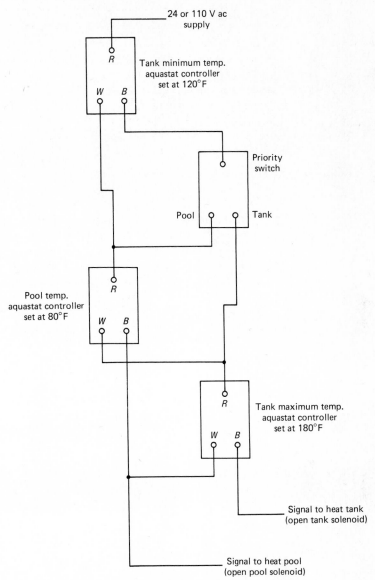

24 or 110 V ac
supply

R

Tank minimum temp.
aquastat controller
set at 120°F

W   B

Priority
switch

Pool   Tank

R

Pool temp.
aquastat controller
set at 80°F

W   B

R

Tank maximum temp.
aquastat controller
set at 180°F

W   B

Signal to heat tank
(open tank solenoid)

Signal to heat pool
(open pool solenoid)

Note: In all three aquastat controllers, *R* and *W* make contact with
a temperature rise to the set point, while *R* and *B* make
contact at temperatures below the set point.

FIGURE 9.11 Wiring schematic for pool/domestic hot water priority.

A residential space heat and domestic hot water system that incorporates the heating capability for an outdoor swimming pool in use during the summer months is illustrated in Figure 9.12. This multipurpose system provides a greater return on investment for the homeowner than just the space heating version since the surplus heat generated by the collectors during the summer months is used to heat the pool rather than going to waste. The more the system output is utilized, the greater will be the return on investment for the system owner.

Open Loop

When unglazed, uninsulated swimming pool collectors are utilized in a pool heating system, an open-loop system design is generally employed. In this type of installation, the ambient air temperature is close to or greater than the pool water temperature. The pool water being

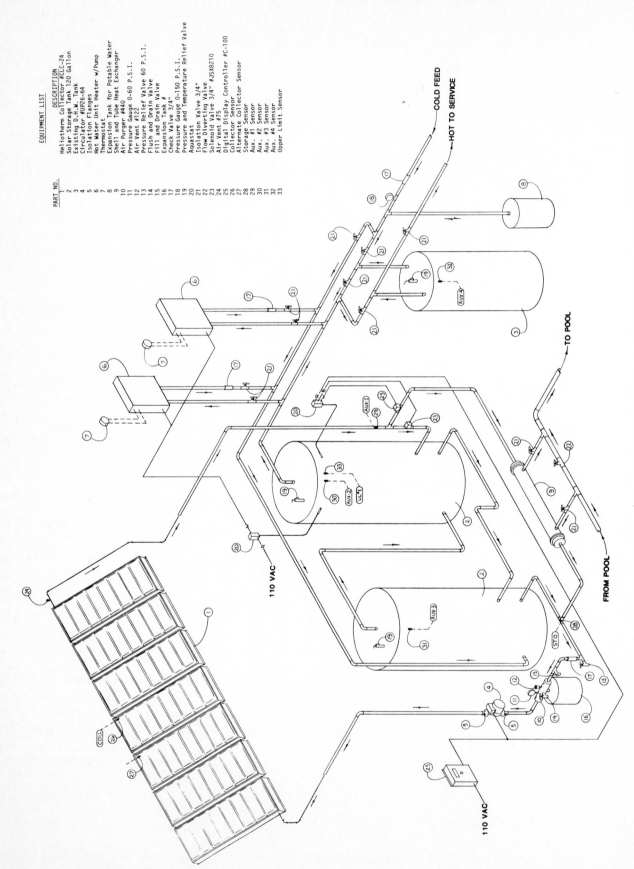

EQUIPMENT LIST

PART NO.	DESCRIPTION
2	Heliotherm Collector #CLC-24
3	Solar Storage Tank 120 Gallon
4	Existing D.H.W. Tank
5	Circulator #UP26-64
6	Isolation Flanges
7	Hot Water Unit Heater w/Pump
8	Thermostat
9	Expansion Tank for Potable Water
10	Shell and Tube Heat Exchanger
11	Air Purger #440
12	Pressure Gauge 0-60 P.S.I.
13	Air Vent #122
14	Pressure Relief Valve 60 P.S.I.
15	Flush and Drain Valve
16	Fill and Drain Valve
17	Expansion Tank #
18	Check Valve 3/4"
19	Pressure Gauge 0-150 P.S.I.
20	Pressure and Temperature Relief Valve
21	Aquastat
22	Isolation Valve 3/4"
23	Flow Diverting Valve
24	Solenoid Valve 3/4" #JSX8210
25	Air Vent #75
26	Digital Display Controller #C-100
27	Collector Sensor
28	Alternate Collector Sensor
29	Storage Sensor
30	Aux. #1 Sensor
31	Aux. #2 Sensor
32	Aux. #3 Sensor
33	Aux. #4 Sensor
	Upper Limit Sensor

COLD FEED

HOT TO SERVICE

TO POOL

FROM POOL

110 VAC

110 VAC

FIGURE 9.12 Residential space heat, domestic hot water, and pool heating system. (Courtesy of Heliotherm.)

circulated by the existing filter pump is diverted either manually or automatically through the collector array when sufficient solar energy is available for heating purposes. When the pump is turned off, the pool water drains from the collectors into the swimming pool. This type of system combines simplicity of design with low cost and reliable operation and is illustrated in Figure 9.13.

Automatic diverting valves and controls

Automatic control of pool water to and from the collector array is accomplished by using a factory-assembled control panel illustrated in Figure 9.14. This panel incorporates a solar differential controller with associated sensors and motorized gate valves to control the direction of pool water flowing within the system. When a positive temperature differential exists between the solar collectors and the pool water, the controller energizes the gate valves to divert the pool water through the collector array and back into the swimming pool. This flow pattern will continue as long as there is sufficient sunlight available to maintain sufficient temperatures within the solar collectors to heat the pool water. When the temperature differential between the collectors and pool water drops to a predetermined low limit, the flow control valves automatically close, isolating the collector array. In this operating mode, the pool water simply bypasses the solar collectors until such time as there is sufficient solar energy available to heat the pool water.

Manual diverting valves

In addition to automatic controls, a simple manually operated valve can be used by the pool owner to divert pool water through the collectors. This manual diverting valve is illustrated in Figure 9.15. If the pool filter pump is timed to operate only during the hours of

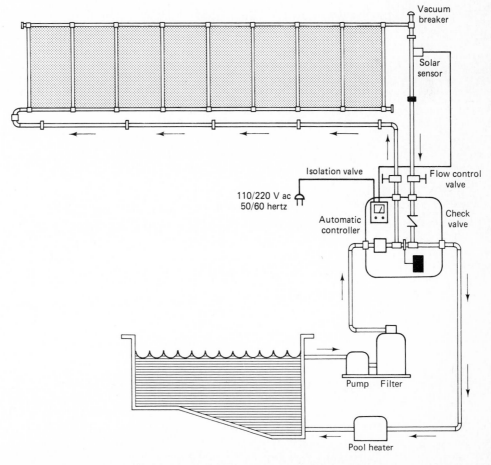

FIGURE 9.13 Open-loop pool heating system. (Courtesy of Sun-Glo Solar Ltd.)

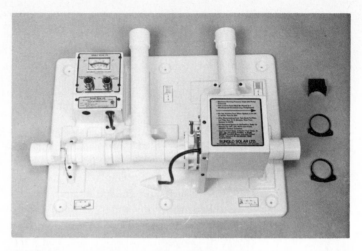

FIGURE 9.14   Automatic control panel for open-loop pool heating systems. (Courtesy of Sun-Glo Solar Ltd.)

strong sunlight, the pool owner need only isolate the collector array on cloudy days. Manual operation in this instance is limited and of little concern.

Prior to system installation, the capacity of the existing circulating pump must be examined carefully to prevent improper flow rate through the pool collectors. Too slow a flow rate through the collectors will result in overheating the collectors, reducing their operating efficiency. Excess pump capacity, on the other hand, results in pressure buildup within the

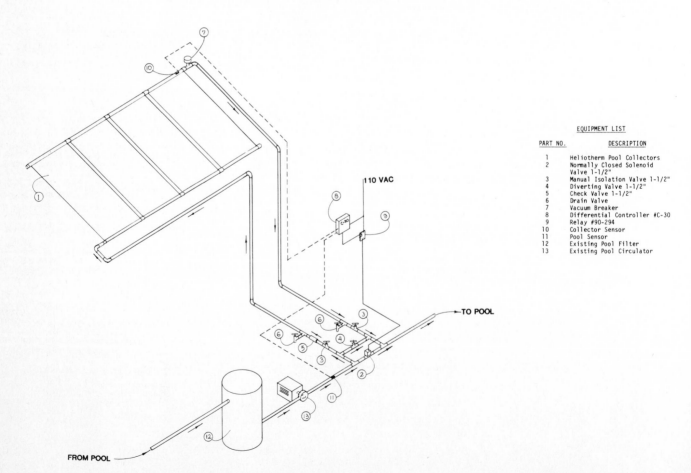

EQUIPMENT LIST

PART NO.	DESCRIPTION
1	Heliotherm Pool Collectors
2	Normally Closed Solenoid Valve 1-1/2"
3	Manual Isolation Valve 1-1/2"
4	Diverting Valve 1-1/2"
5	Check Valve 1-1/2"
6	Drain Valve
7	Vacuum Breaker
8	Differential Controller #C-30
9	Relay #90-294
10	Collector Sensor
11	Pool Sensor
12	Existing Pool Filter
13	Existing Pool Circulator

FIGURE 9.15   Manual diverting valve for open-loop pool heating systems. (Courtesy of Heliotherm.)

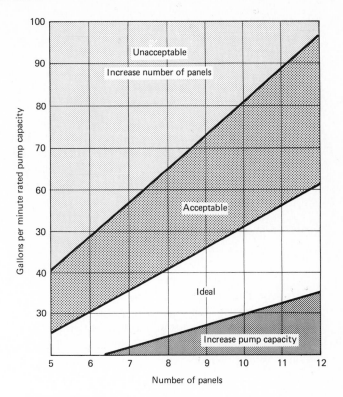

FIGURE 9.16 Proper flow versus number of collectors. (Courtesy of Sun-Glo Solar Ltd.)

collector array, which can cause premature collector failure. Figure 9.16 illustrates proper flow rates for a typical 4 ft $\times$ 10 ft pool collector used in different-sized collector arrays. High efficiency within the collector requires proper flow rate and system pressures.

SYSTEM INSTALLATION PROCEDURES

The installation procedures for solar pool heating systems are similar in many respects to those procedures involved in installing domestic hot water and space heating applications. Open-loop pool systems can be installed quickly and easily, while the installation of closed-loop pool systems is the same as the installation of any closed-loop domestic hot water system. In most instances, only a few standard hand tools are required. Each collector manufacturer has its own set of installation procedures, which should be obtained prior to undertaking system installation.

Site Characteristics

The installation site must be carefully scrutinized prior to making any decisions affecting placement or location of the solar collectors. A successful installation is a combination of knowledgeable installation procedures, the wishes of the pool owner, and common sense. After the collector array has been sized, the number of collectors must be compared with the flow rate of the filter pump to assure a proper flow rate throughout the collector array. A location must then be found that will orient and incline the collectors properly. Also, the height of the collectors above the pool pump should be kept to a minimum to prevent excessive back pressure on the pump and filter system. The collectors normally require at least a 15° inclination angle from horizontal to assure complete drainage.

Where possible, the collectors should be mounted as close to the existing pool pump and filter to reduce pressure drop through the system piping. As in any solar system, shading of the collector array should be avoided. However, in pool systems this factor can become even more critical since with the use of pool collectors that are uninsulated and unglazed, heat loss will take place at an accelerated rate.

Racking Arrangements

Most swimming pool collectors other than conventional flat-plate collectors are not structurally self-supporting but rather, sheets of rubber or copolymer plastic manufactured with integral flow channels in semiflexible states. To prevent damage due to high winds or due to collapsing from excessive weight, collectors must be securely fastened to a supporting surface. If mounting the array on a roof is not possible, a support rack should be constructed as illustrated in Figure 9.17. This type of rack will provide ample collector support and prevent wind damage as well. Note that the structural frame is sheathed with exterior-grade plywood or flakeboard to provide an uninterrupted flat surface behind the collectors. The entire rack should be fabricated from pressure-treated lumber to withstand continuous outdoor exposure. The rack frame should be securely fastened to concrete footings that extend below the frost line. While racks of the type illustrated above add to the cost of system installation, this additional expense is well justified in ensuring successful long-term system operation. Damage to either collectors or property by inadequately designed or constructed support racks are far more costly in the long run than proper design and construction in the initial phases of system installation. Figure 9.18 illustrates a typical installation of swimming pool collectors on an existing flat surface.

Piping Procedures

To assure equal flow rate, hence balanced heating through the individual collectors in the array, reverse return piping must be used when plumbing the system, as illustrated in Figure 9.19. If the collectors were to be piped as illustrated in the upper left arrangement, the tendency would be for the water in the array to short-cycle through the collectors on the right side, causing the collectors farthest to the left to operate at much higher temperatures than those on the right. Uninsulated pool collectors drop sharply in efficiency when their temperature is increased due to poor fluid flow rates. The proper arrangement, shown on the right of Figure 9.19, assures an equal flow rate throughout, eliminating the possibilities of short cycling of the pool water.

A combination air vent/vacuum breaker is required at the high point of the system to facilitate filling and draining of the solar collectors (Figure 9.20). All system piping should

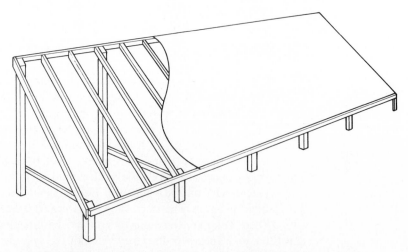

FIGURE 9.17 Racking structure for swimming pool collectors.

FIGURE 9.18 Swimming pool collectors installed on existing roof. (Courtesy of Bio-Energy Systems (Beiscorp).)

be pitched back toward the swimming pool to enable the collectors to drain completely when the system is turned off or bypassed.

During installation, every effort should be made to minimize system pressure drop. For example, sharp elbows can cause significant restriction to flow, so 45° fittings should be used wherever possible, especially for compound angles (Figure 9.21). These procedures will en-

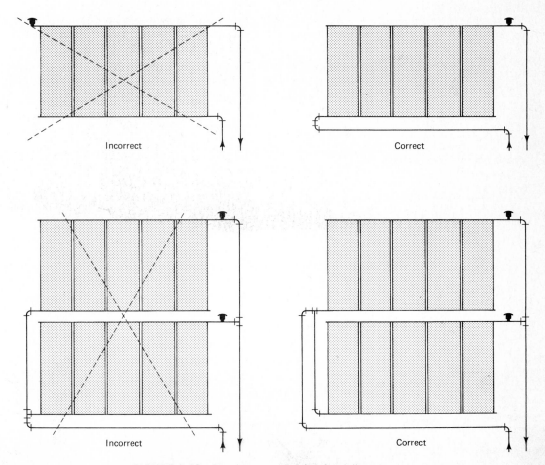

FIGURE 9.19 Reverse return piping for collector array.

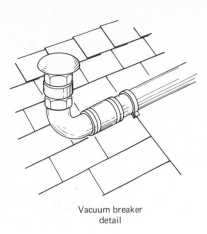

Vacuum breaker
detail

FIGURE 9.20   Vacuum breaker.

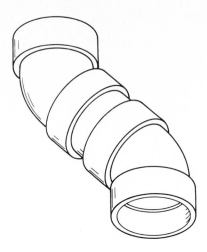

FIGURE 9.21   Compound angles to reduce flow restriction.

able the existing pump and filter system to operate at peak efficiency with a minimum of excess power consumption.

Most piping in a residential system will require the use of white 1½-in.-diameter schedule 40 PVC pipe. If the distance from the filter pump to the collector array exceeds 50 ft, 2-in.-diameter pipe should be installed.

A gate valve should be installed on both collector feed and return pipes. The feed pipe valve is used to isolate the collectors from the pool filter system when necessary. The valve on the collector return pipe is used as a flow control valve to assure that the entire collector array is full of water and operating at maximum efficiency. Installation of the flow control valve is necessary since there is a tendency for pool water to empty from the collectors more quickly than it can be replaced by the pump. During normal operation, this valve is closed only enough to maintain the collector array at full capacity with the maximum flow rate possible. If the return flow control valve is excessively restricted, too much back pressure can damage both the filter pump and the solar collectors.

WEATHERPROOFING AND WINTERIZATION

The same weatherproofing procedures outlined in Chapter 5 must be used for all screw holes and roof penetrations made during collector installation.

Winterization procedures for closed-loop systems are the same as for any antifreeze system: checks of the level of corrosion inhibitors and freeze protection level of the antifreeze at prescribed service intervals. All joints should be checked for integrity, and pipe insulation should be kept in top condition.

When winterizing an open-loop pool system, it is necessary to ensure that the collector array is completely drained. Many system installers provide break-apart unions on the collector feed and return lines which are opened when the pool is taken out of service for the winter. Also, compressed air should be used to blow out all piping to ensure absence of water. End plugs with air-hose adapters that fit tightly into the plastic pipe can be purchased or fabricated to allow sufficient buildup of air pressure within the system to ensure the adequacy of this procedure. Portable compressors rated up to 40 psi are ideal for this purpose.

If automatic control circuitry is employed for water diversion in the system, procedures should be made to protect the controls from the winter weather, or to remove the control panel physically from the system piping for indoor storage.

Following these procedures, a solar pool heating system will give many years of cost-effective and reliable service.

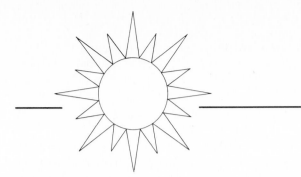

10
Energy
Conservation
Techniques
for the Homeowner

A thoughtful program of energy conservation is an integral part of any successful solar system installation, and one of the most cost-effective efforts a homeowner can make. The primary objective of any energy conservation scheme is to reduce energy consumption while maximizing energy benefits. Such practices reduce the size and hence the investment required for a solar system to perform well. For example, an average well-insulated house can offer heating loads up to 30% lower than a similar house with little or no insulation.

Thus a solar system for the well-insulated home will be designed for a heating load 30% lower than its noninsulated counterpart, resulting in less expense. Additionally, insulation, caulking, and weatherstripping are far less expensive than the purchase of a solar system capable of providing larger heating loads than would otherwise be necessary. Implementation of a well-designed energy conservation plan requires little in the way of formal education. Rather, it requires that the homeowner carefully examine all energy-consuming appliances and associated systems, and determine ways in which each of these can be utilized more efficiently and effectively. Energy conservation techniques have been divided into the following categories: insulation, plumbing and heating, electrical consumption, and caulking and weather stripping.

INSULATION

Insulation refers to any material that restricts or prevents the flow of heat from one surface or area to another. For information relating to the principals of heat transfer and the R values of basic insulating materials, the reader is referred to Chapter 1.

The major portion of lost heat in a building is undoubtedly due to inadequate insulation that results in heating and air-conditioning costs that are much higher than need be. One of the most effective investments that can be made by the homeowner is to upgrade existing insulation characteristics. A number of techniques are available to accomplish this procedure.

Walls

Wall cavities should contain a minimum insulating value of R-11. If the existing R values fall below this level, additional insulation can be added in a number of different ways.

Method 1:   Professional contractors can be employed to blow insulation into the walls of the home. This procedure consists of drilling a series of holes along each wall, in between each stud section, and blowing insulation into the wall cavities under pressure until the spaces are filled. Prices for this service vary, and estimates are made on an individual basis. References should be requested from professional insulating contractors to help ascertain prior customer satisfaction.

Method 2:   Do-it-yourself insulation kits are available on a rental basis from home improvement centers and lumberyards for installing blow-in insulation. In most instances, the customer purchases the insulating material and rents the equipment necessary for blowing it in the walls. In many instances, this equipment is provided free of charge by the company that supplies the insulating material.

Method 3:   Removal of interior walls or exterior building sheathing to install the insulation material. This method is usually undertaken only if the building is to be renovated, since most situations would preclude the removal of the interior walls. When this procedure is employed, fiberglass batts or rigid insulating batts are commonly utilized.

There are many different types of insulating materials available. Some of these have been shown to pose varying degrees of health hazards. The purchaser should thoroughly investigate the type of material to be used prior to undertaking the insulating job. In certain localities, building inspectors should be consulted to determine recommended materials.

Ceilings

The ceiling in the top level of the home, directly below the attic, is the area of most concern. This area should be insulated to a minimum *R* value of 19, since so much heat in the home is lost through this barrier, and inadequate insulation at this point so profoundly increases overall heat loss. Two common types of installation techniques are employed in this area:

Method 1:   Blown-in insulation. This method is similar to the method described for blowing insulation into wall cavities, and utilizes the same equipment. One major difference, however, is that rather than drilling holes through the sheathing of the building, all holes are drilled in the floor boards of the attic, and the insulation is blown in from the sides toward the center.

Method 2:   Removal of floor boards. After the floor boards have been removed, loose mica pellets, sold under the generic name of "vermiculite," are poured in between the joists and leveled with a trowel. This type of insulation is sold in bags that range in volume from 2 to 6 ft^3. After all the insulation has been installed, the floor boards can easily be replaced. In addition to loose fill, fiberglass batts can be inserted between the floor joists with equally satisfactory results.

Foundation Walls

The foundation walls of most homes and commercial buildings are rarely insulated properly. If slab construction has been employed in a building without a basement area, the foundation walls are those which become the first story of the house. These walls should be insulated as exterior side walls. If, on the other hand, the foundation walls of the home form the basement area and are uninsulated, significant heat loss in the lower level of the structure results. This condition accounts for significant amounts of air infiltration, in addition to conducted heat loss during the winter months. Interior masonry walls can be insulated by adhering furring strips to the interior wall surface and placing insulation material between the furring strips.

Since these walls are sometimes subject to periods of high moisture, insulating materials such as Styrofoam, rigid board, or other moisture-resistant approved and fire-rated insulation is recommended. Table 10.1 lists the most common types of insulating materials

Table 10.1 PROPERTIES OF BUILDING INSULATION

Property	Material					
	Cellulose	Vermiculite	Fiberglass	Rock Wool	Urethane Foam	Urea Formaldehyde
Density	2.2–3.0	4–10	0.6–1.0	1.5–2.5	2.0	0.6–1.0
Thermal conductivity (K)	0.27–0.31	0.30–0.41	Varies	0.27–0.31	0.13–0.17	0.23
R value 1 in. at 75°F	3.2–3.7	2.4–3.0	3.16	3.2–3.7	5.8–7.7	4.3
Water absorption (wt %)	5–20	None	1%	2	Negligible	1.6 (volume)
Fire resistance	Combustible	Non-combustible	Non-combustible	Non-combustible	Combustible	Combustibility varies
Flame spread	15–40	None	15–20	15	30–50	15
Fuel contribution	0–40	None	5–15	15	10–25	15
Smoke developed	0–45	None	0–20	None	155–500	0
Toxicity	CO	None	Of binder	None	CO	Varies
Effect of Age						
Dimensional stability	Settles to 20%	None	None	None	0–12% change	Shrinkage varies
Thermal performance	Unknown	None	None	None	0.17 aged	Unknown
Fire resistance	Varies	None	None	None	None	Unknown
Degradation due to:						
Temperature	Unknown	Not below 1200°F	Not below 180°F	None	Above 250°F	Unknown
Cycling	Unknown	None	None	None	Unknown	Unknown
Animal	Unknown	None	None	None	None	None
Moisture	Not severe	None	None	Transient	Limited information	Unknown
Fungal/bacterial	May support	None	None	None	None	None
Weathering	Unknown	None	None	None	None	None
Corrosiveness	Possible	None	None	None	None	Unknown
Odor	None	None	None	None	None	Varies

SOURCE: Adapted from The Insulator's Estimating Handbook, *Frank R. Walker Publishing Co. Chicago, 1980.*

together with their prominent characteristics. Local building codes should be consulted prior to undertaking any purchases to determine if restrictions concerning the use of specific insulation materials exist.

Crawl Spaces and Closet Walls

Crawl spaces and walls behind closets are sometimes uninsulated due to their inaccessibility. Walls behind closets can often be insulated in a manner similar to insulating exterior building walls, where small holes are drilled in between interior stud sections and the insulation injected into the wall cavities under pressure.

Crawl spaces are one of the most difficult areas in a home to insulate properly. Due to low head room and unfinished and damp ground surfaces, working in and around crawl spaces can be difficult. However, if the nature of these working conditions is taken into account when planning the job, the insulation can be satisfactorily installed. The first step is to install a vapor barrier directly below the flooring surface. Since the adhering surface and built-in vapor barrier of the insulation faces away from the living area rather than toward it in a crawl space, a layer of polyethylene sheeting is usually employed. After the vapor barrier has been fastened in place, fiberglass batt insulation is fastened underneath the polyethylene and the butting joints are sealed with duct tape or similar moisture resistant-adhesive to prevent air infiltration between seams. This results in an effective insulated seal below the living space and will prevent condensation of moisture from coming into contact with the wood flooring above.

Windows

Windows are often overlooked when it comes to upgrading existing insulation within the home. However, windows account for relatively large heat losses and should be prime candidates in any energy conservation scheme. One of three techniques for accomplishing this are employed: replacement of old, drafty window units with double- or triple-glazed replacement units; addition of interior or exterior storm windows to existing window units; or the installation of movable insulating window shades. Any of these procedures will effectively reduce heat loss and air infiltration through the window area.

Replacement units

If the window sashes are drafty or deteriorated but the frame of the window is in good condition, the window sashes can be replaced while leaving the original frame intact. In this type of upgrading, the homeowner pays only for the new sash and jamb liners, not for the whole window. This type of assembly is illustrated in Figure 10.1.

If the existing windows are old and in a general state of disrepair, installation of new double- or triple-glazed windows might be necessary. This is economically justifiable if the condition of the existing windows is such that upgrading them is not feasible due to advanced deterioration. Double- or triple-glazed replacement units, such as illustrated in Figure 10.2, are recommended over the single-glazed variety. The additional expense involved for the extra layer or two of glazing will pay for itself many times over in reduced heat loss over the lifetime of the window unit.

Storm windows

If the existing window is to be retained, a cost-effective measure for reducing the heat loss through the window is to install either an interior or exterior storm window. Both interior and exterior storm windows are easily installed. The exterior unit fits over the outside of

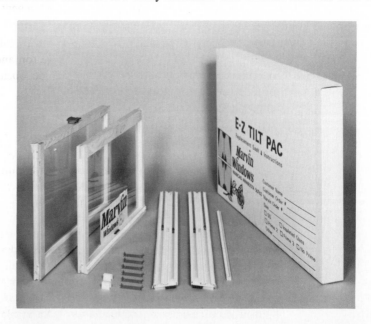

FIGURE 10.1   Sash replacement kits. (Courtesy of Marvin Windows.)

FIGURE 10.2 Double- and triple-glazed replacement window units. (Courtesy of Marvin Windows.)

the window frame and it should be sealed with caulking prior to installation with noncorrosive fasteners. Most of these windows have top and bottom levels of movable glass together with a screen, so that the window can be opened during the summer months with the screened insect barrier in place.

An interior storm window (Figure 10.3) is available with a magnetic seal to hold it in place. The storm window is cut to size, and incorporates steel strips along the perimeter of the glazing. Magnetic strips are adhered around the window frame. To install, the window is merely snapped into place. The magnetic seals of these storm windows are extremely tight, which result in almost totally eliminating air infiltration. The glazing used in most of these windows is ultraviolet-light-stabilized acrylic plastic.

Thermal insulating shades

Another highly effective method used to increase the insulating value of a typical window is the use of movable insulated window shades. Fabricated from a flexible insulating material, these shades are similar in appearance and operating characteristics to ordinary window shades. R values of these shades are in excess of 4.0 when combined with double-glazed windows or combinations of standard and storm windows. Thus they are highly effective in reducing window heat losses. An insulated window shade is illustrated in Figure 10.4.

FIGURE 10.3 Magnetic interior storm window. (Courtesy of Magnetite Corp.)

FIGURE 10.4 Insulating window shades. (Courtesy of Appropriate Technology Corp.)

Weather Stripping and Caulking

Weather stripping and caulking are two techniques that have been developed to reduce air infiltration through cracks in the building structure. As a home "settles" over the years, once airtight doors and windows develop cracks. Also, the framing members of the structure begin to settle, resulting in cracks in the exterior sheathing and sill plates, leading to increased air infiltration over the years.

Controlling air infiltration is inexpensive and easily accomplished by the use of either weather stripping or caulking compounds. Locating areas of air infiltration can be simplified by the use of solid-state electronic devices which are extremely sensitive to temperature changes (Figure 10.5).

A distinction exists between weather stripping and caulking: Weather stripping employs techniques and materials designed to stop leaks between two moving surfaces; caulking involves materials and techniques designed to stop air leaks between nonmoving surfaces.

Weather Stripping

The type of weather stripping to be used in a particular situation is determined by the type of joint that is causing the air leak and is classified either as a compression joint or a friction joint.

Compression joints

Compression joint air leaks are due, for example, to doors that do not close completely or windows that do not seal properly. To remedy this situation there are a variety of weather stripping materials available (Figure 10.6). Most of these materials feature a self-adhesive or mechanically fastened strip of foam, felt, or rubber beading. Foam strips work well on uneven surfaces since the foam can be compressed to form a uniform seal around windows and doors (Figure 10.7). Caution should be exercised in weather stripping a compression joint, since excessive amounts of weather stripping can cause warping in the moving part of the joint.

FIGURE 10.5 Electronic air leak detector.
(Courtesy of Controlled Energy Systems Corp.)

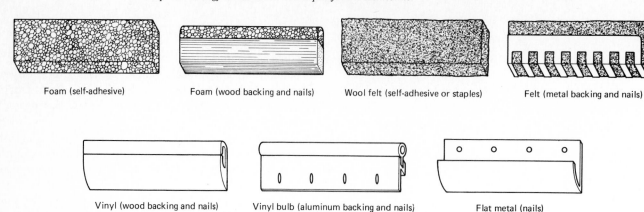

Foam (self-adhesive)   Foam (wood backing and nails)   Wool felt (self-adhesive or staples)   Felt (metal backing and nails)

Vinyl (wood backing and nails)   Vinyl bulb (aluminum backing and nails)   Flat metal (nails)

FIGURE 10.6   Weather stripping materials.

In situations where the joint is a combination of compression in addition to possessing some friction characteristics, vinyl bead strips can be mechanically fastened to the stationary member of the joint for an effective seal (Figure 10.8).

Friction joints

When the joint exhibits a maximum amount of friction, as in a window frame, the weather stripping material must be securely fastened to the stationary window frame and should be made from corrosion-resistant spring-type metal strips (Figure 10.9). In many instances, the backing material of the weather stripping is just as important as the weather stripping itself. For example, felt strips are ideal for narrow, uniform gaps such as those found in most doors of the home. However, if the felt strip is installed in a flexible metal material, it will last much longer than if it is used without any backing material. Vinyl and foam weather stripping are usually embedded in either a wood or a metal strip with slotted holes to aid in installation and pressure adjustment. Given the proper selection of the weather stripping material, the installation should result in a long-lasting and effective seal against air infiltration.

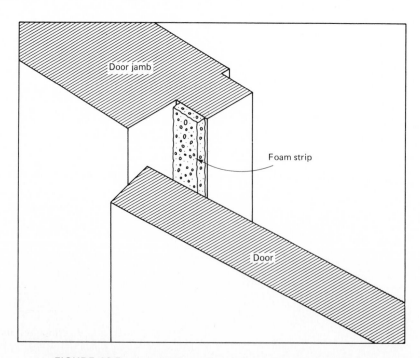

FIGURE 10.7   Use of foam strip on door compression joint.

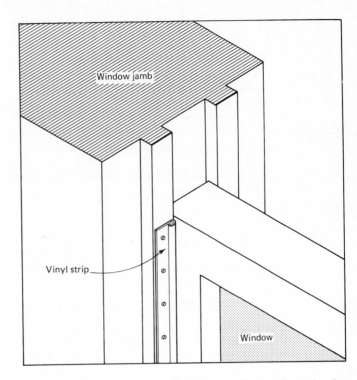

FIGURE 10.8 Use of vinyl bead with mechanical fastening for window joint.

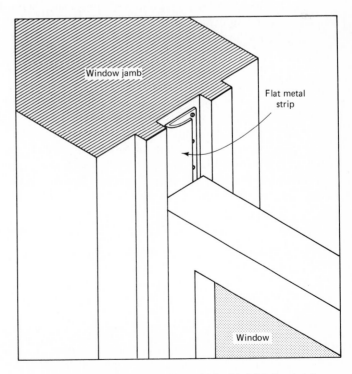

FIGURE 10.9 Metal spring strip for high-friction joints.

Table 10.2   CAULKING MATERIALS

Caulking Materials	Life Span/Performance
Silicone rubber, acrylic latex, urethane foam[a]	10–20 years/excellent
Butyl rubber, latex[b]	3–10 years/good
Oil-based caulks[c]	2–4 years/fair

[a]*Highly flexible; requires solvents for cleaning; cannot be easily painted.*

[b]*Flexible; good adherence to most surfaces; can be painted; requires solvents for cleaning.*

[c]*Good adherence to most surfaces; does not retain flexibility when dry; requires solvents for cleaning.*

Caulking

Caulking materials are used for the repair of cracks and leaks that occur between two nonmoving surfaces. There are a variety of caulking materials on the market, and there are two principal criteria for the selection of the proper one: durability of the bond between the caulked surfaces and the expected lifetime of the caulking material. The characteristics of the most commonly used caulking materials are listed in Table 10.2.

Silicone rubber, acrylic latex, and urethane foam caulks are the highest-quality caulking materials. Highly moisture resistant, silicone caulking can be used to repair leaks in nonpressurized water containers such as drain pipes, shower stalls, and tubs in addition to caulking applications on exterior wood and masonry. Silicone rubber is also a durable adhesive. Acrylic latex and urethane foam caulks weather very well. All these caulks can be painted with either acrylic latex or oil-based house paints.

Butyl rubber and acrylic latex caulks are among the most widely used caulking materials due to their medium price range and good durability characteristics. These caulks will bond to almost any building material but are especially well suited to use in masonry and metal joint applications. They are easily painted and will last approximately 10 years.

Oil-based caulks are the least expensive of the caulking materials. They have a limited life span, generally in the neighborhood of 3 to 4 years, and should not be used to fill cracks greater than ⅜ in. wide.

Most caulks are available in different colors to blend into a wide variety of interior and exterior applications. The acrylic latex and silicone caulks have a strong odor when applied and therefore are used most widely in exterior applications.

The primary rule of thumb for weather stripping and caulking materials is to consider the application prior to purchasing. It is not always necessary to purchase the most expensive material, for even less expensive products will significantly reduce air infiltration and save money if proper installation procedures are followed.

PLUMBING AND HEATING

Many energy conservation techniques related to plumbing and heating have already been discussed as they affect the efficient functioning of fossil-fuel, solid-fuel, and solar energy systems. What follows is a summary of pertinent measures that relate to plumbing and heating appliances in the home.

1. Lowered thermostat settings in unoccupied areas of the home or building can result in significant savings over the course of a heating season. For most effective operation, doors in these areas should be closed to the rest of the house.
2. For maximum performance, room thermostats should be located on interior walls, out of the path of direct sunlight. Avoid placement of thermostats on chimney walls, or over radiators, baseboard heaters, or hot air ducts. The space surrounding the thermostats should be unrestricted to allow for sufficient airflow over the unit.

3. Insulate all hot and cold water piping in the home. Even piping in the heated areas of the home can lose significant amounts of heat if not properly insulated.

4. When using baseboard hot water heating systems, aquastat settings should be lowered during the fall and spring months. Low-limit settings of between 120 and 140°F and high-limit settings between 140 and 160°F will be satisfactory for space-heating purposes prior to the onset of winter. These lowered settings can result in significant savings of fuel.

5. Bottom air openings of baseboard hot water radiation units should not be blocked by carpeting or furniture. Restricting airflow in this manner will significantly reduce the operating efficiency of the heating unit.

6. If the heating boiler also furnishes domestic hot water, boiler water temperature should be kept as low as possible consistent with the hot water requirements of the household.

7. If a water storage tank is used for domestic hot water, it should be well insulated. The use of auxiliary insulation blankets (Figure 10.10) greatly reduces heat loss from the tank, thereby reducing energy input. Also, the tank should be flushed periodically to prevent the buildup of sediment and sludge that can otherwise impair the efficiency of heat transfer.

8. All leaking faucets and water fixtures should be repaired. The use of restricted-flow aerators and shower heads can significantly reduce hot water consumption. The consumption of a standard showerhead is approximately 6 to 8 gallons of water per minute. Assuming a 50/50 mixture of hot and cold water, a 10-minute shower can consume upward of 40 gallons of hot water. Installing a restricted-flow showerhead (Figure 10.11) can reduce total water consumption to between 3 and 5 gallons per minute with a total hot water consumption for the same shower of only 15 gallons.

9. The central heating plant in any building should be tuned on an annual basis to enable the unit to perform as efficiently as possible and to assure the integrity of all peripheral components. Significant advances have taken place in central heating technology during the past several years. If the heating unit is more than 10 years old, an updating and modernization of equipment should be considered. Many states have energy tax credits as well as federal tax incentives for undertaking modernization of such equipment.

10. Equal comfort levels can be achieved at lower temperature levels with the use of a power humidifier. These units are most easily installed in conjunction with forced hot air heating systems, although portable units are available that perform the same function for hot water and steam heating systems.

11. Baseboard heating units should be periodically cleaned to prevent a buildup of dust and lint, which in excessive amounts insulates the radiating fins of the unit, preventing efficient heat transfer.

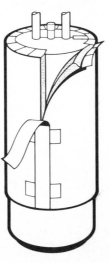

FIGURE 10.10 Add-on water storage tank insulation. (Courtesy of Conserv, Inc.)

FIGURE   10.11 Restricted-flow   showerhead. (Courtesy of Whedon Products.)

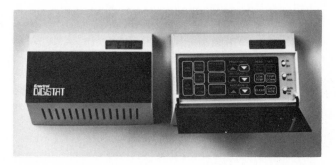

FIGURE 10.12 Solid-state set-back thermostat. (Courtesy of American Stabilis.)

12. Automatic setback thermostats that feature 24-hour timing clocks that allow for lowered thermostat settings at various times during the day or evening can reduce annual energy requirements for space heating by up to 15% (Figure 10.12).
13. All furnace air filters should be changed at regular intervals during the heating season.
14. The use of whole-house cooling fans consume less energy than room or central air-conditioning systems and provide a great deal of comfort. When air-conditioning systems are in use, higher thermostat settings on the unit results in reduced energy consumption.
15. Hot air heating systems can often be used for whole-house cooling by removing the furnace cover over the blower motor when the furnace is located in a cool basement area. Operating in this mode, the furnace will bypass the cold air return ducts and instead pull in cool air from the surrounding basement area and send this air throughout the house.

CONSUMPTION OF ELECTRICAL ENERGY

The use and cost of electrical energy within the home has risen dramatically during the past 30 years. With the advent of relatively inexpensive crude oil during the early nineteenth century, the price of utility-produced electrical power continued to decrease as electrical consumption increased. This trend continued into the 1960s. As utility prices began their dramatic increase due to the rising prices of crude oil and natural gas, energy consumption began to level off during the late 1970s. In spite of this, however, utility bills are presently many times higher than in the recent past, due to kilowatt charges that remain at historically high levels and continue to rise.

The first step in any electrical conservation program is to determine points of consumption and then decide which areas can be reduced, eliminated, or replaced with more efficient appliances. Table 10.3 lists the average electrical consumption, in watts per hour, of a wide variety of standard household appliances. To determine the monthly operating costs for each, multiply the number of hours each month that the appliance is in use by the power consumption of the appliance and divide this figure by 1000. This yields the total number of kilowatthours per month consumed by each appliance. Multiply this last figure by your utility cost per kilowatthour to determine monthly operating costs. It should be noted that the utility cost used in these calculations should include all fuel adjustment charges and applicable local taxes. Fuel adjustment charges can be obtained from either the utility bill or the local utility information service. Using this procedure, let's examine the operating costs for a 100-watt light bulb for a month. If the bulb is in use 8 hours per day, the daily electrical consumption is 800 watt hours. On a monthly basis, this totals 24 kilowatt hours (800 × 30 days ÷ 1,000 = 24 kWh). If the total cost of electricity from the utility company is 9.5 cents/kWh, monthly operating cost for this light bulb is $2.28. On a yearly basis, this figure totals $27.36. If it were possible to substitute a 75-watt bulb for the 100-watt bulb, the yearly operating costs would then be reduced to $20.52, a savings of 25%. If similar savings can be made for other devices, the reduction in annual utility costs can be substantial. Some suggestions for achieving these savings are offered below.

1. Change light bulbs to a lower wattage where reduced illumination is not a hazard.
2. Install draft sealers on all electrical outlet and switch boxes. These kits are available at most home energy centers for little cost and will reduce air infiltration.

Table 10.3 ENERGY REQUIREMENTS OF COMMON HOUSEHOLD ELECTRICAL APPLIANCES[a]

	Average Wattage	Est. kWh Annually		Average Wattage	Est. kWh Annually
Air cleaner	50	200	Radio/record player	100	100
Air conditioner (room)	1,500	1,250	Range with oven	12,500	1,200
Blender	375	15	Range without oven	12,500	1,250
Broiler	1,250	100	Refrigerator (12 ft^3)	250	750
Carving knife	100	5	Frostless 12 ft^3	350	1,200
Clock	2	15	Refrigerator/freezer		
Clothes dryer	4,500	1,000	14 ft^3	350	1,200
Coffee maker	900	100	Frostless 14 ft^3	600	1,800
Deep fryer	1,500	75	Roaster	1,250	200
Dehumidifier	240	375	Sandwich grill	1,000	25
Dishwasher	1,200	350	Sewing machine	75	10
Fan (circulating)	100	45	Shaver	15–25	2
Fan (whole house)	300–500	300	Sunlamp	275	15
Fan (window)	200	175	Television		
Freezer 15 ft^3	350	2,000	Black and white		
Frostless 15 ft^3	450	1,750	tube	150	350
Frying pan	1,200	150	Solid state	50	125
Hair dryer	175	15	Color tube	300	700
Heater (portable)	1,000–2,500	150	Solid state	200	450
Heating pad	35–100	10	Toaster	1,000	35
Heat lamp	250	15	Toothbrush	7–10	1
Hot plate	1,250	75	Trash compactor	400–500	50
Humidifier	175	175	Vacuum cleaner	600	50
Iron	1,000	150	Video recorder	35	50
Mixer	125	15	Waffle iron	1,000–1,200	20
Oven (microwave)	1,500	200	Washing machine	500	100
Radio	75	75–100	Waste diposer	400–600	30
			Water heater	4,500	5,000

[a]*Figures are approximations and may vary from one manufacturer to the other.*

3. Use a clothes dryer only during the cooler hours of the day when possible. Also, deflector kits are available that are installed in the dryer vent pipe to funnel heat from the unit into the basement or utility room during the winter, and allow for normal outdoor exhaust during the summer months.

4. When using a dishwasher, allow the dishes to air-dry rather than using the heating and drying cycle of the machine. The dishes will dry quickly due to the heat of the water, and the electrical consumption of the process will be greatly reduced.

5. Turn electric hot water heater thermostats down to the lowest temperature consistent with adequate hot water requirements. This setting is generally 120°F.

6. Installation of a 24-hour timing switch on the electric hot water heater, available at most hardware and electrical supply houses, can reduce electrical consumption of hot water heaters by almost 40%. The hot water heater will still furnish ample amounts of heated water.

7. The Energy Efficiency Rating (EER) of all electrical appliances should be considered before making a purchase. This rating is determined by dividing the number of Btu delivered by the appliance by its electrical consumption. An EER of 7.5 or more is considered desirable. Thus a unit with an EER of 8.0 will deliver more usable energy at a lower electrical consumption than a corresponding unit with an EER of 6.0.

8. Replace incandescent light fixtures with fluorescent light fixtures where convenient. Fluorescent fixtures use less electricity than conventional incandescent fixtures to deliver equivalent amounts of light.

PASSIVE SOLAR ADDITIONS

The sunroom, an updated version of the passive solar attached greenhouse, is a feature incorporated in many newly constructed houses and as retrofits in others. These room additions come equipped with a variety of features and options that create, in effect, an area in

FIGURE 10.13 Passive solar addition (exterior). (Courtesy of Garden Way, Inc.)

FIGURE 10.14   Passive solar addition (interior). (Courtesy of Garden
Way, Inc.)

the home which not only can be enjoyed on a year-round basis but can supply limited amounts of solar heat to the dwelling as well. Typical of these structures are sun-space additions such as those illustrated in Figures 10.13 and 10.14.

A complete site analysis should be undertaken prior to the installation of any of these kits. Since tax codes vary state by state, inquiries will need to be made for the applicability of any of these structures for existing energy tax credits.

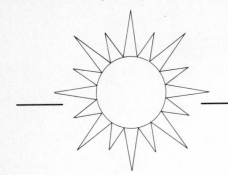

Appendix 1

Units of Measurement and Conversions

Solar Constant:
   428 Btu/hr-ft² = 1350.2 watts/meter

Area:
   1 ft² = 144 in.² = 0.092903 meter²

Volume:
   1 ft³ = 1728 in.³ = 7.48 gallons
   1 gallon = 3.785 liter
   1 barrel, oil = 5.615 ft³ = 42 gallons

Mass:
   1 gallon water = 8.345 pounds = 3.784 kilograms

Temperature:
   F = (C × 1.8) + 32
   C = (F − 32) × 0.55556

Energy:
   1 Btu = 1.05506 kilojoules = 252 calories
   1 kWh = 3413 Btu

Note: For conversion factors and Btu equivalents of common fuels, see Table 1.2.

Appendix 2

Average Temperature
of Selected Cities
in the United States

From U.S. Department of Commerce, National Oceanic and Atmospheric Administration, Environmental Data Service.

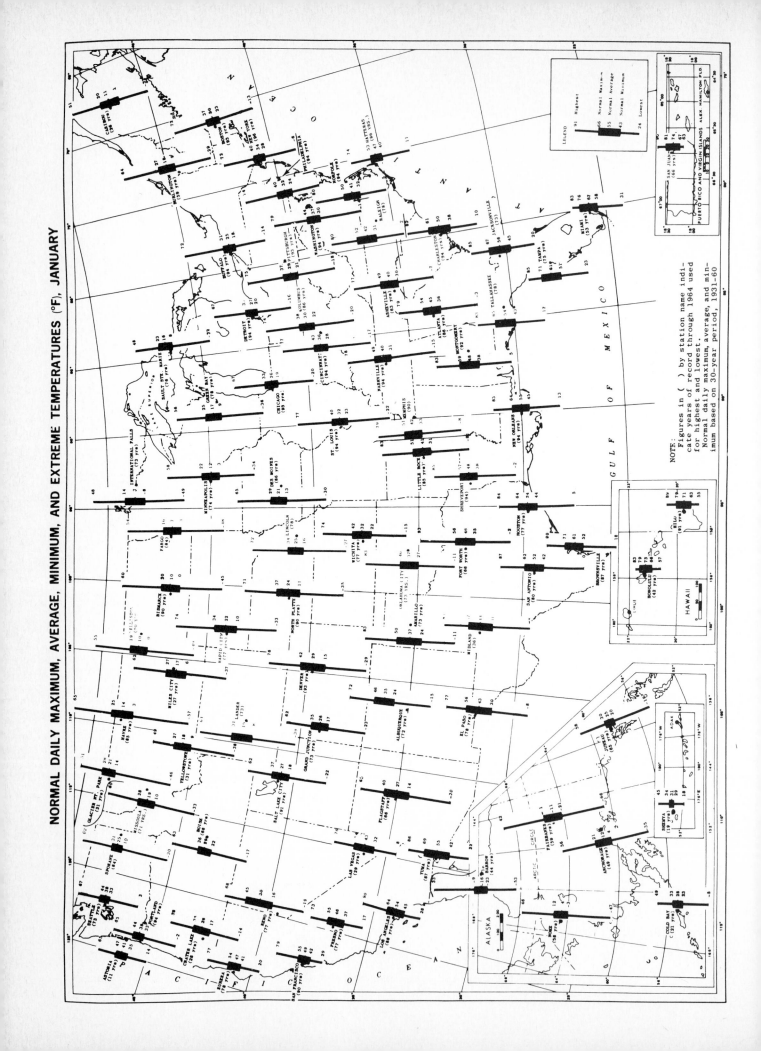

NORMAL DAILY MAXIMUM, AVERAGE, MINIMUM, AND EXTREME TEMPERATURES (°F), JANUARY

NOTE: Figures in ( ) by station name indicate years of record through 1964 used
for highest and lowest.
Normal daily maximum, average, and minimum based on 30-year period, 1931–60

NORMAL DAILY MAXIMUM, AVERAGE, MINIMUM, AND EXTREME TEMPERATURES (°F), FEBRUARY

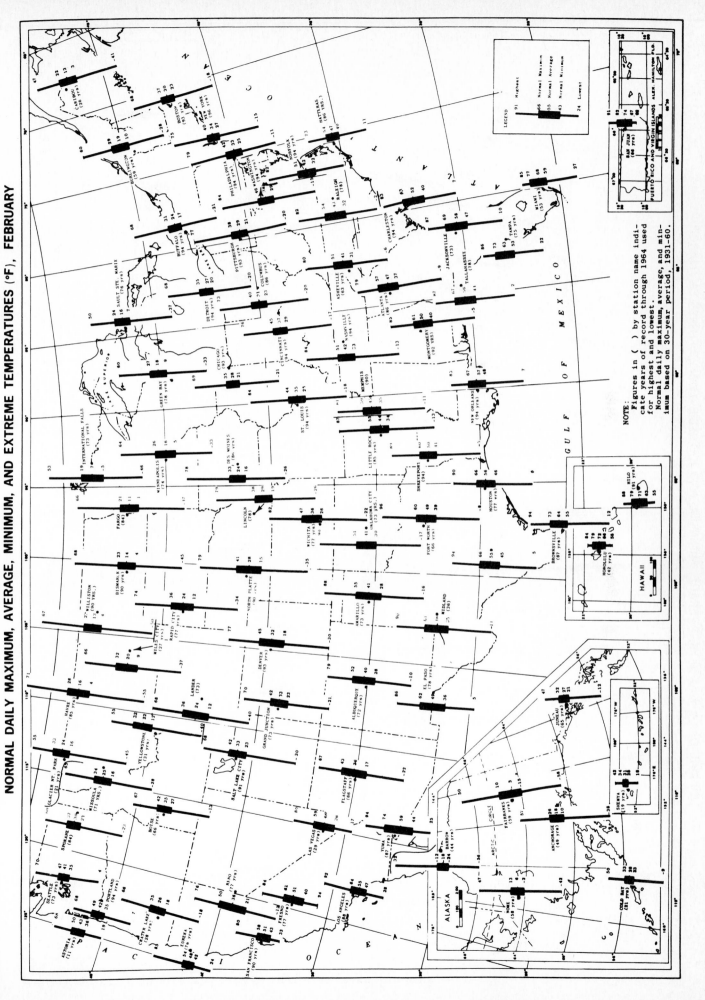

NOTE: Figures in ( ) by station name indicate years of record through 1964 used for highest and lowest. Normal daily maximum, average, and minimum based on 30-year period, 1931-60.

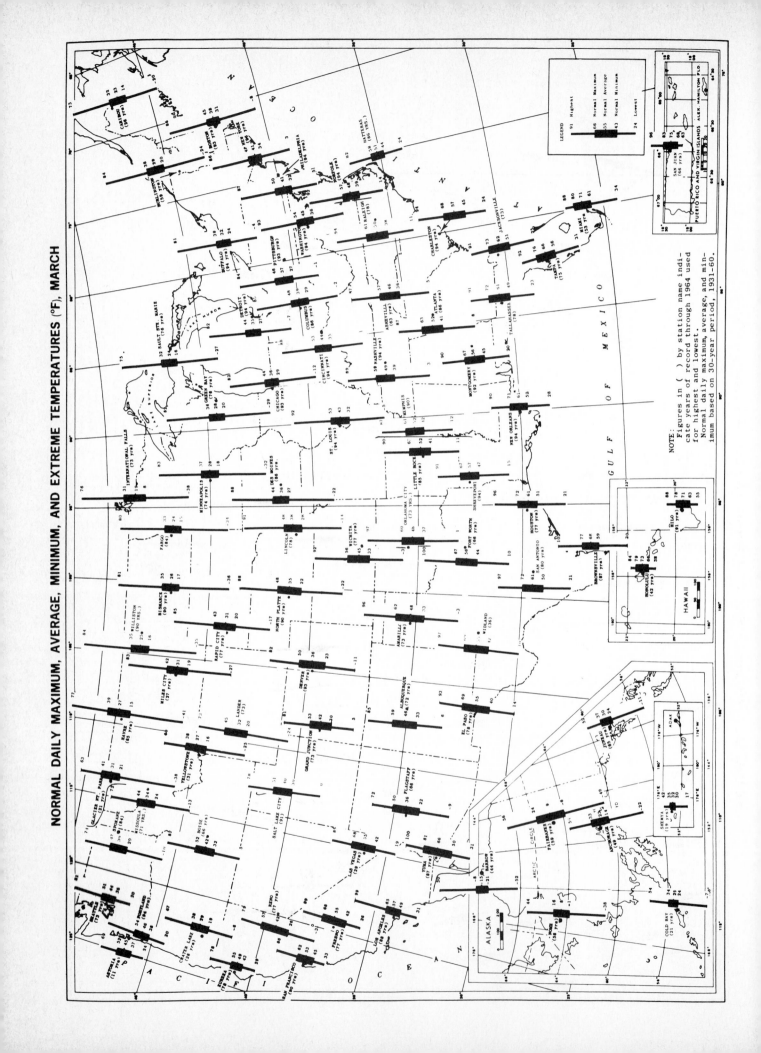

NORMAL DAILY MAXIMUM, AVERAGE, MINIMUM, AND EXTREME TEMPERATURES (°F), MARCH

NOTE: Figures in ( ) by station name indicate years of record through 1964 used for highest and lowest. Normal daily maximum, average, and minimum based on 30-year period, 1931-60.

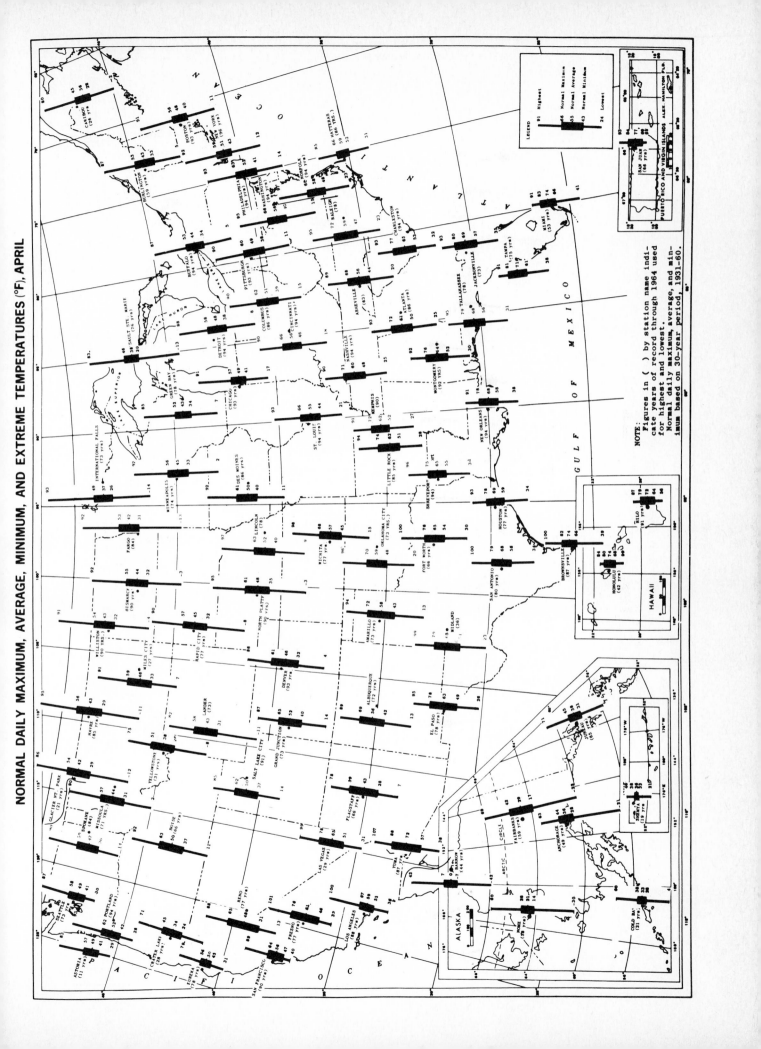

NORMAL DAILY MAXIMUM, AVERAGE, MINIMUM, AND EXTREME TEMPERATURES (°F), APRIL

NORMAL DAILY MAXIMUM, AVERAGE, MINIMUM, AND EXTREME TEMPERATURES (°F), MAY

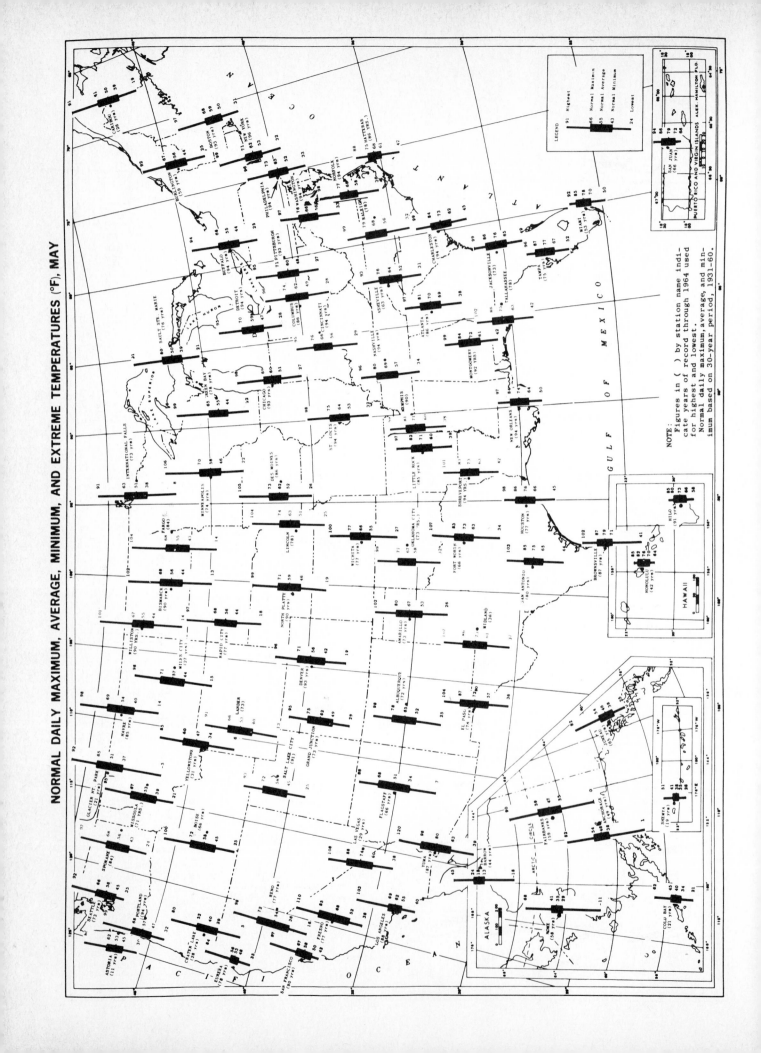

LEGEND

91   Highest
66   Normal Maximum
55   Normal Average
43   Normal Minimum
24   Lowest

NOTE:
Figures in ( ) by station name indicate years of record through 1964 used for highest and lowest. Normal daily maximum, average, and minimum based on 30-year period, 1931–60.

NORMAL DAILY MAXIMUM, AVERAGE, MINIMUM, AND EXTREME TEMPERATURES (°F), JUNE

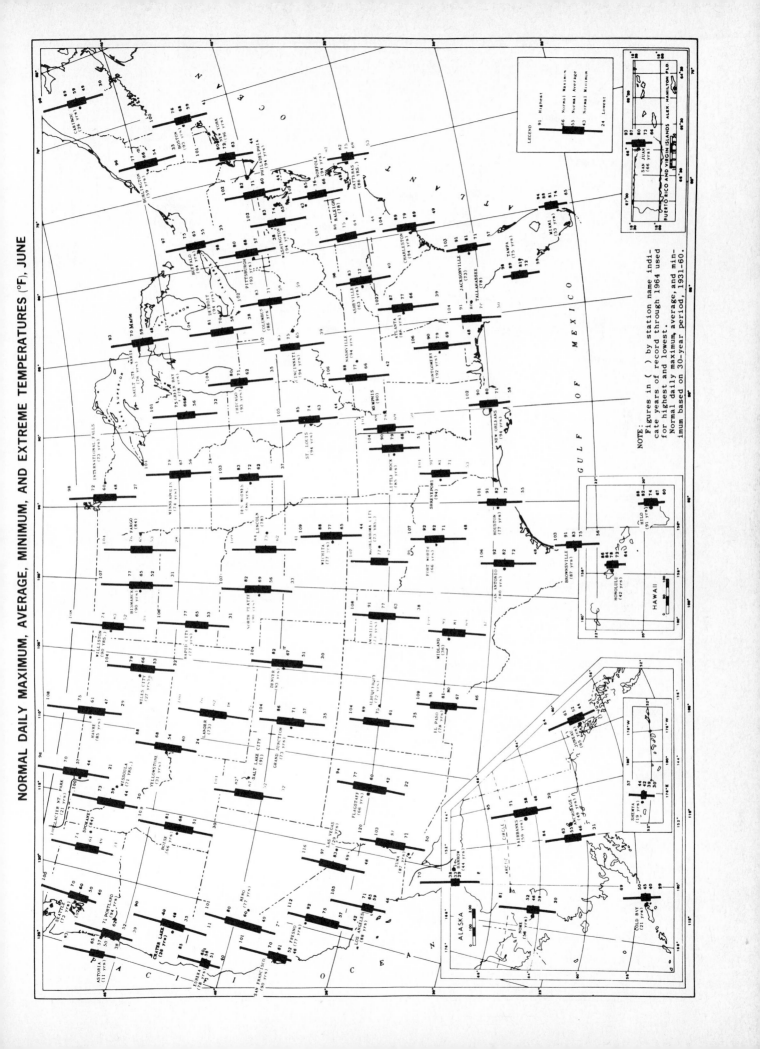

NOTE:

Figures in ( ) by station name indicate years of record through 1964 used for highest and lowest. Normal daily maximum, average, and minimum based on 30-year period, 1931–60.

NORMAL DAILY MAXIMUM, AVERAGE, MINIMUM, AND EXTREME TEMPERATURES (°F), JULY

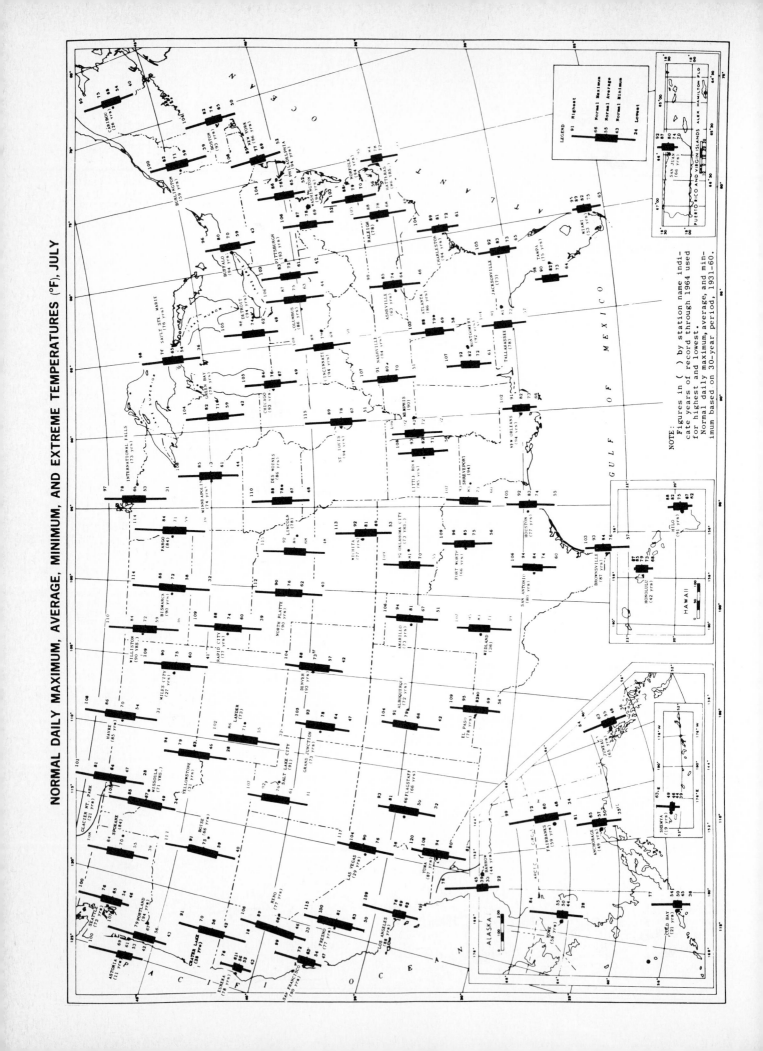

NOTE:
Figures in ( ) by station name indicate years of record through 1964 used
for highest and lowest. Normal daily maximum, average, and min-
imum based on 30-year period, 1931–60.

NORMAL DAILY MAXIMUM, AVERAGE, MINIMUM, AND EXTREME TEMPERATURES (°F), AUGUST

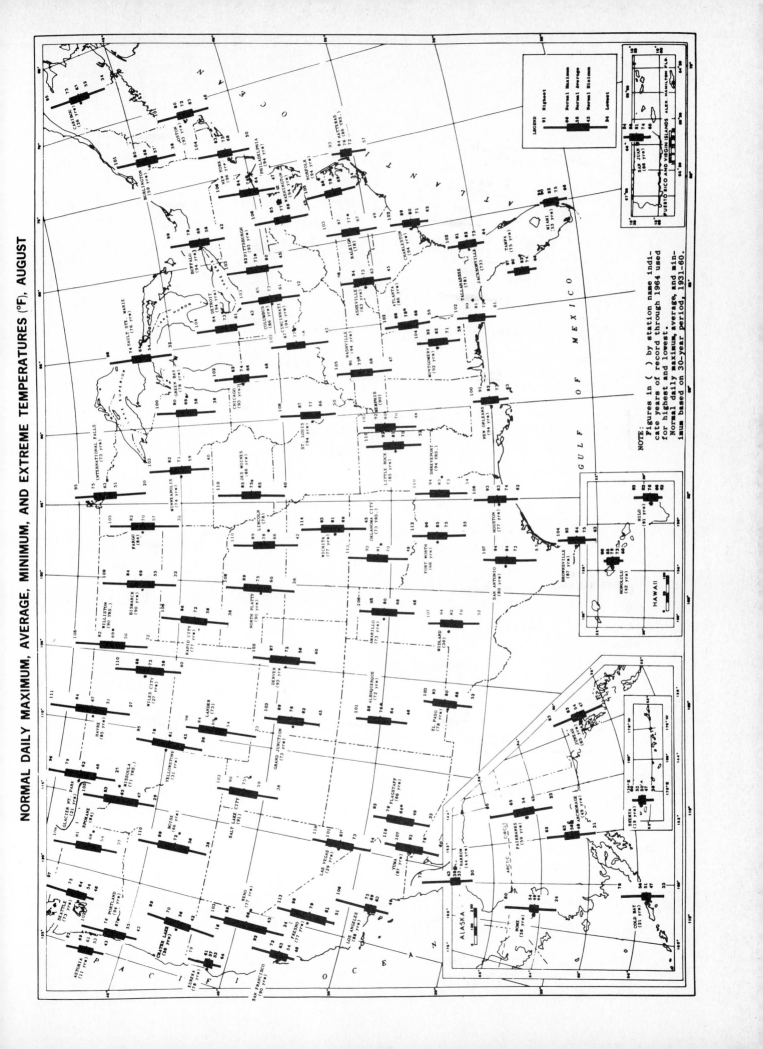

NORMAL DAILY MAXIMUM, AVERAGE, MINIMUM, AND EXTREME TEMPERATURES (°F), SEPTEMBER

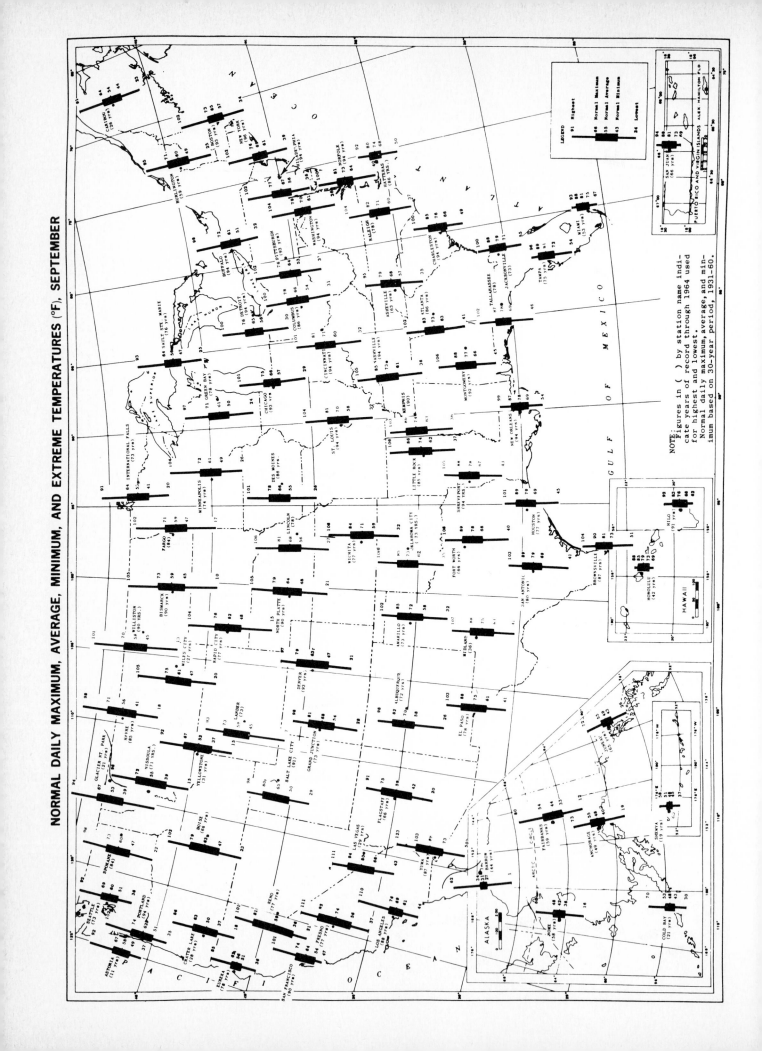

NOTE: Figures in ( ) by station name indicate years of record through 1964 used for highest and lowest. Normal daily maximum, average, and minimum based on 30-year period, 1931-60.

NORMAL DAILY MAXIMUM, AVERAGE, MINIMUM, AND EXTREME TEMPERATURES (°F), OCTOBER

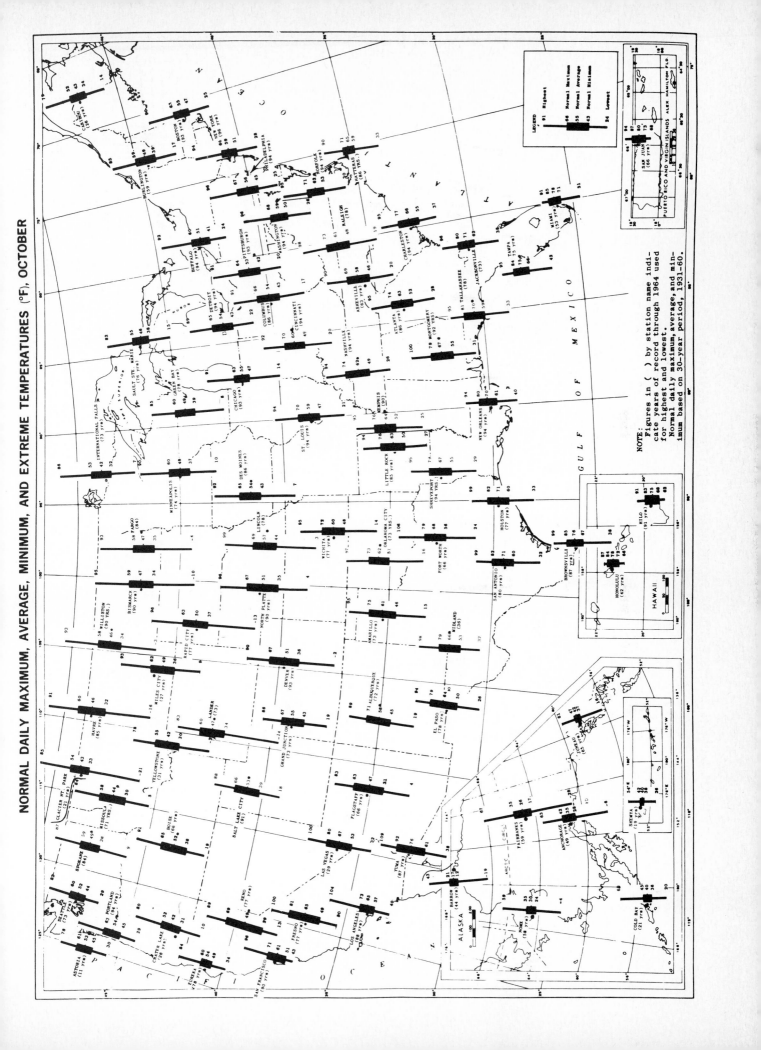

NOTE: Figures in ( ) by station name indicate years of record through 1964 used for highest and lowest. Normal daily maximum, average, and minimum based on 30-year period, 1931–60.

NORMAL DAILY MAXIMUM, AVERAGE, MINIMUM, AND EXTREME TEMPERATURES (°F), NOVEMBER

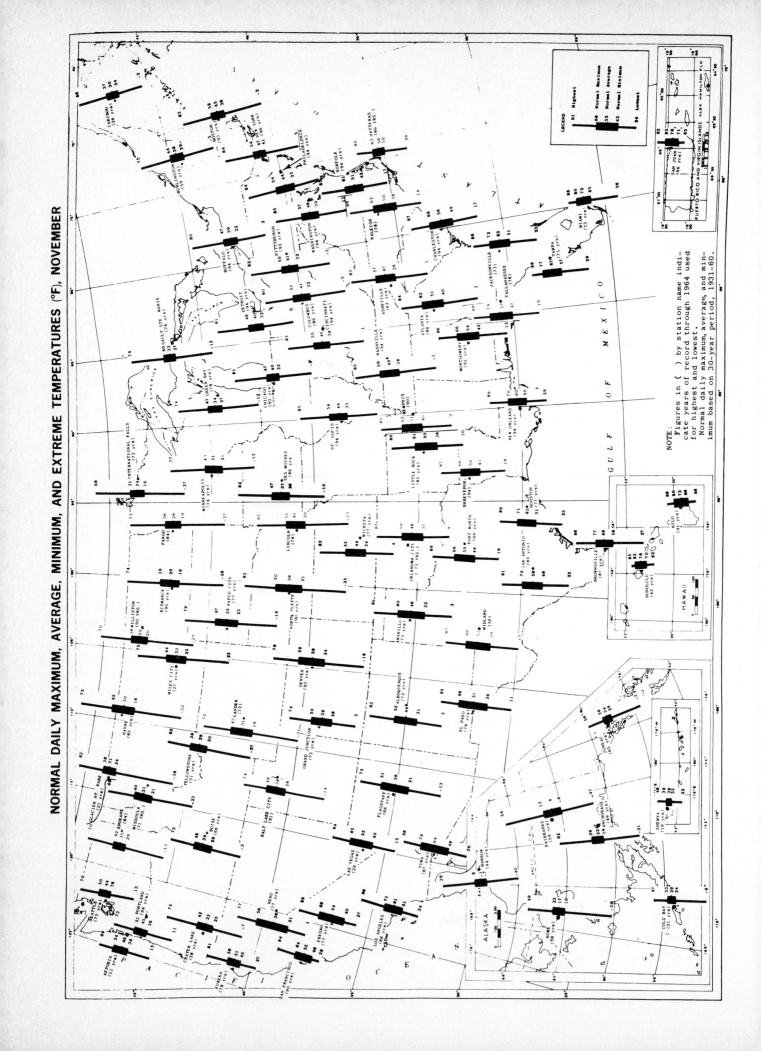

NOTE:
Figures in ( ) by station name indicate years of record through 1964 used for highest and lowest. Normal daily maximum, average, and minimum based on 30-year period, 1931-60.

NORMAL DAILY MAXIMUM, AVERAGE, MINIMUM, AND EXTREME TEMPERATURES (°F), DECEMBER

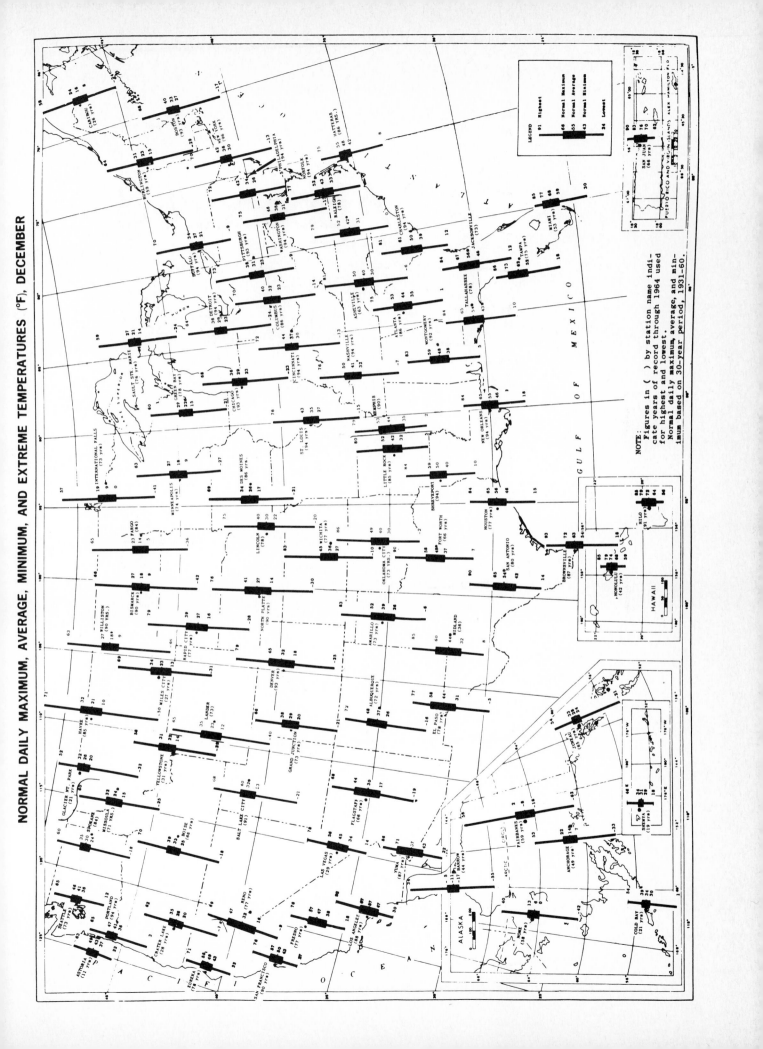

NOTE: Figures in ( ) by station name indicate years of record through 1964 used for highest and lowest.
Normal daily maximum, average, and minimum based on 30-year period, 1931-60.

Appendix 3 —

Degree–Days
of Selected Cities
in the United States

From 1981 Ashrae Fundamentals Handbook.

AVERAGE MONTHLY AND YEARLY DEGREE DAYS FOR CITIES IN THE UNITED STATES AND CANADA[a,b,c] (BASE 65 F)

State	Station	Avg. Winter Temp[d]	July	Aug.	Sept.	Oct.	Nov.	Dec.	Jan.	Feb.	Mar.	Apr.	May	June	Yearly Total
Ala.	Birmingham .... A	54.2	0	0	6	93	363	555	592	462	363	108	9	0	2551
	Huntsville .... A	51.3	0	0	12	127	426	663	694	557	434	138	19	0	3070
	Mobile .... A	59.9	0	0	0	22	213	357	415	300	211	42	0	0	1560
	Montgomery .... A	55.4	0	0	0	68	330	527	543	417	316	90	0	0	2291
Alaska	Anchorage .... A	23.0	245	291	516	930	1284	1572	1631	1316	1293	879	592	315	10864
	Fairbanks .... A	6.7	171	332	642	1203	1833	2254	2359	1901	1739	1068	555	222	14279
	Juneau .... A	32.1	301	338	483	725	921	1135	1237	1070	1073	810	601	381	9075
	Nome .... A	13.1	481	496	693	1094	1455	1820	1879	1666	1770	1314	930	573	14171
Ariz.	Flagstaff .... A	35.6	46	68	201	558	867	1073	1169	991	911	651	437	180	7152
	Phoenix .... A	58.5	0	0	0	22	234	415	474	328	217	75	0	0	1765
	Tucson .... A	58.1	0	0	0	25	231	406	471	344	242	75	6	0	1800
	Winslow .... A	43.0	0	0	6	245	711	1008	1054	770	601	291	96	0	4782
	Yuma .... A	64.2	0	0	0	0	108	264	307	190	90	15	0	0	974
Ark.	Fort Smith .... A	50.3	0	0	12	127	450	704	781	596	456	144	22	0	3292
	Little Rock .... A	50.5	0	0	9	127	465	716	756	577	434	126	9	0	3219
	Texarkana .... A	54.2	0	0	0	78	345	561	626	468	350	105	0	0	2533
Calif.	Bakersfield .... A	55.4	0	0	0	37	282	502	546	364	267	105	19	0	2122
	Bishop .... A	46.0	0	0	48	260	576	797	874	680	555	306	143	36	4275
	Blue Canyon .... A	42.2	28	37	108	347	594	781	896	795	806	597	412	195	5596
	Burbank .... A	58.6	0	0	6	43	177	301	366	277	239	138	81	18	1646
	Eureka C	49.9	270	257	258	329	414	499	546	470	505	438	372	285	4643
	Fresno .... A	53.3	0	0	0	84	354	577	605	426	335	162	62	6	2611
	Long Beach .... A	57.8	0	0	9	47	171	316	397	311	264	171	93	24	1803
	Los Angeles .... A	57.4	28	28	42	78	180	291	372	302	288	219	158	81	2061
	Los Angeles C	60.3	0	0	6	31	132	229	310	230	202	123	68	18	1349
	Mt. Shasta C	41.2	25	34	123	406	696	902	983	784	738	525	347	159	5722
	Oakland .... A	53.5	53	50	45	127	309	481	527	400	353	255	180	90	2870
	Red Bluff .... A	53.8	0	0	0	53	318	555	605	428	341	168	47	0	2515
	Sacramento .... A	53.9	0	0	0	56	321	546	583	414	332	178	72	0	2502
	Sacramento C	54.4	0	0	0	62	312	533	561	392	310	173	76	0	2419
	Sandberg C	46.8	0	0	30	202	480	691	778	661	620	426	264	57	4209
	San Diego .... A	59.5	9	0	21	43	135	236	298	235	214	135	90	42	1458
	San Francisco .... A	53.4	81	78	60	143	306	462	508	395	363	279	214	126	3015
	San Francisco C	55.1	192	174	102	118	231	388	443	336	319	279	239	180	3001
	Santa Maria .... A	54.3	99	93	96	146	270	391	459	370	363	282	233	165	2967
Colo.	Alamosa .... A	29.7	65	99	279	639	1065	1420	1476	1162	1020	696	440	168	8529
	Colorado Springs .... A	37.3	9	25	132	456	825	1032	1128	938	893	582	319	84	6423
	Denver .... A	37.6	6	9	117	428	819	1035	1132	938	887	558	288	66	6283
	Denver C	40.8	0	0	90	366	714	905	1004	851	800	492	254	48	5524
	Grand Junction .... A	39.3	0	0	30	313	786	1113	1209	907	729	387	146	21	5641
	Pueblo .... A	40.4	0	0	54	326	750	986	1085	871	772	429	174	15	5462
Conn.	Bridgeport .... A	39.9	0	0	66	307	615	986	1079	966	853	510	208	27	5617
	Hartford .... A	37.3	0	12	117	394	714	1101	1190	1042	908	519	205	33	6235
	New Haven .... A	39.0	0	12	87	347	648	1011	1097	991	871	543	245	45	5897
Del.	Wilmington .... A	42.5	0	0	51	270	588	927	980	874	735	387	112	6	4930
D.C.	Washington .... A	45.7	0	0	33	217	519	834	871	762	626	288	74	0	4224
Fla.	Apalachicola C	61.2	0	0	0	16	153	319	347	260	180	33	0	0	1308
	Daytona Beach .... A	64.5	0	0	0	0	75	211	248	190	140	15	0	0	879
	Fort Myers .... A	68.6	0	0	0	0	24	109	146	101	62	0	0	0	442
	Jacksonville .... A	61.9	0	0	0	12	144	310	332	246	174	21	0	0	1239
	Key West .... A	73.1	0	0	0	0	0	28	40	31	9	0	0	0	108
	Lakeland C	66.7	0	0	0	0	57	164	195	146	99	0	0	0	661
	Miami .... A	71.1	0	0	0	0	0	65	74	56	19	0	0	0	214

[a] Data for United States cities from a publication of the United States Weather Bureau, *Monthly Normals of Temperature, Precipitation and Heating Degree Days,* 1962, are for the period 1931 to 1960 inclusive. These data also include information from the 1963 revisions to this publication, where available.
[b] Data for airport stations, A, and city stations, C, are both given where available.
[c] Data for Canadian cities were computed by the Climatology Division, Department of Transport from normal monthly mean temperatures, and the monthly values of heating degree days data were obtained using the National Research Council computer and a method devised by H. C. S. Thom of the United States Weather Bureau. The heating degree days are based on the period from 1931 to 1960.
[d] For period October to April, inclusive.

Appendix 3 Degree-Days of Selected Cities in the United States

AVERAGE MONTHLY AND YEARLY DEGREE DAYS FOR CITIES
IN THE UNITED STATES AND CANADA (BASE 65 F)

State	Station	Avg. Winter Temp[d]	July	Aug.	Sept.	Oct.	Nov.	Dec.	Jan.	Feb.	Mar.	Apr.	May	June	Yearly Total
Fla.	Miami Beach C	72.5	0	0	0	0	0	40	56	36	9	0	0	0	141
(Cont'd)	Orlando .... A	65.7	0	0	0	0	72	198	220	165	105	6	0	0	766
	Pensacola .... A	60.4	0	0	0	19	195	353	400	277	183	36	0	0	1463
	Tallahassee .... A	60.1	0	0	0	28	198	360	375	286	202	36	0	0	1485
	Tampa............................ A	66.4	0	0	0	0	60	171	202	148	102	0	0	0	683
	West Palm Beach.......... A	68.4	0	0	0	0	6	65	87	64	31	0	0	0	253
Ga.	Athens .... A	51.8	0	0	12	115	405	632	642	529	431	141	22	0	2929
	Atlanta .... A	51.7	0	0	18	124	417	648	636	518	428	147	25	0	2961
	Augusta .... A	54.5	0	0	0	78	333	552	549	445	350	90	0	0	2397
	Columbus .... A	54.8	0	0	0	87	333	543	552	434	338	96	0	0	2383
	Macon............................ A	56.2	0	0	0	71	297	502	505	403	295	63	0	0	2136
	Rome .... A	49.9	0	0	24	161	474	701	710	577	468	177	34	0	3326
	Savannah .... A	57.8	0	0	0	47	246	437	437	353	254	45	0	0	1819
	Thomasville C	60.0	0	0	0	25	198	366	394	305	208	33	0	0	1529
Hawaii	Lihue.............................. A	72.7	0	0	0	0	0	0	0	0	0	0	0	0	0
	Honolulu .... A	74.2	0	0	0	0	0	0	0	0	0	0	0	0	0
	Hilo................................ A	71.9	0	0	0	0	0	0	0	0	0	0	0	0	0
Idaho	Boise .... A	39.7	0	0	132	415	792	1017	1113	854	722	438	245	81	5809
	Lewiston......................... A	41.0	0	0	123	403	756	933	1063	815	694	426	239	90	5542
	Pocatello........................ A	34.8	0	0	172	493	900	1166	1324	1058	905	555	319	141	7033
Ill.	Cairo C	47.9	0	0	36	164	513	791	856	680	539	195	47	0	3821
	Chicago (O'Hare).......... A	35.8	0	12	117	381	807	1166	1265	1086	939	534	260	72	6639
	Chicago (Midway) .... A	37.5	0	0	81	326	753	1113	1209	1044	890	480	211	48	6155
	Chicago.......................... C	38.9	0	0	66	279	705	1051	1150	1000	868	489	226	48	5882
	Moline .... A	36.4	0	9	99	335	774	1181	1314	1100	918	450	189	39	6408
	Peoria .... A	38.1	0	6	87	326	759	1113	1218	1025	849	426	183	33	6025
	Rockford......................... A	34.8	6	9	114	400	837	1221	1333	1137	961	516	236	60	6830
	Springfield...................... A	40.6	0	0	72	291	696	1023	1135	935	769	354	136	18	5429
Ind.	Evansville .... A	45.0	0	0	66	220	606	896	955	767	620	237	68	0	4435
	Fort Wayne .... A	37.3	0	9	105	378	783	1135	1178	1028	890	471	189	39	6205
	Indianapolis .... A	39.6	0	0	90	316	723	1051	1113	949	809	432	177	39	5699
	South Bend .... A	36.6	0	6	111	372	777	1125	1221	1070	933	525	239	60	6439
Iowa	Burlington .... A	37.6	0	0	93	322	768	1135	1259	1042	859	426	177	33	6114
	Des Moines .... A	35.5	0	6	96	363	828	1225	1370	1137	915	438	180	30	6588
	Dubuque......................... A	32.7	12	31	156	450	906	1287	1420	1204	1026	546	260	78	7376
	Sioux City...................... A	34.0	0	9	108	369	867	1240	1435	1198	989	483	214	39	6951
	Waterloo........................ A	32.6	12	19	138	428	909	1296	1460	1221	1023	531	229	54	7320
Kans.	Concordia .... A	40.4	0	0	57	276	705	1023	1163	935	781	372	149	18	5479
	Dodge City..................... A	42.5	0	0	33	251	666	939	1051	840	719	354	124	9	4986
	Goodland .... A	37.8	0	6	81	381	810	1073	1166	955	884	507	236	42	6141
	Topeka .... A	41.7	0	0	57	270	672	980	1122	893	722	330	124	12	5182
	Wichita .... A	44.2	0	0	33	229	618	905	1023	804	645	270	87	6	4620
Ky.	Covington .... A	41.4	0	0	75	291	669	983	1035	893	756	390	149	24	5265
	Lexington .... A	43.8	0	0	54	239	609	902	946	818	685	325	105	0	4683
	Louisville .... A	44.0	0	0	54	248	609	890	930	818	682	315	105	9	4660
La.	Alexandria .... A	57.5	0	0	0	56	273	431	471	361	260	69	0	0	1921
	Baton Rouge .... A	59.8	0	0	0	31	216	369	409	294	208	33	0	0	1560
	Lake Charles .... A	60.5	0	0	0	19	210	341	381	274	195	39	0	0	1459
	New Orleans .... A	61.0	0	0	0	19	192	322	363	258	192	39	0	0	1385
	New Orleans.................. C	61.8	0	0	0	12	165	291	344	241	177	24	0	0	1254
	Shreveport .... A	56.2	0	0	0	47	297	477	552	426	304	81	0	0	2184
Me.	Caribou.......................... A	24.4	78	115	336	682	1044	1535	1690	1470	1308	858	468	183	9767
	Portland .... A	33.0	12	53	195	508	807	1215	1339	1182	1042	675	372	111	7511
Md.	Baltimore .... A	43.7	0	0	48	264	585	905	936	820	679	327	90	0	4654
	Baltimore C	46.2	0	0	27	189	486	806	859	762	629	288	65	0	4111
	Frederick........................ A	42.0	0	0	66	307	624	955	995	876	741	384	127	12	5087
Mass.	Boston............................ A	40.0	0	9	60	316	603	983	1088	972	846	513	208	36	5634
	Nantucket....................... A	40.2	12	22	93	332	573	896	992	941	896	621	384	129	5891
	Pittsfield .... A	32.6	25	59	219	524	831	1231	1339	1196	1063	660	326	105	7578
	Worcester .... A	34.7	6	34	147	450	774	1172	1271	1123	998	612	304	78	6969

AVERAGE MONTHLY AND YEARLY DEGREE DAYS FOR CITIES IN THE UNITED STATES AND CANADA (BASE 65 F)

State	Station	Avg. Winter Temp[d]	July	Aug.	Sept.	Oct.	Nov.	Dec.	Jan.	Feb.	Mar.	Apr.	May	June	Yearly Total
Mich.	Alpena .... A	29.7	68	105	273	580	912	1268	1404	1299	1218	777	446	156	8506
	Detroit (City) .... A	37.2	0	0	87	360	738	1088	1181	1058	936	522	220	42	6232
	Detroit (Wayne) .... A	37.1	0	0	96	353	738	1088	1194	1061	933	534	239	57	6293
	Detroit (Willow Run) .... A	37.2	0	0	90	357	750	1104	1190	1053	921	519	229	45	6258
	Escanaba C	29.6	59	87	243	539	924	1293	1445	1296	1203	777	456	159	8481
	Flint .... A	33.1	16	40	159	465	843	1212	1330	1198	1066	639	319	90	7377
	Grand Rapids................. A	34.9	9	28	135	434	804	1147	1259	1134	1011	579	279	75	6894
	Lansing .... A	34.8	6	22	138	431	813	1163	1262	1142	1011	579	273	69	6909
	Marquette C	30.2	59	81	240	527	936	1268	1411	1268	1187	771	468	177	8393
	Muskegon .... A	36.0	12	28	120	400	762	1088	1209	1100	995	594	310	78	6696
	Sault Ste. Marie .... A	27.7	96	105	279	580	951	1367	1525	1380	1277	810	477	201	9048
Minn.	Duluth .... A	23.4	71	109	330	632	1131	1581	1745	1518	1355	840	490	198	10000
	Minneapolis .... A	28.3	22	31	189	505	1014	1454	1631	1380.	1166	621	288	81	8382
	Rochester .... A	28.8	25	34	186	474	1005	1438	1593	1366	1150	630	301	93	8295
Miss.	Jackson .... A	55.7	0	0	0	65	315	502	546	414	310	87	0	0	2239
	Meridian .... A	55.4	0	0	0	81	339	518	543	417	310	81	0	0	2289
	Vicksburg C	56.9	0	0	0	53	279	462	512	384	282	69	0	0	2041
Mo.	Columbia .... A	42.3	0	0	54	251	651	967	1076	874	716	324	121	12	5046
	Kansas City .... A	43.9	0	0	39	220	612	905	1032	818	682	294	109	0	4711
	St. Joseph .... A	40.3	0	6	60	285	708	1039	1172	949	769	348	133	15	5484
	St. Louis .... A	43.1	0	0	60	251	627	936	1026	848	704	312	121	15	4900
	St. Louis C	44.8	0	0	36	202	576	884	977	801	651	270	87	0	4484
	Springfield.................. A	44.5	0	0	45	223	600	877	973	781	660	291	105	6	4900
Mont.	Billings .... A	34.5	6	15	186	487	897	1135	1296	1100	970	570	285	102	7049
	Glasgow .... A	26.4	31	47	270	608	1104	1466	1711	1439	1187	648	335	150	8996
	Great Falls .... A	32.8	28	53	258	543	921	1169	1349	1154	1063	642	384	186	7750
	Havre .... A	28.1	28	53	306	595	1065	1367	1584	1364	1181	657	338	162	8700
	Havre C	29.8	19	37	252	539	1014	1321	1528	1305	1116	612	304	135	8182
	Helena .... A	31.1	31	59	294	601	1002	1265	1438	1170	1042	651	381	195	8129
	Kalispell .... A	31.4	50	99	321	654	1020	1240	1401	1134	1029	639	397	207	8191
	Miles City .... A	31.2	6	6	174	502	972	1296	1504	1252	1057	579	276	99	7723
	Missoula .... A	31.5	34	74	303	651	1035	1287	1420	1120	970	621	391	219	8125
Neb.	Grand Island .... A	36.0	0	6	108	381	834	1172	1314	1089	908	462	211	45	6530
	Lincoln C	38.8	0	6	75	301	726	1066	1237	1016	834	402	171	30	5864
	Norfolk .... A	34.0	9	0	111	397	873	1234	1414	1179	983	498	233	48	6979
	North Platte................. A	35.5	0	6	123	440	885	1166	1271	1039	930	519	248	57	6684
	Omaha .... A	35.6	0	12	105	357	828	1175	1355	1126	939	465	208	42	6612
	Scottsbluff.................. A	35.9	0	0	138	459	876	1128	1231	1008	921	552	285	75	6673
	Valentine.................... A	32.6	9	12	165	493	942	1237	1395	1176	1045	579	288	84	7425
Nev.	Elko .... A	34.0	9	34	225	561	924	1197	1314	1036	911	621	409	192	7433
	Ely .... A	33.1	28	43	234	592	939	1184	1308	1075	977	672	456	225	7733
	Las Vegas .... A	53.5	0	0	0	78	387	617	688	487	335	111	6	0	2709
	Reno .... A	39.3	43	87	204	490	801	1026	1073	823	729	510	357	189	6332
	Winnemucca.................... A	36.7	0	34	210	536	876	1091	1172	916	837	573	363	153	6761
N.H.	Concord .... A	33.0	6	50	177	505	822	1240	1358	1184	1032	636	298	75	7383
	Mt. Washington Obsv...........	15.2	493	536	720	1057	1341	1742	1820	1663	1652	1260	930	603	13817
N.J.	Atlantic City .... A	43.2	0	0	39	251	549	880	936	848	741	420	133	15	4812
	Newark .... A	42.8	0	0	30	248	573	921	983	876	729	381	118	0	4589
	Trenton C	42.4	0	0	57	264	576	924	989	885	753	399	121	12	4980
N. M.	Albuquerque.................. A	45.0	0	0	12	229	642	868	930	703	595	288	81	0	4348
	Clayton .... A	42.0	0	6	66	310	699	899	986	812	747	429	183	21	5158
	Raton .... A	38.1	9	28	126	431	825	1048	1116	904	834	543	301	63	6228
	Roswell .... A	47.5	0	0	18	202	573	806	840	641	481	201	31	0	3793
	Silver City .... A	48.0	0	0	6	183	525	729	791	605	518	261	87	0	3705
N.Y.	Albany .... A	34.6	0	19	138	440	777	1194	1311	1156	992	564	239	45	6875
	Albany C	37.2	0	9	102	375	699	1104	1218	1072	908	498	186	30	6201
	Binghamton .... A	33.9	22	65	201	471	810	1184	1277	1154	1045	645	313	99	7286
	Binghamton C	36.6	0	28	141	406	732	1107	1190	1081	949	543	229	45	6451
	Buffalo .... A	34.5	19	37	141	440	777	1156	1256	1145	1039	645	329	78	7062
	New York (Cent. Park)........ C	42.8	0	0	30	233	540	902	986	885	760	408	118	9	4871
	New York (La Guardia) .... A	43.1	0	0	27	223	528	887	973	879	750	414	124	6	4811

AVERAGE MONTHLY AND YEARLY DEGREE DAYS FOR CITIES
IN THE UNITED STATES AND CANADA (BASE 65 F)

State	Station	Avg. Winter Temp[d]	July	Aug.	Sept.	Oct.	Nov.	Dec.	Jan.	Feb.	Mar.	Apr.	May	June	Yearly Total
	New York (Kennedy) .... A	41.4	0	0	36	248	564	933	1029	935	815	480	167	12	5219
	Rochester .... A	35.4	9	31	126	415	747	1125	1234	1123	1014	597	279	48	6748
	Schenectady C	35.4	0	22	123	422	756	1159	1283	1131	970	543	211	30	6650
	Syracuse .... A	35.2	6	28	132	415	744	1153	1271	1140	1004	570	248	45	6756
N. C.	Asheville............................ C	46.7	0	0	48	245	555	775	784	683	592	273	87	0	4042
	Cape Hatteras	53.3	0	0	0	78	273	521	580	518	440	177	25	0	2612
	Charlotte........................... A	50.4	0	0	6	124	438	691	691	582	481	156	22	0	3191
	Greensboro........................ A	47.5	0	0	33	192	513	778	784	672	552	234	47	0	3805
	Raleigh.............................. A	49.4	0	0	21	164	450	716	725	616	487	180	34	0	3393
	Wilmington........................ A	54.6	0	0	0	74	291	521	546	462	357	96	0	0	2347
	Winston-Salem.................... A	48.4	0	0	21	171	483	747	753	652	524	207	37	0	3595
N. D.	Bismarck............................ A	26.6	34	28	222	577	1083	1463	1708	1442	1203	645	329	117	8851
	Devils Lake C	22.4	40	53	273	642	1191	1634	1872	1579	1345	753	381	138	9901
	Fargo................................ A	24.8	28	37	219	574	1107	1569	1789	1520	1262	690	332	99	9226
	Williston............................ A	25.2	31	43	261	601	1122	1513	1758	1473	1262	681	357	141	9243
Ohio	Akron-Canton .... A	38.1	0	9	96	381	726	1070	1138	1016	871	489	202	39	6037
	Cincinnati C	45.1	0	0	39	208	558	862	915	790	642	294	96	6	4410
	Cleveland .... A	37.2	9	25	105	384	738	1088	1159	1047	918	552	260	66	6351
	Columbus .... A	39.7	0	6	84	347	714	1039	1088	949	809	426	171	27	5660
	Columbus C	41.5	0	0	57	285	651	977	1032	902	760	396	136	15	5211
	Dayton.............................. A	39.8	0	6	78	310	696	1045	1097	955	809	429	167	30	5622
	Mansfield .... A	36.9	9	22	114	397	768	1110	1169	1042	924	543	245	60	6403
	Sandusky C	39.1	0	6	66	313	684	1032	1107	991	868	495	198	36	5796
	Toledo............................... A	36.4	0	16	117	406	792	1138	1200	1056	924	543	242	60	6494
	Youngstown .... A	36.8	6	19	120	412	771	1104	1169	1047	921	540	248	60	6417
Okla.	Oklahoma City.................... A	48.3	0	0	15	164	498	766	868	664	527	189	34	0	3725
	Tulsa................................. A	47.7	0	0	18	158	522	787	893	683	539	213	47	0	3860
Ore.	Astoria .... A	45.6	146	130	210	375	561	679	753	622	636	480	363	231	5186
	Burns C	35.9	12	37	210	515	867	1113	1246	988	856	570	366	177	6957
	Eugene .... A	45.6	34	34	129	366	585	719	803	627	589	426	279	135	4726
	Meacham .... A	34.2	84	124	288	580	918	1091	1209	1005	983	726	527	339	7874
	Medford............................ A	43.2	0	0	78	372	678	871	918	697	642	432	242	78	5008
	Pendleton .... A	42.6	0	0	111	350	711	884	1017	773	617	396	205	63	5127
	Portland............................ A	45.6	25	28	114	335	597	735	825	644	586	396	245	105	4635
	Portland............................ C	47.4	12	16	75	267	534	679	769	594	536	351	198	78	4109
	Roseburg........................... A	46.3	22	16	105	329	567	713	766	608	570	405	267	123	4491
	Salem .... A	45.4	37	31	111	338	594	729	822	647	611	417	273	144	4754
Pa.	Allentown .... A	38.9	0	0	90	353	693	1045	1116	1002	849	471	167	24	5810
	Erie.................................. A	36.8	0	25	102	391	714	1063	1169	1081	973	585	288	60	6451
	Harrisburg .... A	41.2	0	0	63	298	648	992	1045	907	766	396	124	12	5251
	Philadelphia .... A	41.8	0	0	60	297	620	965	1016	889	747	392	118	40	5144
	Philadelphia C	44.5	0	0	30	205	513	856	924	823	691	351	93	0	4486
	Pittsburgh.......................... A	38.4	0	9	105	375	726	1063	1119	1002	874	480	195	39	5987
	Pittsburgh.......................... C	42.2	0	0	60	291	615	930	983	885	763	390	124	12	5053
	Reading............................. C	42.4	0	0	54	257	597	939	1001	885	735	372	105	0	4945
	Scranton............................ A	37.2	0	19	132	434	762	1104	1156	1028	893	498	195	33	6254
	Williamsport....................... A	38.5	0	9	111	375	717	1073	1122	1002	856	468	177	24	5934
R. I.	Block Island........................ A	40.1	0	16	78	307	594	902	1020	955	877	612	344	99	5804
	Providence......................... A	38.8	0	16	96	372	660	1023	1110	988	868	534	236	51	5954
S. C.	Charleston .... A	56.4	0	0	0	59	282	471	487	389	291	54	0	0	2033
	Charleston C	57.9	0	0	0	34	210	425	443	367	273	42	0	0	1794
	Columbia .... A	54.0	0	0	0	84	345	577	570	470	357	81	0	0	2484
	Florence .... A	54.5	0	0	0	78	315	552	552	459	347	84	0	0	2387
	Greenville-Spartenburg .... A	51.6	0	0	6	121	399	651	660	546	446	132	19	0	2980
S. D.	Huron .... A	28.8	9	12	165	508	1014	1432	1628	1355	1125	600	288	87	8223
	Rapid City .... A	33.4	22	12	165	481	897	1172	1333	1145	1051	615	326	126	7345
	Sioux Falls .... A	30.6	19	25	168	462	972	1361	1544	1285	1082	573	270	78	7839
Tenn.	Bristol............................... A	46.2	0	0	51	236	573	828	828	700	598	261	68	0	4143
	Chattanooga .... A	50.3	0	0	18	143	468	698	722	577	453	150	25	0	3254
	Knoxville .... A	49.2	0	0	30	171	489	725	732	613	493	198	43	0	3494
	Memphis............................ A	50.5	0	0	18	130	447	698	729	585	456	147	22	0	3232

AVERAGE MONTHLY AND YEARLY DEGREE DAYS FOR CITIES
IN THE UNITED STATES AND CANADA (BASE 65 F)

State or Prov.	Station	Avg. Winter Temp^d	July	Aug.	Sept.	Oct.	Nov.	Dec.	Jan.	Feb.	Mar.	Apr.	May	June	Yearly Total
	Memphis C	51.6	0	0	12	102	396	648	710	568	434	129	16	0	3015
	Nashville .... A	48.9	0	0	30	158	495	732	778	644	512	189	40	0	3578
	Oak Ridge C	47.7	0	0	39	192	531	772	778	669	552	228	56	0	3817
Tex.	Abilene .... A	53.9	0	0	0	99	366	586	642	470	347	114	0	0	2624
	Amarillo .... A	47.0	0	0	18	205	570	797	877	664	546	252	56	0	3985
	Austin .... A	59.1	0	0	0	31	225	388	468	325	223	51	0	0	1711
	Brownsville .... A	67.7	0	0	0	0	66	149	205	106	74	0	0	0	600
	Corpus Christi .... A	64.6	0	0	0	0	120	220	291	174	109	0	0	0	914
	Dallas .... A	55.3	0	0	0	62	321	524	601	440	319	90	6	0	2363
	El Paso .... A	52.9	0	0	0	84	414	648	685	445	319	105	0	0	2700
	Fort Worth .... A	55.1	0	0	0	65	324	536	614	448	319	99	0	0	2405
	Galveston .... A	62.2	0	0	0	6	147	276	360	263	189	33	0	0	1274
	Galveston C	62.0	0	0	0	0	138	270	350	258	189	30	0	0	1235
	Houston .... A	61.0	0	0	0	6	183	307	384	288	192	36	0	0	1396
	Houston C	62.0	0	0	0	0	165	288	363	258	174	30	0	0	1278
	Laredo .... A	66.0	0	0	0	0	105	217	267	134	74	0	0	0	797
	Lubbock .... A	48.8	0	0	18	174	513	744	800	613	484	201	31	0	3578
	Midland .... A	53.8	0	0	0	87	381	592	651	468	322	90	0	0	2591
	Port Arthur .... A	60.5	0	0	0	22	207	329	384	274	192	39	0	0	2255
	San Angelo .... A	56.0	0	0	0	68	318	536	567	412	288	66	0	0	1546
	San Antonio .... A	60.1	0	0	0	31	204	363	428	286	195	39	0	0	1546
	Victoria .... A	62.7	0	0	0	6	150	270	344	230	152	21	0	0	1173
	Waco .... A	57.2	0	0	0	43	270	456	536	389	270	66	0	0	2030
	Wichita Falls .... A	53.0	0	0	0	99	381	632	698	518	378	120	6	0	2832
Utah	Milford .... A	36.5	0	0	99	443	867	1141	1252	988	822	519	279	87	6497
	Salt Lake City .... A	38.4	0	0	81	419	849	1082	1172	910	763	459	233	84	6052
	Wendover .... A	39.1	0	0	48	372	822	1091	1178	902	729	408	177	51	5778
Vt.	Burlington .... A	29.4	28	65	207	539	891	1349	1513	1333	1187	714	353	90	8269
Va.	Cape Henry C	50.0	0	0	0	112	360	645	694	633	536	246	53	0	3279
	Lynchburg .... A	46.0	0	0	51	223	540	822	849	731	605	267	78	0	4166
	Norfolk .... A	49.2	0	0	0	136	408	698	738	655	533	216	37	0	3421
	Richmond .... A	47.3	0	0	36	214	495	784	815	703	546	219	53	0	3865
	Roanoke .... A	46.1	0	0	51	229	549	825	834	722	614	261	65	0	4150
Wash.	Olympia .... A	44.2	68	71	198	422	636	753	834	675	645	450	307	177	5236
	Seattle-Tacoma .... A	44.2	56	62	162	391	633	750	828	678	657	474	295	159	5145
	Seattle C	46.9	50	47	129	329	543	657	738	599	577	396	242	117	4424
	Spokane .... A	36.5	9	25	168	493	879	1082	1231	980	834	531	288	135	6655
	Walla Walla C	43.8	0	0	87	310	681	843	986	745	589	342	177	45	4805
	Yakima .... A	39.1	0	12	144	450	828	1039	1163	868	713	435	220	69	5941
W. Va.	Charleston .... A	44.8	0	0	63	254	591	865	880	770	648	300	96	9	4476
	Elkins .... A	40.1	9	25	135	400	729	992	1008	896	791	444	198	48	5675
	Huntington .... A	45.0	0	0	63	257	585	856	880	764	636	294	99	12	4446
	Parkersburg C	43.5	0	0	60	264	606	905	942	826	691	339	115	6	4754
Wisc.	Green Bay .... A	30.3	28	50	174	484	924	1333	1494	1313	1141	654	335	99	8029
	La Crosse .... A	31.5	12	19	153	437	924	1339	1504	1277	1070	540	245	69	7589
	Madison .... A	30.9	25	40	174	474	930	1330	1473	1274	1113	618	310	102	7863
	Milwaukee .... A	32.6	43	47	174	471	876	1252	1376	1193	1054	642	372	135	7635
Wyo.	Casper .... A	33.4	6	16	192	524	942	1169	1290	1084	1020	657	381	129	7410
	Cheyenne .... A	34.2	28	37	219	543	909	1085	1212	1042	1026	702	428	150	7381
	Lander .... A	31.4	6	19	204	555	1020	1299	1417	1145	1017	654	381	153	7870
	Sheridan .... A	32.5	25	31	219	539	948	1200	1355	1154	1051	642	366	150	7680
Alta.	Banff C	—	220	295	498	797	1185	1485	1624	1364	1237	855	589	402	10551
	Calgary .... A	—	109	186	402	719	1110	1389	1575	1379	1268	798	477	291	9703
	Edmonton .... A	—	74	180	411	738	1215	1603	1810	1520	1330	765	400	222	10268
	Lethbridge .... A	—	56	112	318	611	1011	1277	1497	1291	1159	696	403	213	8644
B. C.	Kamloops .... A	—	22	40	189	546	894	1138	1314	1057	818	462	217	102	6799
	Prince George* .... A	—	236	251	444	747	1110	1420	1612	1319	1122	747	468	279	9755
	Prince Rupert C	—	273	248	339	539	708	868	936	808	812	648	493	357	7029
	Vancouver* .... A	—	81	87	219	456	657	787	862	723	676	501	310	156	5515
	Victoria* .... A	—	136	140	225	462	663	775	840	718	691	504	341	204	5699
	Victoria C	—	172	184	243	426	607	723	805	668	660	487	354	250	5579

Appendix 3 Degree-Days of Selected Cities in the United States

AVERAGE MONTHLY AND YEARLY DEGREE DAYS FOR CITIES
IN THE UNITED STATES AND CANADA (BASE 65 F)

State or Prov.	Station		Avg. Winter Temp	July	Aug.	Sept.	Oct.	Nov.	Dec.	Jan.	Feb.	Mar.	Apr.	May	June	Yearly Total
Man.	Brandon*	A	—	47	90	357	747	1290	1792	2034	1737	1476	837	431	198	11036
	Churchill	A	—	360	375	681	1082	1620	2248	2558	2277	2130	1569	1153	675	16728
	The Pas	C	—	59	127	429	831	1440	1981	2232	1853	1624	969	508	228	12281
	Winnipeg	A	—	38	71	322	683	1251	1757	2008	1719	1465	813	405	147	10679
N. B.	Fredericton*	A	—	78	68	234	592	915	1392	1541	1379	1172	753	406	141	8671
	Moncton	C	—	62	105	276	611	891	1342	1482	1336	1194	789	468	171	8727
	St. John	C	—	109	102	246	527	807	1194	1370	1229	1097	756	490	249	8219
Nfld.	Argentia	A	—	260	167	294	564	750	1001	1159	1085	1091	879	707	483	8440
	Corner Brook	C	—	102	133	324	642	873	1194	1358	1283	1212	885	639	333	8978
	Gander	A	—	121	152	330	670	909	1231	1370	1266	1243	939	657	366	9254
	Goose*	A	—	130	205	444	843	1227	1745	1947	1689	1494	1074	741	348	11887
	St. John's*	A	—	186	180	342	651	831	1113	1262	1170	1187	927	710	432	8991
N. W. T.	Aklavik	C	—	273	459	807	1414	2064	2530	2632	2336	2282	1674	1063	483	18017
	Fort Norman	C	—	164	341	666	1234	1959	2474	2592	2209	2058	1386	732	294	16109
	Resolution Island	C	—	843	831	900	1113	1311	1724	2021	1850	1817	1488	1181	942	16021
N. S.	Halifax	C	—	58	51	180	457	710	1074	1213	1122	1030	742	487	237	7361
	Sydney	A	—	62	71	219	518	765	1113	1262	1206	1150	840	567	276	8049
	Yarmouth	A	—	102	115	225	471	696	1029	1156	1065	1004	726	493	258	7340
Ont.	Cochrane	C	—	96	180	405	760	1233	1776	1978	1701	1528	963	570	222	11412
	Fort William	A	—	90	133	366	694	1140	1597	1792	1557	1380	876	543	237	10405
	Kapuskasing	C	—	74	171	405	756	1245	1807	2037	1735	1562	978	580	222	11572
	Kitchener	C	—	16	59	177	505	855	1234	1342	1226	1101	663	322	66	7566
	London	A	—	12	43	159	477	837	1206	1305	1198	1066	648	332	66	7349
	North Bay	C	—	37	90	267	608	990	1507	1680	1463	1277	780	400	120	9219
	Ottawa	C	—	25	81	222	567	936	1469	1624	1441	1231	708	341	90	8735
	Toronto	C	—	7	18	151	439	760	1111	1233	1119	1013	616	298	62	6827
P.E.I.	Charlottetown	C	—	40	53	198	518	804	1215	1380	1274	1169	813	496	204	8164
	Summerside	C	—	47	84	216	546	840	1246	1438	1291	1206	841	518	216	8488
Que.	Arvida	C	—	102	136	327	682	1074	1659	1879	1619	1407	891	521	231	10528
	Montreal*	A	—	9	43	165	521	882	1392	1566	1381	1175	684	316	69	8203
	Montreal	C	—	16	28	165	496	864	1355	1510	1328	1138	657	288	54	7899
	Quebec*	A	—	56	84	273	636	996	1516	1665	1477	1296	819	428	126	9372
	Quebec	C	—	40	68	243	592	972	1473	1612	1418	1228	780	400	111	8937
Sasks	Prince Albert	A	—	81	136	414	797	1368	1872	2108	1763	1559	867	446	219	11630
	Regina	A	—	78	93	360	741	1284	1711	1965	1687	1473	804	409	201	10806
	Saskatoon	C	—	56	87	372	750	1302	1758	2006	1689	1463	798	403	186	10870
Y. T.	Dawson	C	—	164	326	645	1197	1875	2415	2561	2150	1838	1068	570	258	15067
	Mayo Landing	C	—	208	366	648	1135	1794	2325	2427	1992	1665	1020	580	294	14454

*The data for these normals were from the full ten-year period 1951–1960, adjusted to the standard normal period 1931–1960.

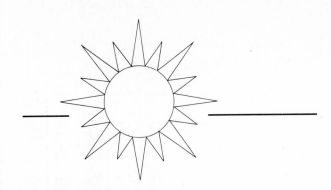

Appendix 4

Solar Insolation
for Various Latitudes

SOLAR POSITION AND INSOLATION VALUES FOR 24 DEGREES NORTH LATITUDE

DATE	SOLAR TIME AM	SOLAR TIME PM	SOLAR POSITION ALT	SOLAR POSITION AZM	BTUH/SQ. FT. TOTAL INSOLATION ON SURFACES NORMAL	HORIZ.	SOUTH FACING SURFACE ANGLE WITH HORIZ. 14	24	34	54	90
JAN 21	7	5	4.8	65.6	71	10	17	21	25	28	31
	8	4	16.9	58.3	239	83	110	126	137	145	127
	9	3	27.9	48.8	288	151	188	207	221	228	176
	10	2	37.2	36.1	308	204	246	268	282	287	207
	11	1	43.6	19.6	317	237	283	306	319	324	226
	12		46.0	0.0	320	249	296	319	332	336	232
	SURFACE DAILY TOTALS				2766	1622	1984	2174	2300	2360	1766
FEB 21	7	5	9.3	74.6	158	35	44	49	53	56	46
	8	4	22.3	67.2	263	116	135	145	150	151	102
	9	3	34.4	57.6	298	187	213	225	230	228	141
	10	2	45.1	44.2	314	241	273	286	291	287	168
	11	1	53.0	25.0	321	276	310	324	328	323	185
	12		56.0	0.0	324	288	323	337	341	335	191
	SURFACE DAILY TOTALS				3036	1998	2276	2396	2446	2424	1476
MAR 21	7	5	13.7	83.8	194	60	63	64	62	59	27
	8	4	27.2	76.8	267	141	150	152	149	142	64
	9	3	40.2	67.9	295	212	226	229	225	214	95
	10	2	52.3	54.8	309	266	285	288	283	270	120
	11	1	61.9	33.4	315	300	322	326	320	305	135
	12		66.0	0.0	317	312	334	339	333	317	140
	SURFACE DAILY TOTALS				3078	2270	2428	2456	2412	2298	1022
APR 21	6	6	4.7	100.6	40	7	5	4	4	3	2
	7	5	18.3	94.9	203	83	77	70	62	51	10
	8	4	32.0	89.0	256	160	157	149	137	122	15
	9	3	45.6	81.9	280	227	227	220	206	186	41
	10	2	59.0	71.8	292	278	282	275	259	237	61
	11	1	71.1	51.6	298	310	316	309	293	269	74
	12		77.6	0.0	299	321	328	321	305	280	79
	SURFACE DAILY TOTALS				3036	2454	2458	2374	2228	2016	488
MAY 21	6	6	8.0	108.4	86	22	15	10	9	9	5
	7	5	21.2	103.2	203	98	85	73	59	44	12
	8	4	34.6	98.5	248	171	159	145	127	106	15
	9	3	48.3	93.6	269	233	224	210	190	165	16
	10	2	62.0	87.7	280	281	275	261	239	211	22
	11	1	75.5	76.9	286	311	307	293	270	240	34
	12		86.0	0.0	288	322	317	304	281	250	37
	SURFACE DAILY TOTALS				3032	2556	2447	2286	2072	1800	246
JUN 21	6	6	9.3	111.6	97	29	20	12	12	11	7
	7	5	22.3	106.8	201	103	87	73	58	41	13
	8	4	35.5	102.6	242	173	158	142	122	99	16
	9	3	49.0	98.7	263	234	221	204	182	155	18
	10	2	62.6	95.0	274	280	269	253	229	199	18
	11	1	76.3	90.8	279	309	300	283	259	227	19
	12		89.4	0.0	281	319	310	294	269	236	22
	SURFACE DAILY TOTALS				2994	2574	2422	2230	1992	1700	204
JUL 21	6	6	8.2	109.0	81	23	16	11	10	9	6
	7	5	21.4	103.8	195	98	85	73	59	44	13
	8	4	34.8	99.2	239	169	157	143	125	104	16
	9	3	48.4	94.5	261	231	221	207	187	161	18
	10	2	62.1	89.0	272	278	270	256	235	206	21
	11	1	75.7	79.2	278	307	302	287	265	235	32
	12		86.6	0.0	280	317	312	298	275	245	36
	SURFACE DAILY TOTALS				2932	2526	2412	2250	2036	1766	246
AUG 21	6	6	5.0	101.3	35	7	5	4	4	4	2
	7	5	18.5	95.6	186	82	76	69	60	50	11
	8	4	32.2	89.7	241	158	154	146	134	118	16
	9	3	45.9	82.9	265	223	222	214	200	181	39
	10	2	59.3	73.0	278	273	275	268	252	230	58
	11	1	71.6	53.2	284	304	309	301	285	261	71
	12		78.3	0.0	286	315	320	313	296	272	75
	SURFACE DAILY TOTALS				2864	2408	2402	2316	2168	1958	470
SEP 21	7	5	13.7	83.8	173	57	60	60	59	56	26
	8	4	27.2	76.8	248	136	144	146	143	136	62
	9	3	40.2	67.9	278	205	218	221	217	206	93
	10	2	52.3	54.8	292	258	275	278	273	261	116
	11	1	61.9	33.4	299	291	311	315	309	295	131
	12		66.0	0.0	301	302	323	327	321	306	136
	SURFACE DAILY TOTALS				2878	2194	2342	2366	2322	2212	992
OCT 21	7	5	9.1	74.1	138	32	40	45	48	50	42
	8	4	22.0	66.7	247	111	129	139	144	145	99
	9	3	34.1	57.1	284	180	206	217	223	221	138
	10	2	44.7	43.8	301	234	265	277	282	279	165
	11	1	52.5	24.7	309	268	301	315	319	314	182
	12		55.5	0.0	311	279	314	328	332	327	188
	SURFACE DAILY TOTALS				2868	1928	2198	2314	2364	2346	1442
NOV 21	7	5	4.9	65.8	67	10	16	20	24	27	29
	8	4	17.0	58.4	232	82	108	123	135	142	124
	9	3	28.0	48.9	282	150	186	205	217	224	172
	10	2	37.3	36.3	303	203	244	265	278	283	204
	11	1	43.8	19.7	312	236	280	302	316	320	222
	12		46.2	0.0	315	247	293	315	328	332	228
	SURFACE DAILY TOTALS				2706	1610	1962	2146	2268	2324	1730
DEC 21	7	5	3.2	62.6	30	3	7	9	11	12	14
	8	4	14.9	55.3	225	71	99	116	129	139	130
	9	3	25.5	46.0	281	137	176	198	214	223	184
	10	2	34.3	33.7	304	189	234	258	275	283	217
	11	1	40.4	18.2	314	221	270	295	312	320	236
	12		42.6	0.0	317	232	282	308	325	332	243
	SURFACE DAILY TOTALS				2624	1474	1852	2058	2204	2286	1808

NOTE: 1) BASED ON DATA IN TABLE 1, pp 387 in 1972 ASHRAE HANDBOOK OF FUNDAMENTALS; 0% GROUND REFLECTANCE; 1.0 CLEARNESS FACTOR.
2) SEE FIG. 4, pp 394 in 1972 ASHRAE HANDBOOK OF FUNDAMENTALS FOR TYPICAL REGIONAL CLEARNESS FACTORS.
3) GROUND REFLECTION NOT INCLUDED ON NORMAL OR HORIZONTAL SURFACES.

"Reprinted from ASHRAE TRANSACTIONS 1974, Volume 80, Part II, by permission of the American Society of Heating, Refrigerating and Air-Conditioning Engineers, Inc."

SOLAR POSITION AND INSOLATION VALUES FOR 32 DEGREES NORTH LATITUDE

DATE	SOLAR TIME AM	SOLAR TIME PM	SOLAR POSITION ALT	SOLAR POSITION AZM	BTUH/SQ. FT. TOTAL INSOLATION ON SURFACES NORMAL	HORIZ.	SOUTH FACING SURFACE ANGLE WITH HORIZ. 22	32	42	52	90
JAN 21	7	5	1.4	65.2	1	0	0	0	0	1	1
	8	4	12.5	56.5	203	56	93	106	116	123	115
	9	3	22.5	46.0	269	118	175	193	206	212	181
	10	2	30.6	33.1	295	167	235	256	269	274	221
	11	1	36.1	17.5	306	198	273	295	308	312	245
	12		38.0	0.0	310	209	285	308	321	324	253
	SURFACE DAILY TOTALS				2458	1288	1839	2008	2118	2166	1779
FEB 21	7	5	7.1	73.5	121	22	34	37	40	42	38
	8	4	19.0	64.4	247	95	127	136	140	141	108
	9	3	29.9	53.4	288	161	206	217	222	220	158
	10	2	39.1	39.4	306	212	266	278	283	279	193
	11	1	45.6	21.4	315	244	304	317	321	315	214
	12		48.0	0.0	317	255	316	330	334	328	222
	SURFACE DAILY TOTALS				2872	1724	2188	2300	2345	2322	1644
MAR 21	7	5	12.7	81.9	185	54	60	60	59	56	32
	8	4	25.1	73.0	260	129	146	147	144	137	78
	9	3	36.8	62.1	290	194	222	224	220	209	119
	10	2	47.3	47.5	304	245	280	283	278	265	150
	11	1	55.0	26.8	311	277	317	321	315	300	170
	12		58.0	0.0	313	287	329	333	327	312	177
	SURFACE DAILY TOTALS				3012	2084	2378	2403	2358	2246	1276
APR 21	6	6	6.1	99.9	66	14	9	6	6	5	3
	7	5	18.8	92.2	206	86	78	71	62	51	10
	8	4	31.5	84.0	255	158	156	148	136	120	35
	9	3	43.9	74.2	278	220	225	217	203	183	68
	10	2	55.7	60.3	290	267	279	272	256	234	95
	11	1	65.4	37.5	295	297	313	306	290	265	112
	12		69.6	0.0	297	307	325	318	301	276	118
	SURFACE DAILY TOTALS				3076	2390	2444	2356	2206	1994	764
MAY 21	6	6	10.4	107.2	119	36	21	13	13	12	7
	7	5	22.8	100.1	211	107	88	75	60	44	13
	8	4	35.4	92.9	250	175	159	145	127	105	15
	9	3	48.1	84.7	269	233	223	209	188	163	33
	10	2	60.6	73.3	280	277	273	259	237	208	56
	11	1	72.0	51.9	285	305	305	290	268	237	72
	12		78.0	0.0	286	315	315	301	278	247	77
	SURFACE DAILY TOTALS				3112	2582	2454	2284	2064	1788	469
JUN 21	6	6	12.2	110.2	131	45	26	16	15	14	9
	7	5	24.3	103.4	210	115	91	76	59	41	14
	8	4	36.9	96.8	245	180	159	143	122	99	16
	9	3	49.6	89.4	264	236	221	204	181	153	19
	10	2	62.2	79.7	274	279	268	251	227	197	41
	11	1	74.2	60.9	279	306	299	282	257	224	56
	12		81.5	0.0	280	315	309	292	267	234	60
	SURFACE DAILY TOTALS				3084	2634	2436	2234	1990	1690	370
JUL 21	6	6	10.7	107.7	113	37	22	14	13	12	8
	7	5	23.1	100.6	203	107	87	75	60	44	14
	8	4	35.7	93.6	241	174	158	143	125	104	16
	9	3	48.4	85.5	261	231	220	205	185	159	31
	10	2	60.9	74.3	271	274	269	254	232	204	54
	11	1	72.4	53.3	277	302	300	285	262	232	69
	12		78.6	0.0	279	311	310	296	273	242	74
	SURFACE DAILY TOTALS				3012	2558	2422	2250	2030	1754	458
AUG 21	6	6	6.5	100.5	59	14	9	7	6	6	4
	7	5	19.1	92.8	190	85	77	69	60	50	12
	8	4	31.8	84.7	240	156	152	144	132	116	33
	9	3	44.3	75.0	263	216	220	212	197	178	65
	10	2	56.1	61.3	276	262	272	264	249	226	91
	11	1	66.0	38.4	282	292	305	298	281	257	107
	12		70.3	0.0	284	302	317	309	292	268	113
	SURFACE DAILY TOTALS				2902	2352	2388	2296	2144	1934	736
SEP 21	7	5	12.7	81.9	163	51	56	56	55	52	30
	8	4	25.1	73.0	240	124	140	141	138	131	75
	9	3	36.8	62.1	272	188	213	215	211	201	114
	10	2	47.3	47.5	287	237	270	273	268	255	145
	11	1	55.0	26.8	294	268	306	309	303	289	164
	12		58.0	0.0	296	278	318	321	315	300	171
	SURFACE DAILY TOTALS				2808	2014	2288	2308	2264	2154	1226
OCT 21	7	5	6.8	73.1	99	19	29	32	34	36	32
	8	4	18.7	64.0	229	90	120	128	133	134	104
	9	3	29.5	53.0	273	155	198	208	213	212	153
	10	2	38.7	39.1	293	204	257	269	273	270	188
	11	1	45.1	21.1	302	236	294	307	311	306	209
	12		47.5	0.0	304	247	306	320	324	318	217
	SURFACE DAILY TOTALS				2696	1654	2100	2208	2252	2232	1588
NOV 21	7	5	1.5	65.4	2	0	0	0	1	1	1
	8	4	12.7	56.6	196	55	91	104	113	119	111
	9	3	22.6	46.1	263	118	173	190	202	208	176
	10	2	30.8	33.2	289	166	233	252	265	270	217
	11	1	36.2	17.6	301	197	270	291	303	307	241
	12		38.2	0.0	304	207	282	304	316	320	249
	SURFACE DAILY TOTALS				2406	1280	1816	1980	2084	2130	1742
DEC 21	8	4	10.3	53.8	176	41	77	90	101	108	107
	9	3	19.8	43.6	257	102	161	180	195	204	183
	10	2	27.6	31.2	288	150	221	244	259	267	226
	11	1	32.7	16.4	301	180	258	282	298	305	251
	12		34.6	0.0	304	190	271	295	311	318	259
	SURFACE DAILY TOTALS				2348	1136	1704	1888	2016	2086	1794

NOTE: 1) BASED ON DATA IN TABLE 1, pp 387 in 1972 ASHRAE HANDBOOK OF FUNDAMENTALS; 0% GROUND REFLECTANCE; 1.0 CLEARNESS FACTOR.
2) SEE FIG. 4, pp 394 in 1972 ASHRAE HANDBOOK OF FUNDAMENTALS FOR TYPICAL REGIONAL CLEARNESS FACTORS.
3) GROUND REFLECTION NOT INCLUDED ON NORMAL OR HORIZONTAL SURFACES

"Reprinted from ASHRAE TRANSACTIONS 1974, Volume 80, Part II, by permission of the American Society of Heating, Refrigerating and Air-Conditioning Engineers, Inc."

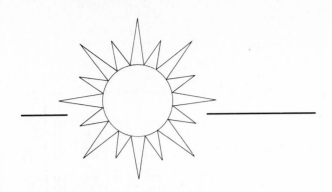

Appendix 4

Solar Insolation
for Various Latitudes

SOLAR POSITION AND INSOLATION VALUES FOR 24 DEGREES NORTH LATITUDE

DATE	SOLAR TIME AM	SOLAR TIME PM	SOLAR POSITION ALT	SOLAR POSITION AZM	BTUH/SQ. FT. TOTAL INSOLATION ON SURFACES NORMAL	HORIZ.	SOUTH FACING SURFACE ANGLE WITH HORIZ. 14	24	34	54	90
JAN 21	7	5	4.8	65.6	71	10	17	21	25	28	31
	8	4	16.9	58.3	239	83	110	126	137	145	127
	9	3	27.9	48.8	288	151	188	207	221	228	176
	10	2	37.2	36.1	308	204	246	268	282	287	207
	11	1	43.6	19.6	317	237	283	306	319	324	226
	12		46.0	0.0	320	249	296	319	332	336	232
	SURFACE DAILY TOTALS				2766	1622	1984	2174	2300	2360	1766
FEB 21	7	5	9.3	74.6	158	35	44	49	53	56	46
	8	4	22.3	67.2	263	116	135	145	150	151	102
	9	3	34.4	57.6	298	187	213	225	230	228	141
	10	2	45.1	44.2	314	241	273	286	291	287	168
	11	1	53.0	25.0	321	276	310	324	328	323	185
	12		56.0	0.0	324	288	323	337	341	335	191
	SURFACE DAILY TOTALS				3036	1998	2276	2396	2446	2424	1476
MAR 21	7	5	13.7	83.8	194	60	63	64	62	59	27
	8	4	27.2	76.8	267	141	150	152	149	142	64
	9	3	40.2	67.9	295	212	226	229	225	214	95
	10	2	52.3	54.8	309	266	285	288	283	270	120
	11	1	61.9	33.4	315	300	322	326	320	305	135
	12		66.0	0.0	317	312	334	339	333	317	140
	SURFACE DAILY TOTALS				3078	2270	2428	2456	2412	2298	1022
APR 21	6	6	4.7	100.6	40	7	5	4	4	3	2
	7	5	18.3	94.9	203	83	77	70	62	51	10
	8	4	32.0	89.0	256	160	157	149	137	122	15
	9	3	45.6	81.9	280	227	227	220	206	186	41
	10	2	59.0	71.8	292	278	282	275	259	237	61
	11	1	71.1	51.6	298	310	316	309	293	269	74
	12		77.6	0.0	299	321	328	321	305	280	79
	SURFACE DAILY TOTALS				3036	2454	2458	2374	2228	2016	488
MAY 21	6	6	8.0	108.4	86	22	15	10	9	9	5
	7	5	21.2	103.2	203	98	85	73	59	44	12
	8	4	34.6	98.5	248	171	159	145	127	106	15
	9	3	48.3	93.6	269	233	224	210	190	165	16
	10	2	62.0	87.7	280	281	275	261	239	211	22
	11	1	75.5	76.9	286	311	307	293	270	240	34
	12		86.0	0.0	288	322	317	304	281	250	37
	SURFACE DAILY TOTALS				3032	2556	2447	2286	2072	1800	246
JUN 21	6	6	9.3	111.6	97	29	20	12	12	11	7
	7	5	22.3	106.8	201	103	87	73	58	41	13
	8	4	35.5	102.6	242	173	158	142	122	99	16
	9	3	49.0	98.7	263	234	221	204	182	155	18
	10	2	62.6	95.0	274	280	269	253	229	199	18
	11	1	76.3	90.8	279	309	300	283	259	227	19
	12		89.4	0.0	281	319	310	294	269	236	22
	SURFACE DAILY TOTALS				2994	2574	2422	2230	1992	1700	204
JUL 21	6	6	8.2	109.0	81	23	16	11	10	9	6
	7	5	21.4	103.8	195	98	85	73	59	44	13
	8	4	34.8	99.2	239	169	157	143	125	104	16
	9	3	48.4	94.5	261	231	221	207	187	161	18
	10	2	62.1	89.0	272	278	270	256	235	206	21
	11	1	75.7	79.2	278	307	302	287	265	235	32
	12		86.6	0.0	280	317	312	298	275	245	36
	SURFACE DAILY TOTALS				2932	2526	2412	2250	2036	1766	246
AUG 21	6	6	5.0	101.3	35	7	5	4	4	4	2
	7	5	18.5	95.6	186	82	76	69	60	50	11
	8	4	32.2	89.7	241	158	154	146	134	118	16
	9	3	45.9	82.9	265	223	222	214	200	181	39
	10	2	59.3	73.0	278	273	275	268	252	230	58
	11	1	71.6	53.2	284	304	309	301	285	261	71
	12		78.3	0.0	286	315	320	313	296	272	75
	SURFACE DAILY TOTALS				2864	2408	2402	2316	2168	1958	470
SEP 21	7	5	13.7	83.8	173	57	60	60	59	56	26
	8	4	27.2	76.8	248	136	144	146	143	136	62
	9	3	40.2	67.9	278	205	218	221	217	206	93
	10	2	52.3	54.8	292	258	275	278	273	261	116
	11	1	61.9	33.4	299	291	311	315	309	295	131
	12		66.0	0.0	301	302	323	327	321	306	136
	SURFACE DAILY TOTALS				2878	2194	2342	2366	2322	2212	992
OCT 21	7	5	9.1	74.1	138	32	40	45	48	50	42
	8	4	22.0	66.7	247	111	129	139	144	145	99
	9	3	34.1	57.1	284	180	206	217	223	221	138
	10	2	44.7	43.8	301	234	265	277	282	279	165
	11	1	52.5	24.7	309	268	301	315	319	314	182
	12		55.5	0.0	311	279	314	328	332	327	188
	SURFACE DAILY TOTALS				2868	1928	2198	2314	2364	2346	1442
NOV 21	7	5	4.9	65.8	67	10	16	20	24	27	29
	8	4	17.0	58.4	232	82	108	123	135	142	124
	9	3	28.0	48.9	282	150	186	205	217	224	172
	10	2	37.3	36.3	303	203	244	265	278	283	204
	11	1	43.8	19.7	312	236	280	302	316	320	222
	12		46.2	0.0	315	247	293	315	328	332	228
	SURFACE DAILY TOTALS				2706	1610	1962	2146	2268	2324	1730
DEC 21	7	5	3.2	62.6	30	3	7	9	11	12	14
	8	4	14.9	55.3	225	71	99	116	129	139	130
	9	3	25.5	46.0	281	137	176	198	214	223	184
	10	2	34.3	33.7	304	189	234	258	275	283	217
	11	1	40.4	18.2	314	221	270	295	312	320	236
	12		42.6	0.0	317	232	282	308	325	332	243
	SURFACE DAILY TOTALS				2624	1474	1852	2058	2204	2286	1808

NOTE: 1) BASED ON DATA IN TABLE 1, pp 387 in 1972 ASHRAE HANDBOOK OF FUNDAMENTALS; 0% GROUND REFLECTANCE; 1.0 CLEARNESS FACTOR.
2) SEE FIG. 4, pp 394 in 1972 ASHRAE HANDBOOK OF FUNDAMENTALS FOR TYPICAL REGIONAL CLEARNESS FACTORS.
3) GROUND REFLECTION NOT INCLUDED ON NORMAL OR HORIZONTAL SURFACES.

"Reprinted from ASHRAE TRANSACTIONS 1974, Volume 80, Part II, by permission of the American Society of Heating, Refrigerating and Air-Conditioning Engineers, Inc."

SOLAR POSITION AND INSOLATION VALUES FOR 32 DEGREES NORTH LATITUDE

DATE	SOLAR TIME AM	SOLAR TIME PM	SOLAR POSITION ALT	SOLAR POSITION AZM	BTUH/SQ. FT. TOTAL INSOLATION ON SURFACES NORMAL	HORIZ.	SOUTH FACING SURFACE ANGLE WITH HORIZ. 22	32	42	52	90
JAN 21	7	5	1.4	65.2	1	0	0	0	0	1	1
	8	4	12.5	56.5	203	56	93	106	116	123	115
	9	3	22.5	46.0	269	118	175	193	206	212	181
	10	2	30.6	33.1	295	167	235	256	269	274	221
	11	1	36.1	17.5	306	198	273	295	308	312	245
	12		38.0	0.0	310	209	285	308	321	324	253
	SURFACE DAILY TOTALS				2458	1288	1839	2008	2118	2166	1779
FEB 21	7	5	7.1	73.5	121	22	34	37	40	42	38
	8	4	19.0	64.4	247	95	127	136	140	141	108
	9	3	29.9	53.4	288	161	206	217	222	220	158
	10	2	39.1	39.4	306	212	266	278	283	279	193
	11	1	45.6	21.4	315	244	304	317	321	315	214
	12		48.0	0.0	317	255	316	330	334	328	222
	SURFACE DAILY TOTALS				2872	1724	2188	2300	2345	2322	1644
MAR 21	7	5	12.7	81.9	185	54	60	60	59	56	32
	8	4	25.1	73.0	260	129	146	147	144	137	78
	9	3	36.8	62.1	290	194	222	224	220	209	119
	10	2	47.3	47.5	304	245	280	283	278	265	150
	11	1	55.0	26.8	311	277	317	321	315	300	170
	12		58.0	0.0	313	287	329	333	327	312	177
	SURFACE DAILY TOTALS				3012	2084	2378	2403	2358	2246	1276
APR 21	6	6	6.1	99.9	66	14	9	6	6	5	3
	7	5	18.8	92.2	206	86	78	71	62	51	10
	8	4	31.5	84.0	255	158	156	148	136	120	35
	9	3	43.9	74.2	278	220	225	217	203	183	68
	10	2	55.7	60.3	290	267	279	272	256	234	95
	11	1	65.4	37.5	295	297	313	306	290	265	112
	12		69.6	0.0	297	307	325	318	301	276	118
	SURFACE DAILY TOTALS				3076	2390	2444	2356	2206	1994	764
MAY 21	6	6	10.4	107.2	119	36	21	13	13	12	7
	7	5	22.8	100.1	211	107	88	75	60	44	13
	8	4	35.4	92.9	250	175	159	145	127	105	15
	9	3	48.1	84.7	269	233	223	209	188	163	33
	10	2	60.6	73.3	280	277	273	259	237	208	56
	11	1	72.0	51.9	285	305	305	290	268	237	72
	12		78.0	0.0	286	315	315	301	278	247	77
	SURFACE DAILY TOTALS				3112	2582	2454	2284	2064	1788	469
JUN 21	6	6	12.2	110.2	131	45	26	16	15	14	9
	7	5	24.3	103.4	210	115	91	76	59	41	14
	8	4	36.9	96.8	245	180	159	143	122	99	16
	9	3	49.6	89.4	264	236	221	204	181	153	19
	10	2	62.2	79.7	274	279	268	251	227	197	41
	11	1	74.2	60.9	279	306	299	282	257	224	56
	12		81.5	0.0	280	315	309	292	267	234	60
	SURFACE DAILY TOTALS				3084	2634	2436	2234	1990	1690	370
JUL 21	6	6	10.7	107.7	113	37	22	14	13	12	8
	7	5	23.1	100.6	203	107	87	75	60	44	14
	8	4	35.7	93.6	241	174	158	143	125	104	16
	9	3	48.4	85.5	261	231	220	205	185	159	31
	10	2	60.9	74.3	271	274	269	254	232	204	54
	11	1	72.4	53.3	277	302	300	285	262	232	69
	12		78.6	0.0	279	311	310	296	273	242	74
	SURFACE DAILY TOTALS				3012	2558	2422	2250	2030	1754	458
AUG 21	6	6	6.5	100.5	59	14	9	7	6	6	4
	7	5	19.1	92.8	190	85	77	69	60	50	12
	8	4	31.8	84.7	240	156	152	144	132	116	33
	9	3	44.3	75.0	263	216	220	212	197	178	65
	10	2	56.1	61.3	276	262	272	264	249	226	91
	11	1	66.0	38.4	282	292	305	298	281	257	107
	12		70.3	0.0	284	302	317	309	292	268	113
	SURFACE DAILY TOTALS				2902	2352	2388	2296	2144	1934	736
SEP 21	7	5	12.7	81.9	163	51	56	56	55	52	30
	8	4	25.1	73.0	240	124	140	141	138	131	75
	9	3	36.8	62.1	272	188	213	215	211	201	114
	10	2	47.3	47.5	287	237	270	273	268	255	145
	11	1	55.0	26.8	294	268	306	309	303	289	164
	12		58.0	0.0	296	278	318	321	315	300	171
	SURFACE DAILY TOTALS				2808	2014	2288	2308	2264	2154	1226
OCT 21	7	5	6.8	73.1	99	19	29	32	34	36	32
	8	4	18.7	64.0	229	90	120	128	133	134	104
	9	3	29.5	53.0	273	155	198	208	213	212	153
	10	2	38.7	39.1	293	204	257	269	273	270	188
	11	1	45.1	21.1	302	236	294	307	311	306	209
	12		47.5	0.0	304	247	306	320	324	318	217
	SURFACE DAILY TOTALS				2696	1654	2100	2208	2252	2232	1588
NOV 21	7	5	1.5	65.4	2	0	0	0	1	1	1
	8	4	12.7	56.6	196	55	91	104	113	119	111
	9	3	22.6	46.1	263	118	173	190	202	208	176
	10	2	30.8	33.2	289	166	233	252	265	270	217
	11	1	36.2	17.6	301	197	270	291	303	307	241
	12		38.2	0.0	304	207	282	304	316	320	249
	SURFACE DAILY TOTALS				2406	1280	1816	1980	2084	2130	1742
DEC 21	8	4	10.3	53.8	176	41	77	90	101	108	107
	9	3	19.8	43.6	257	102	161	180	195	204	183
	10	2	27.6	31.2	288	150	221	244	259	267	226
	11	1	32.7	16.4	301	180	258	282	298	305	251
	12		34.6	0.0	304	190	271	295	311	318	259
	SURFACE DAILY TOTALS				2348	1136	1704	1888	2016	2086	1794

NOTE: 1) BASED ON DATA IN TABLE 1, pp 387 in 1972 ASHRAE HANDBOOK OF FUNDAMENTALS; 0% GROUND REFLECTANCE; 1.0 CLEARNESS FACTOR.
2) SEE FIG. 4, pp 394 in 1972 ASHRAE HANDBOOK OF FUNDAMENTALS FOR TYPICAL REGIONAL CLEARNESS FACTORS.
3) GROUND REFLECTION NOT INCLUDED ON NORMAL OR HORIZONTAL SURFACES

"Reprinted from ASHRAE TRANSACTIONS 1974, Volume 80, Part II, by permission of the American Society of Heating, Refrigerating and Air-Conditioning Engineers, Inc."

SOLAR POSITION AND INSOLATION VALUES FOR 40 DEGREES NORTH LATITUDE

DATE	SOLAR TIME AM	SOLAR TIME PM	SOLAR POSITION ALT	SOLAR POSITION AZM	BTUH/SQ. FT. TOTAL INSOLATION ON SURFACES NORMAL	HORIZ.	SOUTH FACING SURFACE ANGLE WITH HORIZ. 30	40	50	60	90
JAN 21	8	4	8.1	55.3	142	28	65	74	81	85	84
	9	3	16.8	44.0	239	83	155	171	182	187	171
	10	2	23.8	30.9	274	127	218	237	249	254	223
	11	1	28.4	16.0	289	154	257	277	290	293	253
	12		30.0	0.0	294	164	270	291	303	306	263
	SURFACE DAILY TOTALS				2182	948	1660	1810	1906	1944	1726
FEB 21	7	5	4.8	72.7	69	10	19	21	23	24	22
	8	4	15.4	62.2	224	73	114	122	126	127	107
	9	3	25.0	50.2	274	132	195	205	209	208	167
	10	2	32.8	35.9	295	178	256	267	271	267	210
	11	1	38.1	18.9	305	206	293	306	310	304	236
	12		40.0	0.0	308	216	306	319	323	317	245
	SURFACE DAILY TOTALS				2640	1414	2060	2162	2202	2176	1730
MAR 21	7	5	11.4	80.2	171	46	55	55	54	51	35
	8	4	22.5	69.6	250	114	140	141	138	131	89
	9	3	32.8	57.3	282	173	215	217	213	202	138
	10	2	41.6	41.9	297	218	273	276	271	258	176
	11	1	47.7	22.6	305	247	310	313	307	293	200
	12		50.0	0.0	307	257	322	326	320	305	208
	SURFACE DAILY TOTALS				2916	1852	2308	2330	2284	2174	1484
APR 21	6	6	7.4	98.9	89	20	11	8	7	7	4
	7	5	18.9	89.5	206	87	77	70	61	50	12
	8	4	30.3	79.3	252	152	153	145	133	117	53
	9	3	41.3	67.2	274	207	221	213	199	179	93
	10	2	51.2	51.4	286	250	275	267	252	229	126
	11	1	58.7	29.2	292	277	308	301	285	260	147
	12		61.6	0.0	293	287	320	313	296	271	154
	SURFACE DAILY TOTALS				3092	2274	2412	2320	2168	1956	1022
MAY 21	5	7	1.9	114.7	1	0	0	0	0	0	0
	6	6	12.7	105.6	144	49	25	15	14	13	9
	7	5	24.0	96.6	216	214	89	76	60	44	13
	8	4	35.4	87.2	250	175	158	144	125	104	25
	9	3	46.8	76.0	267	227	221	206	186	160	60
	10	2	57.5	60.9	277	267	270	255	233	205	89
	11	1	66.2	37.1	283	293	301	287	264	234	108
	12		70.0	0.0	284	301	312	297	274	243	114
	SURFACE DAILY TOTQLS				3160	2552	2442	2264	2040	1760	724
JUN 21	5	7	4.2	117.3	22	4	3	3	2	2	1
	6	6	14.8	108.4	155	60	30	18	17	16	10
	7	5	26.0	99.7	216	123	92	77	59	41	14
	8	4	37.4	90.7	246	182	159	142	121	97	16
	9	3	48.8	80.2	263	233	219	202	179	151	47
	10	2	59.8	65.8	272	272	266	248	224	194	74
	11	1	69.2	41.9	277	296	296	278	253	221	92
	12		73.5	0.0	279	304	306	289	263	230	98
	SURFACE DAILY TOTALS				3180	2648	2434	2224	1974	1670	610
JUL 21	5	7	2.3	115.2	2	0	0	0	0	0	0
	6	6	13.1	106.1	138	50	26	17	15	14	9
	7	5	24.3	97.2	208	114	89	75	60	44	14
	8	4	35.8	87.8	241	174	157	142	124	102	24
	9	3	47.2	76.7	259	225	218	203	182	157	58
	10	2	57.9	61.7	269	265	266	251	229	200	86
	11	1	66.7	37.9	275	290	296	281	258	228	104
	12		70.6	0.0	276	298	307	292	269	238	111
	SURFACE DAILY TOTALS				3062	2534	2409	2230	2006	1728	702
AUG 21	6	6	7.9	99.5	81	21	12	9	8	7	5
	7	5	19.3	90.0	191	87	76	69	60	49	12
	8	4	30.7	79.9	237	150	150	141	129	113	50
	9	3	41.8	67.9	260	205	216	207	193	173	89
	10	2	51.7	52.1	272	246	267	259	244	221	120
	11	1	59.3	29.7	278	273	300	292	276	252	140
	12		62.3	0.0	280	282	311	303	287	262	147
	SURFACE DAILY TOTALS				2916	2244	2354	2258	2104	1894	978
SEP 21	7	5	11.4	80.2	149	43	51	51	49	47	32
	8	4	22.5	69.6	230	109	133	134	131	124	84
	9	3	32.8	57.3	263	167	206	208	203	193	132
	10	2	41.6	41.9	280	211	262	265	260	247	168
	11	1	47.7	22.6	287	239	298	301	295	281	192
	12		50.0	0.0	290	249	310	313	307	292	200
	SURFACE DAILY TOTALS				2708	1788	2210	2228	2182	2074	1416
OCT 21	7	5	4.5	72.3	48	7	14	15	17	17	16
	8	4	15.0	61.9	204	68	106	113	117	118	100
	9	3	24.5	49.8	257	126	185	195	200	198	160
	10	2	32.4	35.6	280	170	245	257	261	257	203
	11	1	37.6	18.7	291	199	283	295	299	294	229
	12		39.5	0.0	294	208	295	308	312	306	238
	SURFACE DAILY TOTALS				2454	1348	1962	2060	2098	2074	1654
NOV 21	8	4	8.2	55.4	136	28	63	72	78	82	81
	9	3	17.0	44.1	232	82	152	167	178	183	167
	10	2	24.0	31.0	268	126	215	233	245	249	219
	11	1	28.6	16.1	283	153	254	273	285	288	248
	12		30.2	0.0	288	163	267	287	298	301	258
	SURFACE DAILY TOTALS				2128	942	1636	1778	1870	1908	1686
DEC	8	4	5.5	53.0	89	14	39	45	50	54	56
	9	3	14.0	41.9	217	65	135	152	164	171	163
	10	2	20.7	29.4	261	107	200	221	235	242	221
	11	1	25.0	15.2	280	134	239	262	276	283	252
	12		26.6	0.0	285	143	253	275	290	296	263
	SURFACE DAILY TOTALS				1978	782	1480	1634	1740	1796	1646

NOTE: 1) BASED ON DATA IN TABLE 1, pp 387 in 1972 ASHRAE HANDBOOK OF FUNDAMENTALS; 0% GROUND REFLECTANCE; 1.0 CLEARNESS FACTOR.
2) SEE FIG. 4, pp 394 in 1972 ASHRAE HANDBOOK OF FUNDAMENTALS FOR TYPICAL REGIONAL CLEARNESS FACTORS.
3) GROUND REFLECTION NOT INCLUDED ON NORMAL OR HORIZONTAL SURFACES.

SOLAR POSITION AND INSOLATION VALUES FOR 48 DEGREES NORTH LATITUDE

DATE	SOLAR TIME AM	SOLAR TIME PM	SOLAR POSITION ALT	SOLAR POSITION AZM	BTUH/SQ. FT. TOTAL INSOLATION ON SURFACES NORMAL	HORIZ.	SOUTH FACING SURFACE ANGLE WITH HORIZ. 38	48	58	68	90
JAN 21	8	4	3.5	54.6	37	4	17	19	21	22	22
	9	3	11.0	42.6	185	46	120	132	140	145	139
	10	2	16.9	29.4	239	83	190	206	216	220	206
	11	1	20.7	15.1	261	107	231	249	260	263	243
	12		22.0	0.0	267	115	245	264	275	278	255
	SURFACE DAILY TOTALS				1710	596	1360	1478	1550	1578	1478
FEB 21	7	5	2.4	72.2	12	1	3	4	4	4	4
	8	4	11.6	60.5	188	49	95	102	105	106	96
	9	3	19.7	47.7	251	100	178	187	191	190	167
	10	2	26.2	33.3	278	139	240	251	255	251	217
	11	1	30.5	17.2	290	165	278	290	294	288	247
	12		32.0	0.0	293	173	291	304	307	301	258
	SURFACE DAILY TOTALS				2330	1080	1880	1972	2024	1978	1720
MAR 21	7	5	10.0	78.7	153	37	49	49	47	45	35
	8	4	19.5	66.8	236	96	131	132	129	122	96
	9	3	28.2	53.4	270	147	205	207	203	193	152
	10	2	35.4	37.8	287	187	263	266	261	248	195
	11	1	40.3	19.8	295	212	300	303	297	283	223
	12		42.0	0.0	298	220	312	315	309	294	232
	SURFACE DAILY TOTALS				2780	1578	2208	2228	2182	2074	1632
APR 21	6	6	8.6	97.8	108	27	13	9	8	7	5
	7	5	18.6	86.7	205	85	76	69	59	48	21
	8	4	28.5	74.9	247	142	149	141	129	113	69
	9	3	37.8	61.2	268	191	216	208	194	174	115
	10	2	45.8	44.6	280	228	268	260	245	223	152
	11	1	51.5	24.0	286	252	301	294	278	254	177
	12		53.6	0.0	288	260	313	305	289	264	185
	SURFACE DAILY TOTALS				3076	2106	2358	2266	2114	1902	1262
MAY 21	5	7	5.2	114.3	41	9	4	4	4	3	2
	6	6	14.7	103.7	162	61	27	16	15	13	10
	7	5	24.6	93.0	219	118	89	75	60	43	13
	8	4	34.7	81.6	248	171	156	142	123	101	45
	9	3	44.3	68.3	264	217	217	202	182	156	86
	10	2	53.0	51.3	274	252	265	251	229	200	120
	11	1	59.5	28.6	279	274	296	281	258	228	141
	12		62.0	0.0	280	281	306	292	269	238	149
	SURFACE DAILY TOTALS				3254	2482	2418	2234	2010	1728	982
JUN 21	5	7	7.9	116.5	77	21	9	9	8	7	5
	6	6	17.2	106.2	172	74	33	19	18	16	12
	7	5	27.0	95.8	220	129	93	77	59	39	15
	8	4	37.1	84.6	246	181	157	140	119	95	35
	9	3	46.9	71.6	261	225	216	198	175	147	74
	10	2	55.8	54.8	269	259	262	244	220	189	105
	11	1	62.7	31.2	274	280	291	273	248	216	126
	12		65.5	0.0	275	287	301	283	258	225	133
	SURFACE DAILY TOTALS				3312	2626	2420	2204	1950	1644	874
JUL 21	5	7	5.7	114.7	43	10	5	5	4	4	3
	6	6	15.2	104.1	156	62	28	18	16	15	11
	7	5	25.1	93.5	211	118	89	75	59	42	14
	8	4	35.1	82.1	240	171	154	140	121	99	43
	9	3	44.8	68.8	256	215	214	199	178	153	83
	10	2	53.5	51.9	266	250	261	246	224	195	116
	11	1	60.1	29.0	271	272	291	276	253	223	137
	12		62.6	0.0	272	279	301	286	263	232	144
	SURFACE DAILY TOTALS				3158	2474	2386	2200	1974	1694	956
AUG 21	6	6	9.1	98.3	99	28	14	10	9	8	6
	7	5	19.1	87.2	190	85	75	67	58	47	20
	8	4	29.0	75.4	232	141	145	137	125	109	65
	9	3	38.4	61.8	254	189	210	201	187	168	110
	10	2	46.4	45.1	266	225	260	252	237	214	146
	11	1	52.2	24.3	272	248	293	285	268	244	169
	12		54.3	0.0	274	256	304	296	279	255	177
	SURFACE DAILY TOTALS				2898	2086	2300	2200	2046	1836	1208
SEP 21	7	5	10.0	78.7	131	35	44	44	43	40	31
	8	4	19.5	66.8	215	92	124	124	121	115	90
	9	3	28.2	53.4	251	142	196	197	193	183	143
	10	2	35.4	37.8	269	181	251	254	248	236	185
	11	1	40.3	19.8	278	205	287	289	284	269	212
	12		42.0	0.0	280	213	299	302	296	281	221
	SURFACE DAILY TOTALS				2568	1522	2102	2118	2070	1966	1546
OCT 21	7	5	2.0	71.9	4	0	1	1	1	1	1
	8	4	11.2	60.2	165	44	86	91	95	95	87
	9	3	19.3	47.4	233	94	167	176	180	178	157
	10	2	25.7	33.1	262	133	228	239	242	239	207
	11	1	30.0	17.1	274	157	266	277	281	276	237
	12		31.5	0.0	278	166	279	291	294	288	247
	SURFACE DAILY TOTALS				2154	1022	1774	1860	1890	1866	1626
NOV 21	8	4	3.6	54.7	36	5	17	19	21	22	22
	9	3	11.2	42.7	179	46	117	129	137	141	135
	10	2	17.1	29.5	233	83	186	202	212	215	201
	11	1	20.9	15.1	255	107	227	245	255	258	238
	12		22.2	0.0	261	115	241	259	270	272	250
	SURFACE DAILY TOTALS				1668	596	1336	1448	1518	1544	1442
DEC 21	9	3	8.0	40.9	140	27	87	98	105	110	109
	10	2	13.6	28.2	214	63	164	180	192	197	190
	11	1	17.3	14.4	242	86	207	226	239	244	231
	12		18.6	0.0	250	94	222	241	254	260	244
	SURFACE DAILY TOTALS				1444	446	1136	1250	1326	1364	1304

NOTE: 1) BASED ON DATA IN TABLE 1, pp 387 in 1972 ASHRAE HANDBOOK OF FUNDAMENTALS; 0% GROUND REFLECTANCE; 1.0 CLEARNESS FACTOR.
2) SEE FIG. 4, pp 394 in 1972 ASHRAE HANDBOOK OF FUNDAMENTALS FOR TYPICAL REGIONAL CLEARNESS FACTORS.
3) GROUND REFLECTION NOT INCLUDED ON NORMAL OR HORIZONTAL SURFACES.

"Reprinted from ASHRAE TRANSACTIONS 1974, Volume 80, Part II, by permission of the American Society of Heating, Rfrigerating and Air-Conditioning Engineers, Inc."

SOLAR POSITION AND INSOLATION VALUES FOR 56 DEGREES NORTH LATITUDE

TABLE 5.....SOLAR POSITION AND INSOLATION VALUES FOR 56 DEGREES NORTH LATITUDE

DATE	SOLAR TIME		SOLAR POSITION		BTUH/SQ. FT. TOTAL INSOLATION ON SURFACES						
	AM	PM	ALT	AZM	NORMAL	HORIZ.	SOUTH FACING SURFACE ANGLE WITH HORIZ. 46	56	66	76	90
JAN 21	9	3	5.0	41.8	78	11	50	55	59	60	60
	10	2	9.9	28.5	170	39	135	146	154	156	153
	11	1	12.9	14.5	207	58	183	197	206	208	201
	12		14.0	C.0	217	65	198	214	222	225	217
	SURFACE DAILY TOTALS				1126	282	934	1010	1058	1074	1044
FEB 21	8	4	7.6	59.4	129	25	65	69	72	72	69
	9	3	14.2	45.9	214	65	151	159	162	161	151
	10	2	19.4	31.5	250	98	215	225	228	224	208
	11	1	22.8	16.1	266	119	254	265	268	263	243
	12		24.0	0.0	270	126	268	279	282	276	255
	SURFACE DAILY TOTALS				1986	740	1640	1716	1742	1716	1598
MAR 21	7	5	8.3	77.5	128	28	40	40	39	37	32
	8	4	16.2	64.4	215	75	119	120	117	111	97
	9	3	23.3	50.3	253	118	192	193	189	180	154
	10	2	29.0	34.9	272	151	249	251	246	234	205
	11	1	32.7	17.9	282	172	285	288	282	268	236
	12		34.0	0.0	284	179	297	300	294	280	246
	SURFACE DAILY TOTALS				2586	1268	2066	2084	2040	1938	1700
APR 21	5	7	1.4	108.8	0	0	0	0	0	0	0
	6	6	9.6	96.5	122	32	14	9	8	7	6
	7	5	18.0	84.1	201	81	74	66	57	46	29
	8	4	26.1	70.9	239	129	143	135	123	108	82
	9	3	33.6	56.3	260	169	208	200	186	167	133
	10	2	39.9	39.7	272	201	259	251	236	214	174
	11	1	44.1	20.7	278	220	292	284	268	245	200
	12		45.6	0.0	280	227	303	295	279	255	209
	SURFACE DAILY TOTALS				3024	1892	2282	2186	2038	1830	1458
MAY 21	4	8	1.2	125.5	0	0	0	0	0	0	0
	5	7	8.5	113.4	93	25	10	9	8	7	6
	6	6	16.5	101.5	175	71	28	17	15	13	11
	7	5	24.8	89.3	219	119	88	74	58	41	16
	8	4	33.1	76.3	244	163	153	138	119	98	63
	9	3	40.9	61.6	259	201	212	197	176	151	109
	10	2	47.6	44.2	268	231	259	244	222	194	146
	11	1	52.3	23.4	273	249	288	274	251	222	170
	12		54.0	0.0	275	255	299	284	261	231	178
	SURFACE DAILY TOTALS				3340	2374	2374	2188	1962	1682	1218
JUN 21	4	8	4.2	127.2	21	4	2	2	2	2	1
	5	7	11.4	115.3	122	40	14	13	11	10	8
	6	6	19.3	103.6	185	86	34	19	17	15	12
	7	5	27.6	91.7	222	132	92	76	57	38	15
	8	4	35.9	78.8	243	175	154	137	116	92	55
	9	3	43.8	64.1	257	212	211	193	170	143	98
	10	2	50.7	46.4	265	240	255	238	214	184	133
	11	1	55.6	24.9	269	258	284	267	242	210	156
	12		57.5	0.0	271	264	294	276	251	219	164
	SURFACE DAILY TOTALS				3438	2562	2388	2166	1910	1606	1120
JUL 21	4	8	1.7	125.8	0	0	0	0	0	0	0
	5	7	9.0	113.7	91	27	11	10	9	8	6
	6	6	17.0	101.9	169	72	30	18	16	14	12
	7	5	25.3	89.7	212	119	88	74	58	41	15
	8	4	33.6	76.7	237	163	151	136	117	96	61
	9	3	41.4	62.0	252	201	208	193	173	147	106
	10	2	48.2	44.6	261	230	254	239	217	189	142
	11	1	52.9	23.7	265	248	283	268	245	216	165
	12		54.6	0.0	267	254	293	278	255	225	173
	SURFACE DAILY TOTALS				3240	2372	2342	2152	1926	1646	1186
AUG 21	5	7	2.0	109.2	1	0	0	0	0	0	0
	6	6	10.2	97.0	112	34	16	11	10	9	7
	7	5	18.5	84.5	187	82	73	65	56	45	28
	8	4	26.7	71.3	225	128	140	131	119	104	78
	9	3	34.3	56.7	246	168	202	193	179	160	126
	10	2	40.5	40.0	258	199	251	242	227	206	166
	11	1	44.8	20.9	264	218	282	274	258	235	191
	12		46.3	0.0	266	225	293	285	269	245	200
	SURFACE DAILY TOTALS				2850	1884	2218	2118	1966	1760	1392
SEP 21	7	5	8.3	77.5	107	25	36	36	34	32	28
	8	4	16.2	64.4	194	72	111	111	108	102	89
	9	3	23.3	50.3	233	114	181	182	178	168	147
	10	2	29.0	34.9	253	146	236	237	232	221	193
	11	1	32.7	17.9	263	166	271	273	267	254	223
	12		34.0	0.0	266	173	283	285	279	265	233
	SURFACE DAILY TOTALS				2368	1220	1950	1962	1918	1820	1594
OCT 21	8	4	7.1	59.1	104	20	53	57	59	59	57
	9	3	13.8	45.7	193	60	138	145	148	147	138
	10	2	19.0	31.3	231	92	201	210	213	210	195
	11	1	22.3	16.0	248	112	240	250	253	248	230
	12		23.5	0.0	253	119	253	263	266	261	241
	SURFACE DAILY TOTALS				1804	688	1516	1586	1612	1588	1480
NOV 21	9	3	5.2	41.9	76	12	49	54	57	59	58
	10	2	10.0	28.5	165	39	132	143	149	152	148
	11	1	13.1	14.5	201	58	179	193	201	203	196
	12		14.2	0.0	211	65	194	209	217	219	211
	SURFACE DAILY TOTALS				1094	284	914	986	1032	1046	1016
DEC 21	9	3	1.9	40.5	5	0	3	4	4	4	4
	10	2	6.6	27.5	113	19	86	95	101	104	103
	11	1	9.5	13.9	166	37	141	154	163	167	164
	12		10.6	0.0	180	43	159	173	182	186	182
	SURFACE DAILY TOTALS				748	156	620	678	716	734	722

NOTE: 1) BASED ON DATA IN TABLE 1, pp 387 in 1972 ASHRAE HANDBOOK OF FUNDAMENTALS; 0% GROUND REFLECTANCE; 1.0 CLEARNESS FACTOR.
2) SEE FIG. 4, pp 394 in 1972 ASHRAE HANDBOOK OF FUNDAMENTALS FOR TYPICAL REGIONAL CLEARNESS FACTORS.
3) GROUND REFLECTION NOT INCLUDED ON NORMAL OR HORIZONTAL SURFACES.

SOLAR POSITION AND INSOLATION VALUES FOR 64 DEGREES NORTH LATITUDE

DATE	SOLAR TIME		SOLAR POSITION		BTUH/SQ. FT. TOTAL INSOLATION ON SURFACES						
	AM	PM	ALT.	AZM	NORMAL	HORIZ.	SOUTH FACING SURFACE ANGLE WITH HORIZ. 54	64	74	84	90
JAN 21	10	2	2.8	28.1	22	2	17	19	20	20	20
	11	1	5.2	14.1	81	12	72	77	80	81	81
	12		6.0	0.0	100	16	91	98	102	103	103
	SURFACE DAILY TOTALS				306	45	268	290	302	306	304
FEB 21	8	4	3.4	58.7	35	4	17	19	19	19	19
	9	3	8.6	44.8	147	31	103	108	111	110	107
	10	2	12.6	30.3	199	55	170	178	181	178	173
	11	1	15.1	15.3	222	71	212	220	223	219	213
	12		16.0	0.0	228	77	225	235	237	232	226
	SURFACE DAILY TOTALS				1432	400	1230	1286	1302	1282	1252
MAR 21	7	5	6.5	76.5	95	18	30	29	29	27	25
	8	4	20.7	62.6	185	54	101	102	99	94	89
	9	3	18.1	48.1	227	87	171	172	169	160	153
	10	2	22.3	32.7	249	112	227	229	224	213	203
	11	1	25.1	16.6	260	129	262	265	259	246	235
	12		26.0	0.0	263	134	274	277	271	258	246
	SURFACE DAILY TOTALS				2296	932	1856	1870	1830	1736	1656
APR 21	5	7	4.0	108.5	27	5	2	2	2	1	1
	6	6	10.4	95.1	133	37	15	9	8	7	6
	7	5	17.0	81.6	194	76	70	63	54	43	37
	8	4	23.3	67.5	228	112	136	128	116	102	91
	9	3	29.0	52.3	248	144	197	189	176	158	145
	10	2	33.5	36.0	260	169	246	239	224	203	188
	11	1	36.5	18.4	266	184	278	270	255	233	216
	12		97.6	0.0	268	190	289	281	266	243	225
	SURFACE DAILY TOTALS				2982	1644	2176	2082	1936	1736	1594
MAY 21	4	8	5.8	125.1	51	11	5	4	4	3	3
	5	7	11.6	112.1	132	42	13	11	10	9	8
	6	6	17.9	99.1	185	79	29	16	14	12	11
	7	5	24.5	85.7	218	117	86	72	56	39	28
	8	4	30.9	71.5	239	152	148	133	115	94	80
	9	3	36.8	56.1	252	182	204	190	170	145	128
	10	2	41.6	38.9	261	205	249	235	213	186	167
	11	1	44.9	20.1	265	219	278	264	242	213	193
	12		46.0	0.0	267	224	288	274	251	222	201
	SURFACE DAILY TOTALS				3470	2236	2312	2124	1898	1624	1436
JUN 21	3	9	4.2	139.4	21	4	2	2	2	2	1
	4	8	9.0	126.4	93	27	10	9	8	7	6
	5	7	14.7	113.6	154	60	16	15	13	11	10
	6	6	21.0	100.8	194	96	34	19	17	14	13
	7	5	27.5	87.5	221	132	91	74	55	36	23
	8	4	34.0	73.3	239	166	150	133	112	88	73
	9	3	39.9	57.8	251	195	204	187	164	137	119
	10	2	44.9	40.4	258	217	247	230	206	177	157
	11	1	48.3	20.9	262	231	275	258	233	202	181
	12		49.5	0.0	263	235	284	267	242	211	189
	SURFACE DAILY TOTALS				3650	2488	2342	2118	1862	1558	1356
JUL 21	4	8	6.4	125.3	53	13	6	5	5	4	4
	5	7	12.1	112.4	128	44	14	13	11	10	9
	6	6	18.4	99.4	179	81	30	17	16	13	12
	7	5	25.0	86.0	211	118	86	72	56	38	28
	8	4	31.4	71.8	231	152	146	131	113	91	77
	9	3	37.3	56.3	245	182	201	186	166	141	124
	10	2	42.2	39.2	253	204	245	230	208	181	162
	11	1	45.4	20.2	257	218	273	258	236	207	187
	12		46.6	0.0	259	223	282	267	245	216	195
	SURFACE DAILY TOTALS				3372	2248	2280	2090	1864	1588	1400
AUG 21	5	7	4.6	108.8	29	6	3	3	2	2	2
	6	6	11.0	95.5	123	39	16	11	10	8	7
	7	5	17.6	81.9	181	77	69	61	52	42	35
	8	4	23.9	67.8	214	113	132	123	112	97	87
	9	3	29.6	52.6	234	144	190	182	169	150	138
	10	2	34.2	36.2	246	168	237	229	215	194	179
	11	1	37.2	18.5	252	183	268	260	244	222	205
	12		38.3	0.0	254	188	278	270	255	232	215
	SURFACE DAILY TOTALS				2808	1646	2108	1008	1860	1662	1522
SEP 21	7	5	6.5	76.5	77	16	25	25	24	23	21
	8	4	12.7	72.6	163	51	92	92	90	85	81
	9	3	18.1	48.1	206	83	159	159	156	147	141
	10	2	22.3	32.7	229	108	212	213	209	198	189
	11	1	25.1	16.6	240	124	246	248	243	230	220
	12		26.0	0.0	244	129	258	260	254	241	230
	SURFACE DAILY TOTALS				2074	892	1726	1736	1696	1608	1532
OCT 21	8	4	3.0	58.5	17	2	9	9	10	10	10
	9	3	8.1	44.6	122	26	86	91	93	92	90
	10	2	12.1	30.2	176	50	152	159	161	159	155
	11	1	14.6	15.2	201	65	193	201	203	200	295
	12		15.5	0.0	208	71	207	215	217	213	208
	SURFACE DAILY TOTALS				1238	358	1088	1136	1152	1134	1106
NOV 21	10	2	3.0	28.1	23	3	18	20	21	21	21
	11	1	5.4	14.2	79	12	70	76	79	80	79
	12		6.2	0.0	97	17	89	96	100	101	100
	SURFACE DAILY TOTALS				302	46	266	286	298	302	300
DEC 21	11	1	1.8	13.7	4	0	3	4	4	4	4
	12		2.6	0.0	16	2	14	15	16	17	17
	SURFACE DAILY TOTALS				24	2	20	22	24	24	24

NOTE: 1) BASED ON DATE IN TABLE 1, pp 387 in 1972 ASHRAE HANDBOOK OF FUNDAMENTALS; 0% GROUND REFLECTANCE; 1.0 CLEARNESS FACTOR.
2) SEE FIG. 4, pp 394 in 1972 ASHRAE HANDBOOK OF FUNDAMENTALS FOR TYPICAL REGIONAL CLEARNESS FACTORS.
3) GROUND REFLECTION NOT INCLUDED ON NORMAL OR HORIZONTAL SURFACES.

Appendix 5

Pressure Loss
from Pipe Friction

Gallons per Minute	Pipe Size: Inches I.D.	Water	Glycol/Water	Silicone Synthetic Oil
			psi per Foot of Pipe	
1.5	0.50	0.013	0.020	0.055
	0.75	0.004	0.007	0.018
5	0.75	0.029	0.045	0.060
	1.00	0.009	0.009	0.018
10	1.00	0.030	0.045	0.047
	1.25	0.008	0.013	0.016
15	1.25	0.016	0.025	0.025
	1.50	0.008	0.011	0.013
20	1.25	0.028	0.040	0.060
	1.50	0.013	0.018	0.019
30	1.50	0.027	0.040	0.045
	2.00	0.008	0.012	0.016
40	2.00	0.013	0.020	0.024
	2.50	0.006	0.008	0.011
50	2.00	0.020	0.029	0.033
	2.50	0.009	0.012	0.016
100	2.50	0.032	0.043	0.055
	3.00	0.011	0.015	0.019

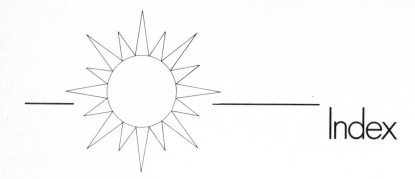

Index